SEVENTH EDITION

HANDBOOK
OF TECHNICAL
WRITING

Gerald J. Alred
University of Wisconsin – Milwaukee

Charles T. Brusaw
NCR Corporation (retired)

Walter E. Oliu
U.S. Nuclear Regulatory Commission

ST. MARTIN'S PRESS New York

FOR BEDFORD/ST.MARTIN'S

Developmental Editor: Sara Eaton
Editorial Assistant: Christine Turnier-Vallecillo
Editorial Assistant, Publishing Services: Maria Teresa Burwell
Production Supervisor: Tina Cameron
Marketing Manager: Richard Cadman
Project Management: Books By Design, Inc.
Text Design: Claire Seng-Niemoeller
Cover Design: Laura Shaw Design
Cover Art: Communication Satellite Receiver Dishes, Brewster, Washington
Copyright 2002 Jeffrey W. Myers/Panoramic Images. All Rights Reserved.
Composition: Pine Tree Composition, Inc.
Printing and Binding: Haddon Craftsman, an R.R. Donnelley &
Sons Company

President: Joan E. Feinberg
Director of Marketing: Karen R. Melton
Editor in Chief: Karen S. Henry
Director of Editing, Design, and Production: Marcia Cohen
Manager, Publishing Services: Emily Berleth

Library of Congress Control Number: 2002101242

8   7   6   5   4   3

f   e   d   c   b

For information, write: Bedford/St. Martin's, 75 Arlington Street,
Boston, MA 02116   (617-399-4000)

ISBN: 0-312-39323-7 (paperback)
          0-312-30923-6 (hardcover)

ACKNOWLEDGMENTS

Figure D–4. Entry for "regard" from *The American Heritage Dictionary of the English
Language,* 3rd Ed. Copyright © 1996 by Houghton Mifflin Company. Reprinted
with the permission of the publisher. All rights reserved.

*Acknowledgments and copyrights are continued at the back of the book on page 590, which
constitutes an extension of the copyright page.*

Contents

Preface

The seventh edition of the *Handbook of Technical Writing*, like previous editions, is a comprehensive resource for both academic and professional audiences. The *Handbook*'s nearly 500 entries cover effective print, oral, and electronic communication in technical writing, as well as grammar, style, and usage. This edition has up-to-date coverage of workplace technologies, library and Internet research, and documenting sources, as well as improved treatment of formal reports, job searches, presentations, and visuals.

This edition of the *Handbook* remains an accessible and easy-to-use guide with entries that have been consolidated and made more concise. In addition, the new companion Web site expands this already thorough reference. Resources available on <www.bedfordstmartins.com/alred> make the *Handbook of Technical Writing* a more effective classroom text and teaching tool while allowing the book to remain the quick reference faithful users have come to appreciate.

How to Use This Book

The New Five-Way Access System. The new five-way access system of the *Handbook of Technical Writing* provides readers with even more ways of retrieving information.

1. *Alphabetically Organized Entries.* The alphabetically organized entries with color tabs enable readers to find information quickly. Within the entries, terms shown as links (underlined and set in an alternate typeface) refer to other entries that contain key definitions of concepts, further information on topics, and additional entries on related subjects.
2. *Topical Key to the Alphabetical Entries.* The Topical Key, on pages xxi–xxiii, groups the entries into categories and serves as a table of contents to all subjects covered in the book. The key can help a writer focusing on a specific task or problem locate helpful entries; it is also useful for instructors who want to correlate the *Handbook* with standard textbooks or their own course materials.
3. *Checklist of the Writing Process.* The checklist, on pages xix–xx, helps readers to reference all writing-related entries.

4. *Comprehensive Index.* The Index lists all the topics covered in the book, including those topics that are not main entries in the alphabetical arrangement.

5. *Topical List of Figures and Model Documents.* A new Topical List of Figures and Model Documents, on pages xxiv–xxvi, makes it easier to find the abundant real-world examples and sample documents throughout the text that provide models for effective technical communication.

Entries Helpful to ESL Writers.   The Topical Key to the Alphabetical Entries includes a list of entries—ESL Trouble Spots—that may be of particular help to ESL writers. This list includes entries that cover persistent problems for ESL writers from a diverse group of languages. For problems not included on that list, ESL writers should check other entries in the *Handbook,* refer to an ESL grammar or reference text, or consult a native speaker of English.

New to This Edition

Those familiar with the *Handbook* will notice the new convention to signal cross-referenced entries—the use of underlined text set in a different typeface—as in, for example, a reference to **audience**. This change as well as many of those listed below reflect new technology and the prevalence of the Web.

- *More concise treatment for a streamlined and comprehensive reference* makes the information even more accessible.

- *A new companion Web site* <www.bedfordstmartins.com/alred> expands this already thorough reference by providing links to online resources, exercises, research and documentation advice, and model documents for technical writing. Web Link boxes throughout the text direct writers to these additional resources available on the companion Web site and beyond.

- *Comprehensive and up-to-date coverage of workplace technology*—in entries such as e-mail, Internet research, writing for the Web, and Web design—focuses on considerations of audience and purpose, the presentation of long documents, and the use of keywords and hyperlinks. The seventh edition offers students expert advice for meeting the demands of online writing.

- *Digital Tip boxes* throughout the text provide practical, concise advice for using software tools for a wide variety of tasks, including creating indexes and outlines, tracking changes, and using collaborative software. The Digital Tips are expanded on the companion Web site, offering more in-depth practical advice for using current workplace technology.

- *Fully revised and expanded coverage of library and Internet research and documenting sources* provides thoroughly updated Internet and library research information, current MLA and APA documentation models, and guidelines for evaluating online sources. The documenting sources entry now includes *Chicago (CMS)* style and is redesigned to make the reasons and rules for documenting sources easier to understand.

- *Improved coverage of formal reports* includes a complete example of a formal report that integrates text and visuals.

- *New and revised entries on brochures, manuals, newsletters, and proposals* focus on audience and purpose and reflect the variety of documents in technical writing.

- *Improved coverage for finding a job* includes new entries on negotiating salaries and writing follow-up letters, as well as updated information on using Web resources and creating electronic résumés.

- *New and revised entries on presentations and visuals* provide more advice on using presentation software and creating and integrating graphics.

Acknowledgments

For their invaluable comments and suggestions for this edition of the *Handbook of Technical Writing,* we thank the following reviewers:

Ralph Batie, Oregon Institute of Technology

LynnDianne Beene, University of New Mexico

Brad Benz, Fort Lewis College

Mary Connerty, Pennsylvania State University–Erie, The Behrend College

Brian Cosbey, University of New Mexico

David L. Dyrud, Oregon Institute of Technology

Marilyn A. Dyrud, Oregon Institute of Technology

David H. Myer, Jacksonville State University

Jackie S. Palmer, Texas A&M University

Cindy Raisor, Texas A&M University

Natalie D. Segal, Ward College of Technology, University of Hartford

Barry Thatcher, New Mexico State University

Valerie J. Vance, Oregon Institute of Technology

For this edition, special thanks are due to Barry Thatcher, New Mexico State University, for his in-depth review of the international communication, global communication, and global graphics entries. We are extremely grateful to Pat Goldstein, University of Wisconsin–Milwaukee, for her expert ESL advice. We wish to thank Valerie Vance, Oregon Institute of Technology, for her detailed review of the Index, and Roger Friedmann, Kansas State University, for his detailed review of the sample résumés. For helping us revise and redesign the documenting sources entry, we wish to thank Mary Connerty, Pennsylvania State University–Erie, The Behrend College; David H. Myer, Jacksonville State University; Cindy Raisor, Texas A&M University; Kris Rose, Davenport University; and Lawrence (Jude) Roy Jr., Madisonville Community College.

We are especially grateful to Rachel Spilka and Nancy Nygaard, both of the University of Wisconsin–Milwaukee, for their contributions to the proposals entry. Dave Clark, University of Wisconsin–Milwaukee, used his impressive knowledge to contribute greatly to the entries related to workplace technology.

We are very much indebted to those who helped us shape the first six editions: Sandra Balzo, First Wisconsin National Bank; Deborah Barrett, Houston Baptist University–Houston; Ralph Batie, Oregon Institute of Technology; Eleanor Berry, University of Wisconsin–Milwaukee; Beatrice Christiana Birchac, University of Houston–Downtown; Shane Borrowman, University of Arizona; James H. Chaffee, Normandale Community College; Darlene Chang, Freelance Graphics Designer; Patricia C. Click, University of Virginia; Mary Coney, University of Washington; Edmund Dandridge, North Carolina State University; Evan Davis, Southwestern Oregon Community College; Pat Dorazio, SUNY Institute of Technology at Utica; Donna Lee Dowdney, De Anza College; Paul Falon, University of Michigan; James P. Farrelly, University of Dayton; M. L. Flynn, South Dakota State University; Pat Goldstein, University of Wisconsin–Milwaukee; Heather Bradie Graves, DePaul University; David Hatfield, Marshall University; Susan Hitchcock, University of Virginia; William H. Holahan, University of Wisconsin–Milwaukee; Suzanne Karberg, Purdue University; Patrick Kelley, New Mexico State University; Kristie Kemper, Floyd College; Wayne Losano, Rensselaer Polytechnic Institute; Carol Mablekos, Temple University; David Mair, University of Oklahoma–Norman; Roger Masse, New Mexico State University; Judy McInish, Hornbake Library, University of Maryland; Susan K. McLaughlin, Business Communications Consultants; Mimi Mejac, Nuclear Regulatory Commission; Mary Mullins, Delux Data Corporation; Brian Murphy, NCR Corporation; Lee Newcomer, University of Wisconsin–Milwaukee; Wendy Osborne, Society of American Foresters; Keith Palmer, Harnischfeger Corporation; Thomas Pearsall, University of Minnesota; Nancy Pointer, Normandale Community College; Alan Rauch, Georgia Institute of Technology; Diana Reep, University of Akron; Lois Rew, San José State University; Rayleona Sanders, Nuclear Regulatory Commission; Peter V. Sands, University of Wisconsin–Milwaukee; Elizabeth A. Schulz, Olympic College; Stuart A. Selber, Pennsylvania State University; Charles Stratton, University of Idaho; Donald R. Swanson, Wright State University; John Taylor, University of Wisconsin–Milwaukee; Erik Thelen, Marquette University; Ann Thomas, Nuclear Regulatory Commission; Mary Thompson, Purdue University; William Van Pelt, University of Wisconsin–Milwaukee; Margaret B. Walters, Kennesaw State University; Thomas Warren, Oklahoma State University; Merrill Whitburn, Rensselaer Polytechnic Institute; Conrad R. Winterhalter, NCR Corporation; Kristin R. Woolever, Northeastern University; and Arthur Young, Michigan Technological University.

We wish to thank Bedford/St. Martin's for supporting this book, especially Joan Feinberg, President, and Karen Henry, Editor in Chief. We are grateful to Emily Berleth, Manager, Publishing Services, at Bedford/St. Martin's, and Herb Nolan of Books By Design for their patience and expert guidance. We are pleased to acknowledge the always enthusiastic assistance of Christine Turnier-Vallecillo, Editorial Assistant at Bedford/St. Martin's. Finally, we wish to thank Sara Eaton, our Developmental Editor at Bedford/St. Martin's, whose editing and energy have helped us meet very tight deadlines.

We offer heartfelt thanks to Barbara Brusaw for her patience and time spent preparing the manuscript for the first five editions. We also gratefully acknowledge the ongoing contributions of many students and instructors at the University of Wisconsin–Milwaukee. For this edition, we owe special thanks to Renee Tegge, whose keen eye and insightful suggestions improved many entries, and to Rosemary Lesnik, who cheerfully proofread multiple versions of many drafts. Finally, special thanks go to Janice Alred for her many hours of substantive assistance and for holding everything together.

G. J. A.
C. T. B.
W. E. O.

Five Steps to Successful Writing

Successful writing on the job is not the product of inspiration, nor is it merely the spoken word converted to print; it is the result of knowing how to structure information using both text and design to achieve an intended purpose for a clearly defined audience. The best way to ensure that your writing will succeed—whether it is in the form of a memo, a résumé, a proposal, or a Web page—is to approach writing using the following steps:

1. Preparation
2. Research
3. Organization
4. Writing
5. Revision

You will very likely need to follow those steps consciously—even self-consciously—at first. The same is true the first time you use new software, interview a candidate for a job, or chair a committee meeting. With practice, the steps become nearly automatic. That is not to suggest that writing becomes easy. It does not. However, the easiest and most efficient way to write effectively is to do it systematically.

As you master the five steps, keep in mind that they are interrelated and often overlap. For example, your readers' needs and your purpose, which you determine in step 1, will affect decisions you make in subsequent steps. You may also need to retrace steps. When you conduct research, for example, you may realize that you need to revise your initial impression of the document's purpose and audience. Similarly, when you begin to organize, you may discover the need to return to the research step to gather more information.

The time required for each step varies with different writing tasks. When writing an informal memo, for example, you might follow the first three steps (preparation, research, and organization) by simply listing the points in the order you want to cover them. In such situations, you gather and organize information mentally as you consider your purpose and audience. For a formal report, the first three steps require well-organized research, careful note-taking, and detailed outlining. For a routine e-mail message to a coworker, the first four steps merge as you type the information on the screen. In short, the five steps expand, contract, and at times must be repeated to fit the complexity or context of the writing task.

Dividing the writing process into steps is especially useful for collaborative writing, in which you typically divide work among team members, keep track of a project, and save time by not duplicating effort. When you collaborate, you can use e-mail to share text and other files, suggest improvements to each other's work, and generally keep everyone informed of your progress as you follow the steps in the writing process.

Preparation

Writing, like most professional tasks, requires solid <u>preparation</u>.* In fact, adequate preparation is as important as <u>writing the draft</u>. In preparation for writing, your goal is to accomplish the following four major tasks:

- Establish your primary purpose.
- Assess your audience (or readers).
- Determine the scope of your coverage.
- Select the appropriate medium.

Establishing Your Purpose. To establish your primary <u>purpose</u> simply ask yourself what you want your readers to know, believe, or be able to do after they have finished reading what you have written. Be precise. Often a writer states a purpose so broadly that it is almost useless. A purpose such as "to report on possible locations for a new facility" is too general. However, "to compare the relative advantages of Paris, Singapore, and San Francisco as possible locations for a new engineering facility so top management can choose the best location" is a purpose statement that can guide you throughout the writing process. In addition to your primary purpose, consider possible secondary purposes for your document. For example, a secondary purpose of the engineering facilities report might be to make corporate executive readers aware of the staffing needs of the new facility so they can ensure its smooth operation in whatever location is selected.

Assessing Your Audience. The next task is to assess your <u>audience</u>. Again, be precise and ask key questions. Who exactly is your reader? Do you have multiple readers? Who needs to see or use the document? What are your readers' needs in relation to your subject?

---

*In this discussion, as elsewhere throughout this book, words and phrases shown as links—underlined and set in an alternate typeface—refer to specific alphabetical entries.

What are your readers' attitudes about the subject? (Skeptical? Supportive? Anxious? Bored?) What do your readers already know about the subject? Should you define basic terminology or will such definitions merely bore, or even impede, your readers? Are you communicating with international readers and therefore dealing with issues inherent in writing <u>international correspondence</u>?

For the engineering facilities report, the readers are described as "top management." But *who* is included in that category? Will one of the people evaluating the report be the Human Resources Manager? If so, that person likely would be interested in the availability of qualified professionals as well as in the presence of training, housing, and perhaps even recreational facilities available to potential employees in each city. The Purchasing Manager would be concerned about available sources for materials needed by the facility. The Marketing Manager would give priority to the facility's proximity to the primary markets for its products and services and the transportation options that are available. The Chief Financial Officer would want to know about land and building costs and about each country's tax structure. The Chief Executive Officer would be interested in all this information and perhaps more.

In addition to knowing the needs and interests of your readers, learn as much as you can about their background knowledge. Have they visited all three cities? Have they already seen other reports on the three cities? Is this the company's first new facility, or has the company chosen locations for new facilities before? As with this example, many workplace documents have audiences composed of multiple readers. You can accommodate their needs through one of a number of approaches described in the entry <u>readers</u>.

(ESL) TIPS FOR CONSIDERING AUDIENCES

In the United States, <u>conciseness</u>, <u>coherence</u>, and <u>clarity</u> characterize good writing. Be brief, make sure readers can follow your writing, be clear, and say only what is necessary to communicate your message. Of course, no writing style is inherently better than another, but to be a successful writer in any language, you must understand the cultural values that underlie the language in which you are writing. See also <u>awkwardness</u>, <u>copyright</u>, <u>global communication</u>, <u>plagiarism</u>, and <u>English as a second language</u>.

Throughout this book we have included ESL Tips boxes like this one with information that may be particularly helpful to nonnative speakers of English. The Topical Key to the Alphabetical Entries on page xxi includes a listing of entries that may be of particular help to ESL writers.

Determining the Scope.   Determining your purpose and assessing your readers will help you decide what to include and what not to include in your writing. Those decisions establish the <u>scope</u> of your writing project. If you do not clearly define the scope, you will spend needless hours on research because you will not be sure what kind of information you need or even how much. Given the purpose and readers established for the report on facility locations, the scope would include such information as land and building costs, available labor force, cultural issues, transportation options, and proximity to suppliers. However, it probably would not include the early history of the cities being considered or their climate and geological features, unless those aspects were directly related to your particular business.

Selecting the Medium.   Finally, you need to determine the most appropriate medium for communicating your message. Professionals on the job face a wide array of options—from <u>e-mail, fax,</u> voice mail, videoconferencing, and Web sites to more traditional means like letters, <u>memos, reports,</u> telephone calls, and face-to-face <u>meetings.</u> See also <u>correspondence</u> and <u>presentations.</u>

The most important considerations in selecting the appropriate medium are the audience and the purpose of the communication. For example, if you need to collaborate with someone to solve a problem or if you need to establish rapport with someone, written exchanges, even by e-mail, could be far less efficient than a phone call or a face-to-face meeting. However, if you need precise wording or you need to provide a record of a complex message, communicate in writing. If you need to make information that is frequently revised accessible to employees at a large company, the best choice might be to place the information on the company's Web site. (See <u>Web design.</u>) If reviewers need to make handwritten comments on a proposal, you may need to provide paper copies that can be faxed. The comparative advantages and primary characteristics of the most typical means of communication are discussed in <u>selecting the medium.</u>

Research

The only way to be sure that you can write about a complex subject is to thoroughly understand it. To do that, you must conduct adequate <u>research,</u> whether that means conducting an extensive investigation for a major proposal—through interviewing, library and Internet research, and careful <u>note-taking</u>—or simply checking a company Web site and jotting down points before you send an e-mail to a colleague.

Methods of Research.   Researchers frequently distinguish between primary and secondary research, depending on the types of sources consulted and the method of gathering information. *Primary research* refers to the gathering of raw data compiled from interviews, direct observation, surveys, experiments, <u>questionnaires,</u> and audio and video recordings, for example. In fact, direct observation and hands-on experience are the only ways to obtain certain kinds of information, such as the behavior of people and animals, certain natural phenomena, mechanical processes, and the operation of systems and equipment. *Secondary research* refers to gathering information that has been analyzed, assessed, evaluated, compiled, or otherwise organized into accessible form. Such forms or sources include books, articles, <u>reports,</u> Web documents, e-mail discussions, business letters, <u>minutes of meetings,</u> operating <u>manuals,</u> and <u>brochures.</u> Use the methods most appropriate to your needs, recognizing that some projects will require several types of research.

Sources of Information.   As you conduct research, numerous sources of information are available to you.

- Your own knowledge and that of your colleagues
- The knowledge of people outside of your workplace, gathered through <u>interviewing for information</u>
- Internet sources, as discussed in <u>Internet research</u>
- Library resources, including databases, as described in <u>library research</u>
- Printed and electronic sources in the workplace, such as brochures, memos and e-mail, and Web documents

Consider all sources of information when you begin your research and use those that are appropriate and useful. The amount of research you will need to do depends on the scope of your project.

Organization

Without <u>organization,</u> the material gathered during your research will be incoherent to your readers. To organize information effectively, you need to determine the best way to structure your ideas; that is, you must choose a primary <u>method of development.</u>

Methods of Development.   An appropriate method of development is the writer's tool for keeping information under control and the readers' means of following the writer's presentation. As you analyze the information you have gathered, choose the method that best suits your subject, your readers' needs, and your purpose. For example, if you were writing instructions for assembling office equipment, you would naturally present the steps of the process in the order readers should perform them: the <u>sequential method of development</u>. If you were writing about the history of an organization, your account would most naturally go from the beginning to the present: the <u>chronological method of development</u>. If your subject naturally lends itself to a certain method of development, use it—do not attempt to impose another method on it.

Sometimes you may need to use combinations of methods of development. For example, a persuasive brochure for a charitable organization might combine a <u>general and specific method of development</u> with a <u>cause-and-effect method of development</u>. That is, you could begin with persuasive case histories of individual people in need and then move to general information about the positive effects of donations on recipients.

Outlining.   Once you have chosen a method of development, you are ready to prepare an outline. <u>Outlining</u> breaks large or complex subjects into manageable parts. It also enables you to emphasize key points by placing them in the positions of greatest importance. Finally, by structuring your thinking at an early stage, a well-developed outline ensures that your document will be complete and logically organized, allowing you to focus exclusively on writing when you begin the rough draft. Even a short letter or memo needs the logic and structure that an outline provides, whether the outline exists in your mind or on-screen or on paper.

At this point, you must begin to consider <u>layout and design</u> elements that will be helpful to your readers and appropriate to your subject and purpose. For example, if <u>visuals</u>, <u>photographs</u>, or <u>tables</u> will be useful, this is a good time to think about where they may be deployed and what kinds of visual elements will be effective, especially if they need to be prepared by someone else while you are writing and revising the draft. The outline can also suggest where <u>headings</u>, <u>lists</u>, and other special design features may be useful.

Writing

When you have established your purpose, your readers' needs, and your scope and have completed your research and your outline, you will be well prepared to write a first draft. Expand your outline into <u>paragraphs</u>, without worrying about <u>grammar</u>, refinements of language <u>usage</u>, or <u>punctuation</u>. Writing and revising are different activities; refinements come with revision.

Write the rough draft, concentrating entirely on converting your outline into sentences and paragraphs. You might try writing as though you were explaining your subject to a reader sitting across from you. Do not worry about a good opening. Just start. There is no need in the rough draft to be concerned about exact <u>word choice</u> unless it comes quickly and easily — concentrate instead on ideas.

Even with good preparation, writing the draft remains a chore for many writers. The most effective way to get started and keep going is to use your outline as a map for your first draft. Do not wait for inspiration — you need to treat writing a draft as you would any on-the-job task. The entry <u>writing a draft</u> describes tactics used by experienced writers — discover which ones are best suited to you and your task.

Consider writing an <u>introduction</u> last because then you will know more precisely what is in the body of the draft. Your opening should announce the subject and give readers essential background information, such as the document's primary purpose. For longer documents, an introduction should serve as a frame into which readers can fit the detailed information that follows.

Finally, you will need to write a <u>conclusion</u> that ties the main ideas together and emphatically makes a final significant point. The final point may be to recommend a course of action, make a prediction or a judgment, or merely summarize your main points — the way you conclude depends on the purpose of your writing and your readers' needs.

Revision

The clearer a finished piece of writing seems to the reader, the more effort the writer has likely put into its <u>revision</u>. If you have followed the steps of the writing process to this point, you will have a rough draft that needs to be revised. Revising, however, requires a different frame of mind than does writing the draft. During revision, be eager to find and correct faults and be honest. Be hard on yourself for the benefit of your readers. Read and evaluate the draft as if you were a reader seeing it for the first time.

Check your draft for accuracy, completeness, and effectiveness in achieving your purpose and meeting your readers' needs and expectations. Trim extraneous information: Your writing should give readers exactly what they need, but it should not burden them with unnecessary information or sidetrack them into loosely related subjects.

Do not try to revise for everything at once. Read your rough draft several times, each time looking for and correcting a different set of problems or errors. Concentrate first on larger issues, such as <u>unity</u> and <u>coherence</u>; save mechanical corrections, like <u>spelling</u> and <u>punctuation</u>, for later reviews. See also <u>ethics in writing</u>.

Finally, for important documents, consider having others review your writing and make suggestions for improvement. Use the Checklist of the Writing Process on page xix to guide you not only as you revise but also throughout the writing process. The checklist refers you to specific entries grouped according to the Five Steps to Successful Writing and can help you diagnose and solve writing problems.

Checklist of the Writing Process

This checklist arranges key entries of *Handbook of Technical Writing* according to the sequence presented in "Five Steps to Successful Writing," which begins on page xi. This checklist is useful both for following the steps and for diagnosing writing problems. The exact titles of the entries are shown as links for quick reference, followed by the page numbers. When you turn to the entries themselves, you will find links to other entries that may be helpful.

You may also wish to refer to the Topical Key to the Alphabetical Entries on pages xxi–xxiii as well as to the Index, which begins on page 591.

☑ Use **quotations** and
paraphrasing   468, 393

☑ Write an
introduction   291

☑ Write a **conclusion**   101

☑ Choose a **title**   549

REVISION   **499**

☑ Check for completeness
(**revision**) and
accuracy   499, 13

☑ Check for **unity** and
coherence   80, 555

conciseness   99

pace   388

transition   548

☑ Check for **sentence**
variety   515

emphasis   178

parallel structure   392

subordination   527

☑ Check for **clarity**   77

ambiguity   36

awkwardness   50

logic errors   329

positive writing   407

voice   569

☑ Check for **ethics in**
writing   187

biased language   54

copyright   105

plagiarism   405

☑ Check for appropriate
word choice   582

abstract/concrete words   8

affectation and
jargon   25, 301

clichés and trite
language   79, 551

connotation/denotation   104

defining terms   126

☑ Eliminate problems
with **grammar**   244

agreement   28

case   69

modifiers   353

pronoun reference   428

sentence faults   513

☑ Review mechanics and
punctuation   457

abbreviations   2

capitalization   66

contractions   105

dates   125

indentation   259

italics   297

numbers   372

proofreading   435

spelling   524

symbols   529

TOPICAL KEY TO THE ALPHABETICAL ENTRIES

This list arranges the alphabetical entries into subject categories. Entries listed under ESL Trouble Spots and marked with an asterisk (*) are entries with ESL Tips boxes that offer additional advice to ESL writers. See the Checklist of the Writing Process on page xix for a list of key entries in a recommended sequence of steps for any writing project.

SENTENCES AND PARAGRAPHS

PARTS OF SPEECH AND GRAMMAR

PUNCTUATION AND MECHANICS

TOPICAL LIST OF FIGURES AND MODEL DOCUMENTS

The *Handbook of Technical Writing* offers abundant figures and examples of technical writing—all of which are listed below, organized into subject categories, for easy access. See also the Topical Key to the Alphabetical Entries on pages xxi–xxiii. For additional model documents and resources for students and instructors, see the companion Web site at <www.bedfordstmartins.com/alred>.

FINDING A JOB

WORKPLACE TECHNOLOGY

PLANNING AND RESEARCH

ORGANIZATION, WRITING, AND REVISION

DESIGN AND VISUALS

ESL TROUBLE SPOTS

STYLE AND LANGUAGE

PARTS OF SPEECH AND GRAMMAR

PUNCTUATION AND MECHANICS

A

a / an

A and *an* are indefinite <u>articles</u>; "indefinite" implies that the <u>noun</u> designated by the article is not a specific person, place, or thing but is one of a group.

- She installed *a* program.
  [not a specific program but an unnamed program]

Use *a* before words beginning with a consonant or consonant sound (including *y* or *w* sounds).

- The year's activities are summarized in *a* one page report.
  [Although the *o* in *one* is a vowel, the first sound is *w,* a consonant sound; hence, the article *a* precedes the word.]
- *A* manual has been written on that subject.
- It was *a* historic event for the laboratory.
- The office manager felt that it was *a* difficult situation.
- We received *a* DNR order.

Use *an* before words beginning with a vowel or vowel sound.

- The report is *an* overview of the year's activities.
- He seems *an* unlikely candidate for the job.
- The interviewer arrived *an* hour early.
  [Although the *h* in *hour* is a consonant, the word begins on a vowel sound; hence, it is preceded by *an*.]
- He bought *an* SLR digital camera.

Be careful not to use unnecessary indefinite articles in a sentence.

- Fill with *a* half *a* pint of fluid.
  [Choose one article and eliminate the other.]

See also <u>adjectives</u>.

a lot

A lot is often incorrectly written as one word (*alot*). Write the phrase as two words: *a lot*. It is, however, an informal phrase that normally should not be used in technical writing. Use *many* or *numerous* instead or give a specific number or amount.

- The peer review group had ~~a lot of~~ objections.

  numerous

abbreviations

DIRECTORY

Abbreviations are shortened versions of words or combinations of the first letters of words, such as Avenue/Ave., Corporation/Corp., and hypertext markup language/HTML. Abbreviations that are formed by combining the first letter or letters of several words are called *acronyms*. Acronyms are pronounced as words and are written without periods, such as *d*isk *o*perating *s*ystem/DOS, *l*ocal *a*rea *n*etwork/LAN, and *s*elf-contained *u*nderwater *b*reathing *a*pparatus/scuba. Abbreviations that are formed by combining the initial letter of each word in a multiword term are called *initialisms*. Initialisms are pronounced as separate letters, such as *f*or *y*our *i*nformation/FYI and *p*ost *m*eridiem/p.m.

Abbreviations, if used appropriately, can be convenient for both the reader and the writer. Like symbols, they can be important space savers in technical writing because it is often necessary to provide the maximum amount of information in a limited amount of space.

Using Abbreviations

In business, industry, and government, specialists and people working together on particular projects often use abbreviations, particularly acronyms and initialisms. Within a group of specialists shortened forms will be easily understood—outside of the group, however, they might be incomprehensible. In fact, abbreviations can be easily overused, either as an **affectation** or in a misguided attempt to make writing concise, especially in **e-mail**. Remember that memos or reports addressed to specific people may be read by other people; you must consider those secondary readers as well. A good rule to follow: When in doubt, spell it out.

Writer's Checklist: Using Abbreviations

Organizations and publishers often establish standards for using abbreviations; however, the following guidelines are common.

☑ Except for commonly used abbreviations (U.S., a.m.), spell out a term to be abbreviated the first time it is used, followed by the abbreviation in parentheses. Thereafter, the abbreviation may be used alone.

☑ In a long document, repeat the full term in parentheses after the abbreviation at regular intervals so readers do not have to search back to the first time the acronym or initialism was used to find its meaning.

   • Remember that the CAR (Capital Appropriations Request) controls the corporate spending.

☑ Do not add an additional period at the end of a sentence that ends with an abbreviation.

   • The official name of the company is DataBase, Inc.

☑ For abbreviations specific to your profession or discipline, use a style guide provided by your professional organization or company. (A listing of style guides appears at the end of **documenting sources**.)

☑ Do not make up your own abbreviations; they will confuse your readers.

☑ Write acronyms in capital letters without periods. The only exceptions are acronyms that have become accepted as common nouns, which are written in lowercase letters, such as *laser* (*l*ight *a*mplification by *s*timulated *e*mission of *r*adiation).

☑ Generally, use periods for lowercase initialisms (a.k.a., e.d.p., p.m.) but not for uppercase ones (GDP, IRA, UFO). Two exceptions are geographic names (U.S., U.K., E.U.) and academic degrees (B.A., M.B.A., Ph.D.).

☑ Form the plural of an acronym or initialism by adding a lowercase *s*. Do not use an apostrophe (CARs, CRTs).

Forming Abbreviations

Measurements.   The following list contains some common abbreviations that are used with units of measurement. Notice that, except for abbreviations that may be confused with words (in. for inch), abbreviations of measurement do not require periods.

amp, ampere	in., inch
atm, atmosphere	kc, kilocycle
Btu, British Thermal Unit	km, kilometer
cal, calorie	lb, pound
cd, candela	lm, lumen
cm, centimeter	min, minute
cos, cosine	oz, ounce
doz or dz, dozen	ppm, parts per million
emf or EMF, electromotive force	qt, quart
F, Fahrenheit	rad, radian
ft, foot (or feet)	rev, revolution
gal., gallon	sec, second or secant
hp, horsepower	tan, tangent
hr, hour	yd, yard
Hz, hertz	yr, year

Abbreviations of units of measure are identical in the singular and plural: 1 *cm* and 15 *cm* (*not* 15 *cms*).

Personal Names and Titles.   Personal names generally should not be abbreviated: Thomas (*not* Thos.) and William (*not* Wm.). An academic, civil, religious, or military title should be spelled out and in lowercase when it does not precede a name.

- The *captain* wanted to check the orders.

When they precede names, some titles are customarily abbreviated (Dr. Smith, Mr. Mills, Ms. Katz). See also <u>Ms./Miss/Mrs.</u>

An abbreviation of a title may follow the name; however, be certain that it does not duplicate a title before the name (Angeline Martinez, Ph.D. *or* Dr. Angeline Martinez). When addressing <u>correspondence</u> and including names in other documents, you normally should spell out titles (The Honorable Mary J. Holt, Professor Charles Matlin). The following is a list of common abbreviations for personal and professional titles.

Atty.	Attorney
B.S.E.E.	Bachelor of Science in Electrical Engineering
Dr.	Doctor [used for anyone with a doctorate]
Drs.	Plural of Dr.
Ed.D.	Doctor of Education
Hon.	Honorable [used with various political and judicial titles]

Jr.	Junior [used when a father with the same name is living]
M.A.	Master of Arts
M.B.A.	Master of Business Administration
M.D.	Doctor of Medicine
Messrs.	Plural of Mr.
Mr.	Mister [spelled out only in the most formal contexts]
Mrs.	Married woman
Ms.	Female equivalent of Mr.
M.S.	Master of Science
Ph.D.	Doctor of Philosophy [for many disciplines]
Rev.	Reverend
Sr.	Senior [used when a son with the same name is living]

Names of Organizations.   A company may include in its name a term such as *Brothers, Incorporated, Corporation,* or *Company.* If the term is abbreviated in the official company name that appears on letterhead stationery or on its Web site, use the abbreviated form: *Bros., Inc., Corp.,* or *Co.* If the term is not abbreviated in the official name, spell it out in writing, except with addresses, footnotes, bibliographies, and <u>lists</u> where abbreviations may be used. A similar guideline applies for use of an <u>**ampersand**</u> (&); that symbol should be used only if it appears in the official company name. For names of divisions within organizations, terms such as *Department* and *Division* should be abbreviated (*Dept.* and *Div.*) only when space is limited.

Dates and Times.   The following are common abbreviations used with dates and times.

A.D.	*anno Domini,* "in the year of the Lord" [placed before the year: A.D. 1790]	Jan.	January
		Feb.	February
B.C.	before Christ [placed after the year: 647 B.C.]	Mar.	March
		Apr.	April
B.C.E.	before the common era [corresponds to B.C.]	Aug.	August
		Sept.	September
C.E.	common era [corresponds to A.D. but placed after the year]	Oct.	October
		Nov.	November
a.m.	*ante meridiem* ["before noon"]	Dec.	December
p.m.	*post meridiem* ["after noon"]		

Months should be spelled out when only the month and the year are given. Standard abbreviations for dates and times appear in most dictionaries, either in regular alphabetical order (by the letters of the abbreviation) or in a separate index.

Common Scholarly Abbreviations.   The following is a partial list of abbreviations commonly used in reference books and for documenting sources in research papers and reports. Other than in formal scholarly work, generally avoid such abbreviations.

anon.	anonymous
assn.	association
bibliog.	<u>bibliography</u>, bibliographer, bibliographic, bibliographical
c., ca.	*circa,* "about" [used with approximate dates: c. 1756]
cf.	*confer,* "compare"
ch., chs.	chapter, chapters
cit.	citation, cited
diss.	dissertation
ed., eds.	edited by, editor(s), edition(s)
e.g.	*exempli gratia,* "for example" (see <u>e.g./i.e.</u>)
enl.	enlarged [as in "rev. and enl. ed."]
esp.	especially
et al.	*et alii,* "and others"
etc.	*et cetera,* "and so forth" (see <u>etc.</u>)
f., ff.	and the following page(s) or line(s)
fwd.	foreword, foreword by
GPO	Government Printing Office, Washington, D.C.
i.e.	*id est,* "that is" (see <u>e.g./i.e.</u>)
l., ll.	line, lines
ms, mss	manuscript, manuscripts
n., nn.	note, notes [used immediately after page number: 56n., 56n.3, 56nn.3–5]
N.B.	*nota bene,* "take notice, mark well"
n.d.	no date (of publication)
n.p.	no place (of publication); no publisher
n. pag.	no pagination
p., pp.	page, pages
pref.	preface, preface by
proc.	proceedings
pseud.	pseudonym
pub (publ.)	published by, publisher, publication
rev.	revised by, revised, revision; review, reviewed by [spell out "review" where "rev." might be ambiguous]
rpt.	reprinted by, reprint
sec., secs.	section, sections
supp.	supplement
trans.	translated by, translator, translation
UP	University Press [used in MLA style of <u>documenting sources</u>, as in Oxford UP]

viz.	*videlicet,* "namely"
vol., vols.	volume, volumes
vs., v.	*versus,* "against" [v. preferred in titles of legal cases]

> **WEB LINK POSTAL ABBREVIATIONS**
>
> The U.S. Postal Service Web site specifies abbreviations for states and pro-tectorates as well as streets and other geographical names. See <*www .bedfordstmartins.com/alred*> and select *Links for Technical Writing.*

above

Avoid using *above* to refer to a preceding passage or <u>visual</u>. Its reference is often vague. The same is true of *aforesaid, aforementioned, the former,* and *the latter.* To refer to something previously mentioned, repeat the <u>noun</u> or <u>pronoun</u>, or construct your <u>paragraph</u> so that your reference is obvious.

- Please fill out and submit ~~the above~~ *your travel voucher* by March 1.

Using such references can also be an <u>affectation</u>. See also <u>former/latter</u>.

absolute words

Absolute words (such as *round, unique, exact,* and *perfect*) are not logically subject to comparison (round*er,* round*est*), especially in technical writing, where accuracy and precision are often crucial. See also <u>adjectives</u>.

- We modified our manual to more ~~exactly~~ *closely* reflect existing specifi-cations.

absolutely

Absolutely means "definitely," "entirely," "completely," or "unquestion-ably." It should not be used as an <u>intensifier</u> to mean "very" or "much."

- Changing the design now is ~~absolutely~~ impossible.

abstract / concrete words

Abstract words refer to general ideas, qualities, conditions, acts, or relationships — intangible things that cannot be detected by the five senses, such as *learning, courage,* and *technology. Concrete words* identify things that can be perceived by the five senses (sight, hearing, touch, taste, and smell), such as *diploma, soldier,* and *keyboard.*

Abstract words must frequently be further defined or explicated.

- The research team needs freedom. *to explore the problem further*

Abstract and concrete words are usually best used together, in support of each other.

- A well-designed desk *chair* [concrete] helps reduce *stress* [abstract].

A word that is abstract in one context can be more concrete in another.

- The gauge measures tensile *stress* in pounds per square inch.

Just how concrete a particular context might require you to be is shown in Figure A–1, which goes from the most abstract, on the left, to the most concrete, on the right.

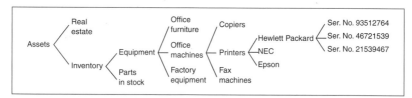

FIGURE A–1. **Abstract-to-Concrete Words**

The example in Figure A–1 represents seven levels of abstraction; the appropriate level depends on your **purpose** in writing and on the context in which you are using the word. See also **word choice**.

abstracts

An *abstract* summarizes and highlights the major points of a **formal report**, **trade journal article**, dissertation, or other work. Its primary purpose is to enable readers to decide whether to read the work in full.

For a discussion of how summaries differ from abstracts, see <u>executive summaries</u>.

Although abstracts, typically 200 to 250 words long, are published with the longer works they condense, they can also be published separately in periodical indexes and by abstracting services (see <u>library research</u>). For this reason, an abstract must be able to stand on its own.

Depending on the kind of information they contain, abstracts are often classified as *descriptive* or *informative*. A *descriptive abstract* includes information about the purpose, scope, and methods used to arrive at the reported findings. It is a slightly expanded table of contents in sentence and paragraph form. A descriptive abstract need not be longer than several sentences. An *informative abstract* is an expanded version of the descriptive abstract. In addition to information about the purpose, scope, and research methods used, the informative abstract includes the results, <u>conclusions</u>, and any recommendations. The informative abstract retains the <u>tone</u> and essential scope of the report, omitting its details. The first two paragraphs of the abstract shown in Figure A–2 (page 10) alone would be descriptive; with the addition of the paragraphs that detail the conclusions of the report, the abstract becomes informative.

The type of abstract you should write depends on your <u>readers</u> and the organization or publication for which you are writing. Informative abstracts work best for wide audiences that need to know conclusions and recommendations; descriptive abstracts work best for compilations, such as proceedings and progress reports, that do not contain conclusions or recommendations.

Writing Style

Write the abstract *after* finishing the report or document. Otherwise, the abstract may not accurately reflect the longer work. Begin with a topic sentence that announces the subject and scope of your report or document. Then, using the major and minor headings of your outline or table of contents to distinguish primary ideas from secondary ones, decide what material is relevant to your abstract. Write with <u>clarity</u> and <u>conciseness</u>, eliminating unnecessary words and ideas. Do not, however, become so terse that you omit articles (*a, an, the*) and important transitional words and phrases (*however, therefore, but, in summary*). Write complete sentences, but avoid stringing together a group of short sentences; instead, combine ideas by using <u>subordination</u> and <u>parallel structure</u>. Spell out most acronyms and all but the most common <u>abbreviations</u>. Typically, an abstract follows the title page and is numbered page iii.

ABSTRACT

Purpose

This report investigates the long-term effects of long-distance running on the bones, joints, and general health of runners aged 50 to 72. The Sports Medicine Institute of Columbia Hospital sponsored this investigation, first to decide whether to add a geriatric unit to the Institute, and second to determine whether physicians should recommend long-distance running for their older patients.

Methods and scope

The investigation is based on recent studies conducted at Stanford University and the University of Florida. The Stanford study tested and compared male and female long-distance runners between 50 and 72 years of age with a control group of runners and nonrunners. The groups were also matched by sex, race, education, and occupation. The Florida study used only male runners who had run at least 20 miles a week for five years and compared them with a group of runners and nonrunners. Both studies based findings on medical histories and on physical and X-ray examinations.

Findings and conclusions

Both studies conclude that long-distance running is not associated with increased degenerative joint disease. Control groups were more prone to spur formation, sclerosis, and joint-space narrowing and showed more joint degeneration than runners. Female long-distance runners exhibited somewhat more sclerosis in knee joints and the lumbar spine area than matched control subjects. Both studies support the role of exercise in retarding bone loss with aging. The investigation concludes that the health risk factors are fewer for long-distance runners than for those less active between the ages of 50 and 72.

Recommendations

The investigation recommends that the Sports Medicine Institute of Columbia Hospital consider the development of a geriatric unit a priority and that it inform physicians that an exercise program that includes long-distance running can be beneficial to their patients' health.

iii

FIGURE A–2. Informative Abstract

accept / except

Accept is a <u>verb</u> meaning "consent to," "agree to take," or "admit willingly." (I *accept* the responsibility.) *Except* is normally used as a preposition meaning "other than" or "excluding." (We agreed on everything *except* the schedule.)

acceptance / refusal letters (for employment)

When you decide to accept a job offer, you can notify your new employer by telephone or in a meeting—but to make your decision official, you need to send an acceptance letter to your new employer. For general advice on letter writing, see **correspondence**.

Figure A–3 shows an example of an acceptance letter written by a college student. Of course, the details you include in your own letter will vary depending on your previous conversations with your new employer. Note that in the first paragraph of Figure A–3, the student identifies the job he is accepting and the salary he has been offered—doing so can avoid any misunderstandings about the job or the salary. In the second paragraph, the student details his plans for moving and reporting for work. Even if the student discussed these arrangements during

2647 Sitwell Road
Charlotte, NC 28210
March 26, 2003

Mr. F. E. Vallone
Manager of Human Resources
Calcutex Industries, Inc.
3275 Commercial Park Drive
Raleigh, NC 27609

Dear Mr. Vallone:

I am pleased to accept your offer of $30,500 per year as an ACR designer in the Calcutex Group.

After graduation, I plan to leave Charlotte on Tuesday, June 17. I should be able to find suitable living accommodations within a few days and be ready to report for work on the following Monday, June 23. Please let me know if this date is satisfactory to you.

I look forward to working with the design team at Calcutex.

Sincerely,

Philip Ming

Philip Ming

FIGURE A–3. Acceptance Letter

earlier conversations, he needs to confirm them, officially, in this letter. The student then concludes with a brief but enthusiastic statement that he looks forward to working for the new employer.

When you decide to reject a job offer, send a job refusal letter to make that decision official, even if you have already notified the employer during a meeting or on the phone. Writing a letter is a gesture that the employer will appreciate. Be especially tactful and courteous — the employer you are refusing has spent time and effort interviewing you and may have counted on your accepting the job. Remember, you may apply for another job at that company in the future. In Figure A–4,

---

2647 Sitwell Road
Charlotte, NC 28210
March 26, 2003

Mr. F. E. Vallone
Manager of Human Resources
Information Systems, Inc.
3275 Commercial Park Drive
Raleigh, NC 27609

Dear Mr. Vallone:

I enjoyed talking with you about your opening for a technical writer, and I was gratified to receive your offer. Although I have given the offer serious thought, I have decided to accept a position as a copywriter with an advertising agency. I feel that the job I have chosen is better suited to my skills and long-term goals.

I appreciate your consideration and the time you spent with me. I wish you the best of luck in filling the position.

Sincerely,

Philip Ming

Philip Ming

FIGURE A–4. Refusal Letter

an example of a job refusal letter, the applicant mentions something positive about his contact with the employer and refers to the specific job offered. He indicates his serious consideration of the offer, provides a logical reason for the refusal, and concludes on a pleasant note. For further strategy on handling refusals and negative messages generally, see refusal letters.

accumulative / cumulative

Accumulative and *cumulative* are synonyms that mean "massed" or "added up over a period of time." *Accumulative* is rarely used except in reference to accumulated property or wealth.

- The Radiology Department warned that *cumulative* doses of X-rays must not exceed the hospital's guidelines.

- The corporation's *accumulative* holdings include 19 real-estate properties.

accuracy / precision

Accuracy means error free and correct. *Precision,* mathematically, refers to the "degree of refinement with which a measurement is made." Thus, a measurement carried to five decimal places (5.28371) is more precise than one carried to two decimal places (5.28), but it is not necessarily more accurate.

acknowledgment letters

One way to build goodwill with colleagues and clients is to let them know that something they sent arrived and to express thanks. A letter that serves this function is called an *acknowledgment letter.* It is usually a short, polite note. If you have established a working relationship with someone, letting that person know by e-mail is appropriate. (See also correspondence.) The example shown in Figure A–5 on page 14 is typical and could be sent as a letter or an e-mail.

Dear Mr. Evans:

I received your report today; it appears to be complete and well done.

When I finish studying it in detail, I'll send you our cost estimate for the installation of the Mark II Energy Saving System.

Again, thanks for your effort.

Regards,

Roger Vonblatz

FIGURE A–5. Acknowledgment Letter or E-mail

acronyms and initialisms

An *acronym* is formed by combining the first letter or letters of several words. Acronyms are pronounced as words and are written without periods (ROM/*r*ead-*o*nly *m*emory). An *initialism* is formed by combining the initial letter of each word in a multiword term. Initialisms are pronounced as separate letters (STC/*S*ociety for *T*echnical *C*ommunication). See the entry **abbreviations** for the appropriate use of acronyms and initialisms.

activate / actuate

Both *activate* and *actuate* mean "make active," although *actuate* is usually applied only to mechanical processes.

* The relay *actuates* the trip hammer. [mechanical process]
* The electrolyte *activates* the battery. [chemical process]
* The governor *activated* the National Guard. [legal process]

active voice (*see* voice)

ad hoc

Ad hoc is Latin for "for this" or "for this particular occasion." An *ad hoc* committee is one set up to consider a particular issue, as opposed to a permanent committee. The term has been fully assimilated into English and thus does not have to be italicized (or underlined). See also <u>affectation</u> and <u>foreign words in English</u>.

adapt / adept / adopt

Adapt is a verb meaning "adjust to a new situation." *Adept* is an adjective meaning "highly skilled." *Adopt* is a verb meaning "take or use as one's own."

- The company will *adopt* a policy of finding engineers who are *adept* managers and who can *adapt* to new situations.

adjectives

DIRECTORY

An *adjective* is any word that modifies a <u>noun</u> or <u>pronoun</u>. An adjective makes the meaning of a noun or a pronoun more specific by highlighting one of its qualities (descriptive adjective) or by imposing boundaries on it (limiting adjective).

- a *hot* surface [descriptive]
- *his three* phone lines [limiting]

Limiting Adjectives

Limiting adjectives include these categories:

Articles (*a, an, the*)
Demonstrative adjectives (*this, that, these, those*)
Possessive adjectives (*my, your, his, her, its, our, their*)
Numeral adjectives (*two, first*)
Indefinite adjectives (*all, none, some, any*)

Articles.   Articles (*a, an, the*) are traditionally classified as adjectives because they modify nouns by either limiting them or making them more precise. See <u>a/an</u> and <u>articles</u>.

Demonstrative Adjectives.   A demonstrative adjective points to the thing it modifies, specifying the object's position in space or time. *This* and *these* specify a closer position; *that* and *those* specify a more remote position.

- *This* report received today is more current than *that* report received earlier.

- *These* samples tested today were easier to process than *those* samples tested last week.

Demonstrative adjectives often cause problems when they modify the nouns *kind, type,* and *sort.* Demonstrative adjectives used with those nouns should agree with them in <u>number</u>.

- *this* kind / *these* kinds; *that* type / *those* types

Confusion often develops when the <u>preposition</u> *of* is added (*this kind of, these kinds of*) and the object of the preposition does not conform in number to the demonstrative adjective and its noun.

- *This kind of* hydraulic ~~cranes~~ crane is best.

- *These kinds of* hydraulic ~~crane is~~ cranes are best.

Using demonstrative adjectives with words like *kind, type,* and *sort* can also easily lead to vagueness. It is better to be more specific. See also <u>kind of/sort of</u>.

Possessive Adjectives.   Because possessive adjectives (*my, your, his, her, its, our, their*) directly modify nouns, they function as adjectives, even though they are pronoun forms. (*My* ideas conflicted with *her* plans for the project.)

Numeral Adjectives.   Numeral adjectives identify quantity, degree, or place in a sequence. They always modify count nouns. Numeral adjectives are divided into two subclasses: cardinal and ordinal. A *cardinal adjective* expresses an exact quantity (*one* pencil, *two* computers); an *ordinal adjective* expresses degree or sequence (*first* quarter, *second* edition).

   In most writing, an ordinal adjective should be spelled out if it is a single word (*tenth*) and written in figures if it is more than one word (*312th*). Ordinal numbers can also function as <u>adverbs</u> ("John arrived *first*"). See also <u>first/firstly</u> and <u>numbers</u>.

Indefinite Adjectives.   Indefinite adjectives do not designate anything specific about the nouns they modify (*some* CD-ROMs, *all* designers). The articles *a* and *an* are included among the indefinite adjectives (*a* chair, *an* application).

Comparison of Adjectives

Most adjectives in the positive form show the comparative form with the <u>suffix</u> *-er* for two items and the superlative form with the suffix *-est* for three or more items.

- The first report is *long*. [positive form]
- The second report is *longer*. [comparative form]
- The third report is *longest*. [superlative form]

Many two-syllable adjectives and most three-syllable adjectives are preceded by the word *more* or *most* to form the comparative or the superlative.

- The new library is *more* impressive than the old one. It is the *most* impressive in the county.

A few adjectives have irregular forms of comparison (*much, more, most; little, less, least*).

Some adjectives (*round, unique, exact, accurate*) are not logically subject to comparison. See also **absolute words**.

Placement of Adjectives

When limiting and descriptive adjectives appear together, the limiting adjectives precede the descriptive adjectives, with the articles usually in the first position.

- *The ten red* cars were parked in a row.
  [article (*The*), limiting adjective (*ten*), descriptive adjective (*red*)]

Within a sentence, adjectives may appear before the nouns they modify (the attributive position) or after the nouns they modify (the predicative position).

- *The small* jobs are given priority. [attributive position]
- The exposure is *brief*. [predicative position]

Use of Adjectives

Because of the need for precise description, it is often necessary to use nouns as adjectives.

- The *focus group's* conclusions resulted in the redesign of the product.

When adjectives modifying the same noun can be reversed and still make sense or when they can be separated by *and* or *or*, they should be separated by commas.

- The company is seeking a *young, energetic, creative* management team.

ESL TIPS FOR USING ADJECTIVES

Unlike in many other languages, adjectives in English have only one form. Do not add -s or -es to an adjective to make it plural.

- the *long* trip
- the *long* trips

Capitalize adjectives of origin (city, state, nation, continent).

- the *Venetian* canals
- the *Texan* hat
- the *French* government
- the *African* deserts

  In English, verbs of feeling (for example, *bore, interest, surprise*) have two adjectival forms: the present participle (-*ing*) and the past participle (-*ed*). Use the present participle to describe what causes the feeling. Use the past participle to describe the person who experiences the feeling.

- We heard the *surprising* election results.
  [The *election results* cause the feeling.]

- Only the losing candidate was *surprised* by the election results.
  [The *candidate* experienced the feeling of surprise.]

  Adjectives follow nouns in English in only two cases: when the adjective functions as a subjective complement

- That project is not *finished*.

and when an adjective phrase or clause modifies the noun

- The project *that was suspended temporarily* . . .

In all other cases, adjectives are placed before the noun.

  When there are multiple adjectives, it is often difficult to know the right order. The guidelines illustrated in the following example would apply in most circumstances, but there are exceptions. (Normally do not use a phrase with so many stacked <u>modifiers</u>.) See also <u>articles</u>.

The six extra large rectangular brown cardboard take-out containers

| determiner | comment | number | size | shape | color | material | qualifier | noun |

Notice in the last example there is no comma after *creative*. Never use a comma between a final adjective and the noun it modifies. When an adjective modifies a phrase, no comma is required.

- We need an *accessible Web page design*.
  [*Accessible* modifies the phrase *Web page design*.]

Writers sometimes string together a series of nouns used as adjectives to form a unit modifier, thereby creating jammed <u>modifiers</u>.

In general, avoid vague (*nice, fine, good*) and trite (a *fond* farewell) adjectives; in fact, it is good practice to question the need for most adjectives in technical writing. If you must use an adjective, try to find one that is precise. See also <u>trite language</u>, <u>vague words</u>, and <u>word choice</u>.

adjustment letters

An adjustment letter is written in response to a complaint and tells the customer what your company intends to do about the complaint (see also <u>complaint letters</u>). You should settle such matters quickly and courteously, and always try to satisfy the customer at a reasonable cost to your company.

Although sent in response to a problem, an adjustment letter actually provides an excellent opportunity to build goodwill for your company. An effective adjustment letter both repairs the damage that has been done and restores the customer's confidence in your company.

Grant adjustments graciously; a settlement made grudgingly will do more harm than good. The <u>tone</u> of your <u>correspondence</u> is crucial. No matter how unreasonable the complaint, your response and tone should be positive and respectful. Avoid emphasizing the unfortunate situation, but do take responsibility for it and focus on what you are doing to correct it. Not only must you be gracious, but you must also acknowledge the error in such a way that the customer will not lose confidence in your company. The adjustment letter in Figure A–6 on page 20, for example, begins by accepting responsibility and offers an apology for the customer's inconvenience (note the use of the pronouns *we* and *us*). The second paragraph expresses a desire to restore goodwill

---

WEB LINK **WRITING AND FORMATTING DOCUMENTS**

For annotated examples of correspondence, reports, and other types of technical writing, see <*www.bedfordstmartins.com/alred*> and select *Model Documents Gallery.*

International Hotels

EXECUTIVE OFFICE

September 26, 2003

Ms. Elizabeth Shapiro
2374 N. Kenwood Ave.
Fresno, CA 93650

Dear Ms. Shapiro:

We are sorry that your and your husband's stay with us did not go smoothly. Providing dependable service is what's expected of us — and when our staff doesn't provide high-quality service, it's easy to understand our guests' disappointment. I truly wish we had performed better and that your vacation plans had not been disrupted.

We are eager to restore your confidence in our ability to provide dependable, high-quality service. Please accept the enclosed certificate for one weekend's stay at any of our 500 hotels worldwide. I hope we will have the pleasure of welcoming you and your husband again soon.

Ms. Shapiro, in addition, we appreciate your taking the time to write. It helps to receive comments such as yours, and we conscientiously follow through to be sure proper procedures are met. I assure you your letter is being put to good use.

Yours truly,

Ms. M. J. Matthews

Ms. M. J. Matthews
Executive Office

Enclosure: Certificate

10113 Executive Drive/Chicago, Illinois 60601
800-964-9400   http://www.interhotel.com

FIGURE A–6. Adjustment Letter (When Company Is at Fault)

and offers compensation. The third paragraph thanks the customer for her letter and assures her that her complaint has been taken seriously.

Writer's Checklist: Writing Adjustment Letters

☑ Open with what the reader will consider good news.

- Grant the adjustment, if appropriate, for uncomplicated situations ("Enclosed is a replacement for the damaged part").

- Reveal that you intend to grant the adjustment by admitting that the customer was right ("Yes, you were incorrectly billed for the delivery"). Then explain the specific details of the adjustment. This method is good for adjustments that require detailed explanations.

- Apologize for the error ("Please accept our apologies for the error in your account"). This method is effective when the customer's inconvenience is as much an issue as money.

- Use a combination of these techniques. Often, situations that require an adjustment also require flexibility.

☑ If an explanation will help restore your reader's confidence or good-will, explain what caused the problem.

☑ Explain specifically how you intend to make the adjustment, if it is not obvious in your opening.

☑ Express appreciation to the customer for calling your attention to the problem. Explain that customer feedback helps your firm keep the quality of its product or service high. Also, point out any steps you may be taking to prevent a recurrence of the problem.

☑ Close pleasantly, looking forward, not back. Avoid recalling the problem in your closing ("Again, we apologize . . .").

Before granting an adjustment to a claim for which your company is at fault, you must determine what happened and what you can do to satisfy the customer. Be certain that you are familiar with your company's adjustment policy. In addition, be careful about your wording; for example, "We have just received your letter of May 7 about our *defective product*" could be ruled in a court of law as an admission that the product is in fact defective. Treat every claim individually, and lean toward giving the customer the benefit of the doubt.

You may sometimes need to grant a partial adjustment—even if a claim is not really justified—to regain the lost goodwill of a customer. In some cases, in which a problem resulted from a customer's incorrect use of a product, for example, you may need to educate your reader about the use of your product or service (Figure A-7). When writing a letter of adjustment in this situation, remember that your customer believes that

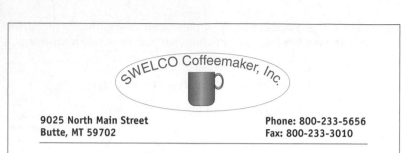

9025 North Main Street
Butte, MT 59702

Phone: 800-233-5656
Fax: 800-233-3010

August 26, 2003

Mr. Carlos Ortiz
638 McSwaney Drive
Butte, MT 59702

Dear Mr. Ortiz:

Enclosed is your SWELCO Coffeemaker, which you shipped to us on
August 18.

In various parts of the country, tap water may contain a high mineral content. If you fill your SWELCO Coffeemaker with water for breakfast coffee
before going to bed, a mineral scale will build up on the inner wall of the
water tube — as explained on page 2 of your SWELCO Instruction Booklet.

We have removed the mineral scale from the water tube of your coffeemaker and thoroughly cleaned the entire unit. To ensure the best service
from your coffeemaker in the future, clean it once a month by operating it
with four ounces of white vinegar and eight cups of water. To rinse out the
vinegar taste, operate the unit twice with clear water.

With proper care, your SWELCO Coffeemaker will serve you well for many
years.

Sincerely,

Helen Upham

Helen Upham
Customer Service

HU/mo
Enclosure

Email: service@swelco.com   Web: http://www.swelco.com

FIGURE A–7. Educational Adjustment Letter (Accompanying a Product)

his or her claim is justified. Therefore, it is wise to give the explanation before granting the claim—otherwise, your reader may never get to the explanation. If your explanation establishes customer responsibility, be sure to do so tactfully. Figure A–7 on page 22 is an example of an educational adjustment letter.

adverbs

An adverb modifies the action or condition expressed by a <u>verb</u>.

- The wrecking ball hit the side of the building *hard*.
  [The adverb tells *how* the wrecking ball hit the building.]

An adverb also can modify an **adjective**, another adverb, or a **clause**.

- The brochure design used *extremely* bright colors.
  [*Extremely* modifies the adjective *bright.*]

- The redesigned brake pad lasted *much* longer.
  [*Much* modifies the adverb *longer.*]

- *Surprisingly,* the machine failed.
  [*Surprisingly* modifies the clause *the machine failed.*]

An adverb answers one of the following questions:

Where? (adverb of place)

- Move the throttle *forward* slightly.

When? (adverb of time)

- Replace the thermostat *immediately.*

How? (adverb of manner)

- Add the solvent *cautiously.*

How much? (adverb of degree)

- The *nearly* completed report was deleted from his disk.

Types of Adverbs

An adverb may be a common, a conjunctive, an interrogative, or a numeric **modifier**. Typical common adverbs are *almost, seldom, down, also, now, ever,* and *always.*

- I *rarely* work on weekends.

A *conjunctive adverb* modifies the clause that it introduces; it operates as a conjunction because it joins two independent clauses. The most common conjunctive adverbs are *however, nevertheless, moreover, therefore, further, then, consequently, besides, accordingly, also,* and *thus.*

- I rarely work on weekends; *however,* this weekend will be an exception.

In this example, note that a semicolon precedes and a comma follows *however.* The conjunctive adverb (*however*) introduces the independent clause (*this weekend will be an exception*) and indicates its relationship to the preceding independent clause (*I rarely work on weekends*). See also **semicolons**.

Interrogative adverbs ask questions. Common interrogative adverbs are *where, when, why,* and *how.*

- *How* many hours did you work last week?

Numeric adverbs tell how often. Typical numeric adverbs are *once* and *twice.*

- I have worked overtime *twice* this week.

Comparison of Adverbs

With most one-syllable adverbs, the suffix *-er* is added to show comparison with one other item, and the suffix *-est* is added to show comparison with two or more items.

- This copier is *fast.* [positive form]
- This copier is *faster* than the old one. [comparative form]
- This copier is the *fastest* of the three tested. [superlative form]

Most adverbs with two or more syllables end in *-ly,* and most adverbs ending in *-ly* are compared by inserting the comparative *more* or *less* or the superlative *most* or *least* in front of them.

- She detected the problem *more quickly* than any other technician.
- *Most surprisingly,* the engine failed during the final test phase.

A few irregular adverbs require a change in form to indicate comparison (*well, better, best; badly, worse, worst; far, farther, farthest*).

- The training program functions *well.*
- Our training program functions *better* than most others in the industry.
- Many consider our training program the *best* in the industry.

Placement of Adverbs

An adverb usually should be placed in front of the verb it modifies.

- The pilot *meticulously* performed the preflight check.

An adverb may, however, follow the verb (or the verb and its object) that it modifies.

- The gauge dipped *suddenly.*
- They replaced the hard drive *quickly.*

An adverb may be placed between a helping verb and a main verb.

- In this temperature range, the pressure will *quickly* drop.

Adverbs such as *only, nearly, almost, just,* and *hardly* should be placed immediately before the words they limit. See also **only** and **modifiers**.

affect / effect

Affect is a **verb** that means "influence."

- The public utility commission's decisions *affect* all state utilities.

Effect can function as a **noun** that means "result."

- The change had a good *effect.*

Effect can also function as a verb that means "bring about" or "cause." It is best, however, to avoid using *effect* as a verb. A less formal word, such as *make,* is usually preferable.

- The new manager will ~~effect~~ several changes that should improve morale.

 make
 ^

affectation

Affectation is the use of language that is more formal, technical, or showy than necessary to communicate information to the reader. Affectation is a widespread writing problem in the workplace because many people feel that affectation lends a degree of authority to their writing. In fact, nothing could be further from the truth. Affectation can alienate customers, clients, and colleagues by placing a barrier between them and the writer.

Affected writing forces readers to work harder to understand the writer's meaning. It typically contains abstract, highly technical, or foreign words and is often liberally sprinkled with trendy words or <u>buzzwords</u>. <u>Jargon</u> and <u>euphemisms</u> can become affectation, especially if their purpose is to hide relevant facts or give a false impression of competence. See also <u>ethics in writing</u>.

Writers are easily lured into affectation through the use of long variants: words created by adding prefixes and suffixes to simpler words (*analyzation* for *analysis*). Unnecessarily formal words (such as *utilization* for *use*) and outdated words (such as *herewith*) can produce affectation. Elegant variation—attempting to avoid repeating a word within a paragraph by substituting a pretentious synonym—is also a form of affectation.

- The use of robotics in the assembly process has increased production . ~~Robotic utilization~~ has cut costs.

  , and it

Another type of affectation is <u>gobbledygook,</u> which is wordy, roundabout writing with many pseudo legal and pseudoscientific terms. See also <u>conciseness/wordiness</u>, <u>clichés</u>, and <u>nominalizations</u>.

Understanding the possible reasons for affectation is the first step toward avoiding it. The following are some causes of affectation.

- *Impression.* Some writers use pretentious language in an attempt to impress the reader with fancy words instead of evidence and logic.

- *Insecurity.* Writers who are insecure about their facts, conclusions, or arguments may try to hide behind a smokescreen of pretentious words.

- *Imitation.* Perhaps unconsciously, some writers imitate poor writing they see around them.

- *Intimidation.* A few writers, consciously or unconsciously, try to intimidate or overwhelm their readers with words, often to protect themselves from criticism.

- *Initiation.* Writers who have just completed their training often feel that one way to prove their professional expertise is to use as much technical terminology and jargon as possible.

- *Imprecision.* Writers who are having trouble being precise sometimes find that an easy solution is to use a vague, trendy, or pretentious word.

Figures A–8 and A–9 illustrate how eliminating affectation leads to
clarity.

In addition to performing interior housekeeping services, the concessionaire shall perform custodial maintenance on the exterior of the facility and
grounds. Where a concessionaire shares a facility with one or more other
concessionaires, exterior custodial maintenance responsibilities will be assigned by Post Exchange management on a fair and equitable basis. In
those instances where the concessionaire's activity is located in a Post
Exchange complex wherein predominant tenancy is by Post Exchange–
operated activities, then Post Exchange management shall be responsible
for exterior custodial maintenance except for those described in 1, 2, 3, and
4 below. The necessary equipment and labor to perform exterior custodial
maintenance, when such a responsibility has been assigned to the concessionaire, shall be furnished by the concessionaire. Exterior custodial maintenance shall include the following tasks:

1. Clean entrance door and exterior of storefront windows daily.
2. Sweep and clean the entrance and customer walks daily.
3. Empty and clean waste and smoking receptacles daily.
4. Check exterior lighting and report failures to the contracting officer's
   representative daily.

FIGURE A–8. Affected Writing

The merchant will, with his or her own equipment, perform the following
duties daily:

1. Maintain a clean and neat appearance inside the store.
2. Clean the entrance door and the outsides of the store windows.
3. Sweep the entrance and the sidewalk.
4. Empty and clean wastebaskets and ashtrays.
5. Check the exterior lighting and report failures to the representative of
   the contracting officer.

The merchant will also maintain the grounds surrounding the store. Where
two or more merchants share the same building, Post Exchange management will assign responsibility for the grounds. Where the merchant's
store is in a building that is occupied predominantly by Post Exchange
operations, Post Exchange management will be responsible for the
grounds.

FIGURE A–9. Affected Writing Revised for Clarity

affinity

Affinity refers to the attraction of two persons or things to each other.

- The *affinity* between these two elements can be explained in terms of their valance electrons.

Affinity should not be used to mean "ability" or "aptitude."

- She has an ~~affinity~~ for problem solving.
  aptitude ^

- She has ~~an affinity~~ for problem solving.
  a talent ^

aforesaid (*see* **above**)

agree to / agree with (*see* **idioms**)

agreement

In grammar, *agreement* means the correspondence in form between different elements of a sentence to indicate <u>number</u>, <u>person</u>, <u>gender</u>, and <u>case</u>.

A subject and its <u>verb</u> must agree in number.

- The *design is* acceptable.
  [The singular subject, *design,* requires the singular verb, *is.*]

- The new *products are* going into production soon.
  [The plural subject, *products,* requires the plural verb, *are.*]

A subject and its verb must agree in person.

- *I am* the designer.
  [The first-person singular subject, *I,* requires the first-person singular verb, *am.*]

- *They are* the designers.
  [The third-person plural subject, *they,* requires the third-person plural verb, *are.*]

A <u>pronoun</u> and its antecedent must agree in person, number, gender, and case.

- The *employees* report that *they* are more efficient in the new facility.
  [The third-person plural subject, *employees,* requires the third-person plural pronoun, *they.*]

- *Kaye McGuire* will meet with the staff on Friday, when *she* will assign duties.
  [The feminine pronoun *she* is in the subjective case because it agrees with its antecedent, *Kaye McGuire,* which is the subject of the sentence.]

See also <u>sentence construction</u>.

Subject-Verb Agreement

Subject-verb agreement is not affected by intervening <u>phrases</u> and <u>clauses</u>.

- *One* in twenty hard drives we receive from our suppliers *is* faulty.
  [The verb, *is,* must agree in number with the subject, *one,* not *hard drives* or *suppliers.*]

The same is true for <u>nouns</u> that fall between a subject and its verb.

- Only *one* of the emergency lights *was* functioning.
  [The subject of the verb is *one,* not *lights.*]

- *Each* of the switches *controls* a separate circuit.
  [The subject of the verb is *each,* not *switches.*]

Note that *one* and *each* are normally singular.

Indefinite pronouns such as *some, none, all, more,* and *most* may be singular or plural, depending on whether they are used with a mass noun ("*Most* of the oil *has* been used") or with a count noun ("*Most* of the drivers *know* why they are here"). Mass nouns are singular, and count nouns are plural. Other words, such as *type, part, series,* and *portion,* take singular verbs even when they precede a phrase containing a plural noun.

- A *series* of meetings *was* held about the best way to market the new product.

- A large *portion* of most industrial annual reports *is* devoted to promoting the corporate image.

Modifying phrases can obscure a simple subject.

- The *advice* of two engineers, one lawyer, and three executives *was* obtained prior to making a commitment.
  [The subject of the verb is *advice*.]

Inverted word order can cause problems with agreement.

- From this work *have come* several important *improvements*.
  [The subject of the verb is *improvements,* not *work*.]

The number of a subjective <u>complement</u> does not affect the number of the verb — the verb must always agree with the subject.

- The *topic* of his report *was* employee benefits.
  [The subject of the sentence is *topic*, not *benefits*.]

A subject that expresses measurement, weight, mass, or total often takes a singular verb even when the subject word is plural in form. Such subjects are treated as a unit.

- *Four years is* the normal duration of the training program.

A verb following the relative pronoun *who* or *that* agrees in number with the noun to which the pronoun refers (its antecedent).

- Steel is one of those *industries* that *are* most affected by energy costs.
  [*That* refers to *industries*.]
- She is one of those *employees* who *are* rarely absent.
  [*Who* refers to *employees*.]

The word *number* sometimes causes confusion. When used to mean a specific number, it is singular.

- *The number* of committee members *was* six.

When used to mean an approximate number, it is plural.

- *A number* of people *were* waiting for the announcement.

Relative pronouns (*who, which, that*) may take either singular or plural verbs, depending on whether the antecedent is singular or plural. See also <u>who/whom</u>.

- He is a bookkeeper *who takes* work home at night.
- He is one of those bookkeepers *who take* work home at night.

Some abstract nouns are singular in meaning but plural in form: *mathematics, news, physics,* and *economics*.

- *Mathematics is* essential in computer science.

Some words, such as the plural *jeans* and *scissors,* cause special problems.

- The *scissors were* ordered last week.
  [The subject is the plural *scissors.*]

- *A pair* of scissors *is* on order.
  [The subject is the singular *pair.*]

A book with a plural title requires a singular verb.

- *Medical Techniques* is an essential resource.

A collective noun (*committee, faculty, class, jury*) used as a subject takes a singular verb when the group is thought of as a unit and a plural verb when the individuals in the group are thought of separately.

- The *committee is* unanimous in its decision.

- The *committee are* returning to their offices.

A clearer way to emphasize the individuals would be to use a phrase.

- The *committee members* are returning to their offices.

Compound Subjects

A *compound subject* is one that is composed of two or more elements joined by a <u>conjunction</u> such as *and, or, nor, either . . . or,* or *neither . . . nor.* Usually, when the elements are connected by *and,* the subject is plural and requires a plural verb.

- *Accounting and chemistry are* both prerequisites for this position.

One exception occurs when the elements connected by *and* form a unit or refer to the same thing. In that case, the subject is regarded as singular and takes a singular verb.

- *Bacon and eggs is* a high-cholesterol meal.

- Our greatest *challenge and business opportunity is* the Internet.

A compound subject with a singular element and a plural element joined by *or* or *nor* requires that the verb agree with the closer element.

- Neither the director nor the *project assistants were* there.

- Neither the project assistants nor the *director was* there.

If *each* or *every* modifies the elements of a compound subject, use the singular verb.

- *Each* manager and supervisor *has* a production goal to meet.

- *Every* manager and supervisor *has* a production goal to meet.

Pronoun-Antecedent Agreement

Every pronoun must have an antecedent—a noun to which it refers. See also <u>pronoun reference</u>.

- When *employees* are hired, *they* must review the policy manual. [The pronoun *they* refers to the antecedent *employees*.]

Gender.   A pronoun must agree in gender with its antecedent.

- *Mr. Swivet* in the accounting department acknowledges *his* share of the responsibility for the misunderstanding, just as *Ms. Barkley* in the research division must acknowledge *hers*.

Traditionally, a masculine, singular pronoun was used to agree with such indefinite antecedents as *anyone* and *person.*

- *Each* may stay or go as *he* chooses.

It is now understood that such usage ignores or excludes women. When alternatives are available, use them. One solution is to use the plural. Another is to use both feminine and masculine pronouns, although that combination is clumsy when used too often.

- ~~Every employee~~ *All employees* must sign ~~his~~ *their* time ~~card.~~ *cards*

- Every employee must sign his *or her* time card.

Do not attempt to avoid expressing gender by resorting to a plural pronoun when the antecedent is singular. You may be able to avoid the pronoun entirely.

- Every employee must sign ~~their~~ *a* time card.

Be careful to avoid gender-related stereotypes in general references, as in "the nurse . . . *she*" or "the doctor . . . *he.*" What if the nurse is male or the doctor is female? See also <u>biased language</u>.

Number.   A pronoun must agree with its antecedent in number. Many problems of agreement are caused by expressions that are not clear in number.

- Although the typical *engine* runs well in moderate temperatures, ~~they~~ *it* often ~~stall~~ *stalls* in extreme cold.

Use singular pronouns with the antecedents *everybody* and *everyone* unless to do so would be illogical because the meaning is obviously plural. See also <u>everybody/everyone</u>.

- *Everyone* pulled *his or her* share of the load.

- *Everyone* thought my plan should be revised, and I really couldn't blame *them*.

Collective nouns may use a singular or plural pronoun, depending on the meaning.

- The *committee* arrived at the recommended solutions only after *it* had deliberated for days. [*committee* thought of as collective]

- The *committee* quit for the day and went to *their* respective homes. [*committee* thought of as plural]

Demonstrative adjectives sometimes cause problems with agreement of number. *This* and *that* are used with singular nouns, and *these* and *those* are used with plural nouns. Demonstrative adjectives often cause problems when they modify the nouns *kind, type,* and *sort.* When used with those nouns, demonstrative adjectives should agree with them in number.

- *this* kind / *these* kinds; *that* type / *those* types

Confusion often develops when the **preposition** *of* is added (*this kind of, these kinds of*) and the object of the preposition does not conform in number to the demonstrative <u>adjective</u> and its <u>noun</u>.

- This kind of retirement ~~plans~~ *plan* is best.

- These kinds of retirement ~~plan~~ *plans* are best.

Avoid that error by remembering to make the demonstrative adjective, the noun, and the object of the preposition — all three — agree in number. The agreement makes the sentence not only correct but also more precise. Keep in mind that using demonstrative adjectives with words like *kind, type,* and *sort* can easily lead to vagueness. See <u>kind of/sort of</u>.

Compound Antecedents.   A compound antecedent joined by *or* or *nor* is singular if both elements are singular and plural if both elements are plural.

- Neither the *engineer* nor the *technician* could do *his* job until *he* understood the new concept.

- Neither the *executives* nor the *directors* were pleased at the performance of *their* company.

When one of the antecedents connected by *or* or *nor* is singular and the other plural, the pronoun agrees with the closer antecedent.

- Either the *computer* or the *printers* should have *their* serial numbers registered.

- Either the *printers* or the *computer* should have *its* serial number registered.

A compound antecedent with its elements joined by *and* requires a plural pronoun.

- *Seon Ju* and *Juanita* took *their* layout drawings with them.

If both elements refer to the same person, however, use the singular pronoun.

- The noted *scientist* and *author* departed from *her* prepared speech.

all right

All right means "all correct," as in "The answers were *all right*." In formal writing, it should not be used to mean "good" or "acceptable." It is always written as two words, with no hyphen; *alright* is nonstandard.

all together / altogether

All together means "all acting together" or "all in one place." (The necessary instruments were *all together* on the tray.) *Altogether* means "entirely" or "completely." (The trip was *altogether* unnecessary.)

allude / elude / refer

Allude means to make an indirect reference to something. (The report simply *alluded* to the problem, rather than stating it explicitly.)
  Elude means to escape notice or detection. (The leak *eluded* the inspectors.) *Refer* is used to indicate a direct reference to something. (She *referred* to the chart three times during her presentation.)

allusions

An *allusion* is an indirect reference to something from past or current events, literature, or other familiar sources. The use of allusion promotes economical writing because it is a shorthand way of referring to a body of material in a few words or of helping to explain a new and unfamiliar process in terms of one that is familiar. In the following example, the writer sums up a description with an allusion to a well-known story. The allusion, with its implicit reference to "right standing up to might," concisely emphasizes the writer's point.

- As it currently exists, the review process involves the consumer's attorney sitting alone, usually without adequate technical assistance, faced by two or three government attorneys, two or three attorneys from CompuSystems, and large teams of experts who support the government and the corporation. The entire proceeding is reminiscent of David versus Goliath.

Be sure, of course, that your reader is familiar with the material to which you allude. Allusions should be used with restraint, especially in international correspondence. If overdone, allusions can lead to affectation or can be viewed merely as cliché. See also technical writing style.

allusion / illusion

An *allusion* is an indirect reference to something not specifically mentioned. (The report made an *allusion* to metal fatigue in the support structures.) An *illusion* is a mistaken perception or a false image. (County officials are under the *illusion* that the landfill will last indefinitely.)

almost / most

Do not use *most* as a colloquial substitute for *almost* in your writing.

- New shipments arrive ~~most~~ *almost* every day.

already / all ready

Already is an <u>adverb</u> that means "before this time" or "previously." *All ready* is a two-word phrase meaning "completely prepared." (They were *all ready* to cancel the order; fortunately, we had *already* corrected the shipments.)

also

Also is an <u>adverb</u> that means "additionally." (Two 5,000-gallon tanks are on-site, and several 2,500-gallon tanks are *also* available.) *Also* should not be used as a connective in the sense of "and."

- He brought the reports, the letters, ~~also~~ *and* the section supervisor's recommendations.

Avoid opening sentences with *also*. It is a weak transitional word that suggests an afterthought rather than planned writing.

- ~~Also,~~ *In addition* he brought statistical data to support his proposal.
- ~~Also, he~~ *He also* brought statistical data to support his proposal.

ambiguity

A word or passage is *ambiguous* when it can be interpreted in two or more ways, yet provides the reader with no certain basis for choosing among the alternatives.

- Mathematics is more valuable to an engineer than a computer. [Does that mean an engineer is more in need of mathematics than a computer is? Or does it mean that mathematics is more valuable to an engineer than a computer is?]

Ambiguity can take many forms, one of which is ambiguous <u>pronoun reference</u>.

AMBIGUOUS	Inadequate quality-control procedures have resulted in more equipment failures. This is our most serious problem at present. [Does *this* refer to *quality-control procedures* or to *equipment failures?*]
SPECIFIC	Inadequate quality-control procedures have resulted in more equipment failures. *These failures* are our most serious problem at present.
SPECIFIC	Inadequate quality-control procedures have resulted in more equipment failures. *Quality control* is our most serious problem at present.

Incomplete <u>comparison</u> and missing or misplaced <u>modifiers</u> (including <u>dangling modifiers</u>) cause ambiguity.

- Ms. Lee values rigid quality-control standards more than Mr.
  does
  Rosenblum ˄ . [Complete the comparison.]

- *also*
  His hobby was cooking. He ˄ was especially fond of cocker spaniels.

  [Add the missing modifier.]

The placement of some modifiers enables them to be interpreted in either of two ways.

- She volunteered *immediately* to deliver the toxic substance.

By moving the word *immediately,* the meaning can be clarified.

- She *immediately* volunteered to deliver the toxic substance.
- She volunteered to deliver *immediately* the toxic substance.

Imprecise <u>word choice</u> (including faulty <u>idioms</u>) can cause ambiguity.

- The general manager has denied reports that the plant's recent
  rescinded
  fuel allocation cut will be ~~restored~~. [inappropriate word choice]
  ˄

Ambiguity also can be caused by various forms of <u>awkwardness</u>.

amount / number

Amount is used with things that are thought of in bulk and that cannot be counted (mass <u>nouns</u>), as in "the *amount* of electricity." *Number* is used with things that can be counted as individual items (count nouns), as in "the *number* of employees."

ampersands

The *ampersand* (&) is a symbol sometimes used to represent the word *and,* especially in the names of organizations ("Kirkwell & Associates"). When you are writing the name of an organization in sentences, addresses, or references, spell out the word *and* unless the ampersand appears in its official name that appears on letterhead stationery or on its Web site. See also <u>documenting sources</u>.

analogies (*see* figures of speech)

and/or

And/or means that either both circumstances are possible or only one of two circumstances is possible. This term is awkward and confusing because it makes the reader stop to puzzle over your distinction.

> **AWKWARD**   Use A *and/or* B.
>
> **IMPROVED**   Use A or B or both.

ante- / anti-

Ante- means "in front of" or "before" (*ante*room: outer room or waiting room; *ante*date: predate or earlier date). *Anti-* means "against" or "opposed to" (*anti*body, *anti*social). See also <u>prefixes</u>.

antonyms

An *antonym* is a word with a meaning opposite that of another word (*good/bad, well/ill, fresh/stale*). Many pairs of words that look as if they are antonyms are not. Be careful not to use words like <u>affect/effect</u> incorrectly. When in doubt, consult a thesaurus or a dictionary. See also <u>synonyms</u>.

apostrophes

An *apostrophe* (') is used to show possession, to indicate the omission of letters, and sometimes to indicate plurals.

Show Possession

An apostrophe is used with an *s* to form the possessive case of some <u>nouns</u> (the *report's* title). For advice on using apostrophes to show possession, see also <u>possessive case</u>.

Indicate Omission

An apostrophe is used to mark the omission of letters or numbers in a <u>contraction</u> or a date (*can't, I'm, I'll;* the class of *'03*).

Form Plurals

The trend for indicating the plural forms of words mentioned as words, of numbers used as nouns, and of abbreviations shown as single or multiple letters is currently to add only *s* rather than *'s*.

When a word (or letter) mentioned as a word is italicized, it is current usage to add *s* in roman type.

- There were five *and*s in his first sentence.

Rather than using italics, you may place a word in quotation marks. If you choose this option, use an apostrophe and *s*.

- There were five "and's" in his first sentence.

To indicate the plural of a number, add *s* (*7s,* the 1990*s*). If the letter and the *s* form a word, you may want to consider using an apostrophe to avoid confusion (*A*'s). Use *s* to pluralize an abbreviation that is in all capital letters or that ends with a capital letter (IOU*s*). However, if the abbreviated term contains periods, some writers use an apostrophe to prevent confusion.

- The university awarded 34 Ph.D.*'s* in information systems last year.

Whatever practice you follow, be consistent.

appendixes

An *appendix,* located at the end of a <u>formal report,</u> a <u>proposal,</u> or another major document, supplements or clarifies the information in the body of the document. Appendixes (or appendices) can provide information that is too detailed or lengthy for the primary audience of the text. For example, an appendix could contain complex <u>graphs</u> and <u>tables,</u> profiles of key personnel involved in a proposed project, or documentation of interest only to secondary <u>readers.</u>

A document may have more than one appendix, with each offering only one type of information. When the document contains more than one appendix, arrange them in the order they are mentioned in the text. Begin each appendix on a new page, and identify each with a letter, starting with the letter *A* ("Appendix A: Sample Questionnaire"). If you have only one appendix, title it simply "Appendix." List the titles and beginning page numbers of the appendixes in the <u>table of contents.</u>

application letters

When applying for a job, you usually need to submit both a <u>résumé</u> and an application letter (also referred to as a cover letter). Unless the prospective employer requests a résumé only, be sure to submit an application letter as well. See also <u>job search.</u>

The application letter is essentially a sales letter in which you market your skills, abilities, and knowledge. Therefore, your application letter must be persuasive. The successful application letter accomplishes four tasks: (1) it catches the reader's attention favorably, (2) it explains which particular job interests you and why, (3) it convinces the reader that you are qualified for the job by drawing your reader's attention to particular elements in your résumé, and (4) it requests an interview. See also <u>interviewing for a job,</u> <u>readers,</u> and <u>salary negotiations.</u>

Opening Paragraph

In the opening paragraph, provide context and show your enthusiasm.

1. Indicate how you heard about the opening. If you have been referred to a company by an employee, a career counselor, a professor, or someone else, be sure to mention this even before you state your job objective ("I recently learned from Jodi Hammel").
2. State your job objective and mention the specific job title ("Karen Jarrett informed me of a possible opening for a district manager"). Those who make hiring decisions review many application letters. To save them time while also calling attention to your strengths as a candidate, state your job objective directly in your first paragraph.
3. Explain why you are interested in the job ("Your firm's buyer training program is considered one of the most effective" or "Your position interests me because I can further develop my skills and talents").

Central Paragraphs

In the second and third paragraphs, show through examples that you are highly qualified for the job. Limit each of these paragraphs to just one basic point that is clearly stated in the topic sentence. (See also **paragraphs.**) For example, your second paragraph might focus on work experience and your third paragraph on educational achievements. Don't just *tell* readers that you're qualified—*show* them by including examples and details. Come across as proud of your achievements and refer to your enclosed résumé. Indicate how (with your talents) you can make valuable contributions to their company, such as "I am confident that my ability to take the initiative would be a valuable asset to your company."

Closing Paragraph

In the final paragraph, request an interview. Let the reader know how to reach you by including your phone number or e-mail address. End with a statement of goodwill, even if it is only "thank you."

Proofread your letter very carefully. Research indicates that if employers notice even one spelling, grammatical, or mechanical error, they often eliminate candidates from consideration immediately. Such errors will give employers the impression that you lack writing skills or that you are generally sloppy and careless in the way you present yourself professionally. See also **proofreading.**

Sample Letters

The four sample application letters shown in Figures A–10 through A–13 follow the application-letter structure described in this entry.

6334 Henry Roberts Street
Butler, PA 16005
March 9, 2003

Bob Lupert
Applied Sciences, Inc.
P.O. Box 9098
883 Four Seasons Road
Butler, PA 16005

Dear Mr. Lupert:

In the February 24, 2003, issue of the *Butler Gazette,* I learned that you have summer technical training internships available. This opportunity interests me because I have the professional and educational background necessary to make positive contributions to your firm.

My professional experiences are representative of my abilities. I am particularly proud of my current project—a computer tutoring system that teaches LISP. This tutor is the first of its kind and is now being sold across the country to various corporations and universities. My responsibilities involve working in a team to solve specific problems and then to test and revise our work until we have solved each problem efficiently. I have also developed leadership and collaborative skills that I could contribute to a summer position at Applied Sciences. For example, as a co-founder of a project to target and tutor high school students with learning disabilities, I was responsible for organizing and implementing with my peers many of the training and tutoring sessions. My ability to take the initiative would be a valuable asset to your company.

Pursuing degrees in Industrial Management and Computer Science has prepared me well to make valuable contributions to your goal of successful implementation of new software. Through my varied courses, which are described in my résumé, I have developed the ability to learn new skills quickly and independently and to interact effectively in a technical environment. I would look forward to applying all these abilities at Applied Sciences, Inc.

I would appreciate the chance to interview with you at your earliest convenience. If you have questions or would like additional information, contact me at (435) 228-3490 any Tuesday or Thursday after 10 a.m. or e-mail me at <sennett@execpr.com>. Thank you for your time.

Sincerely,

Molly Sennett

Molly Sennett

Enclosure: Résumé

FIGURE A–10. Application Letter (College Student Applying for an Internship)

7188 Virginia Avenue
Pittsburgh, PA 15232
27 February 2003

Patrice C. Crandal
Executive Recruiter
Abel's Department Stores, Inc.
599 Seventh Avenue
Pittsburgh, PA 15219

Dear Ms. Crandal:

Recently, I learned that you may be hiring undergraduates for summer internships. Through personal research and sources in the retailing industry, I have discovered that your firm's buyer training program is considered one of the most effective. For this reason, I am interested in your company and I would like to be considered as a possible summer intern.

As indicated in my résumé, I have the professional and analytical qualities necessary to excel at an innovative company such as Abel's. My experiences with the Alumni Relations Program and the University Center Committee have enhanced my communication and persuasive abilities as well as my understanding of compromise and negotiation. For example, in the alumni program, my priority focused on convincing both hostile and friendly alumni to become more involved with the direction of the university. On the University Center Committee, my goal was to balance the students' demands with financial and structural constraints of the administration. In both cases, I succeeded in achieving these important goals through persuasion.

I'd also like to point out that throughout my work experiences and education I have been determined and innovative. My efforts to excel at Abel's will reflect my commitment to these qualities.

I would appreciate the opportunity to meet with you to discuss your summer internship further. If you have questions or would like to speak with me personally, please contact me at (412) 863-2289 any weekday after 3 p.m. Thank you for your time and consideration.

Sincerely,

Marsha S. Parker

Marsha S. Parker

Enclosure: Résumé

FIGURE A–11. Application Letter (College Student Applying for an Internship)

449 Samson Street, Apt. 19
Providence, RI 02906
September 19, 2003

Alice Tobowski
Employee Relations Department
Advertising Media, Inc.
1007 Market Street
Providence, RI 02912

Dear Ms. Tobowski:

I recently learned from Jodi Hammel, a graphic designer at Advertising Media, Inc., and a former colleague, that you are looking for outstanding advertising assistants. Your position interests me greatly, not only because your firm is number one in the region but also because I feel that Advertising Media is, as Jodi and I have discussed, the kind of place where I can further develop my skills and talents.

I understand that you especially need bilingual assistants because of your zone's ethnic diversity. As noted in my enclosed résumé, I speak and write Spanish fluently. I would welcome the chance to apply my language skills at Advertising Media. Ms. Tobowski, I am aware that hundreds of applicants are applying for this position, but I have a combination of qualities probably few can match: in addition to my bilingual skills, I have a degree and experience in advertising, outstanding verbal and written communication skills, an innate ability to work well with colleagues, and the common sense to solve both simple and challenging problems. I have developed my skills by contributing to advertising for Quilted Bear in Providence, where we develop campaigns for diverse audiences. I have also been promoted to leadership positions in my jobs, schools, and community organizations, and I have worked well both individually and in team efforts in each environment.

I would enjoy meeting with you at your convenience to discuss this career opportunity further. Also, I have many references that I encourage you to contact. Feel free to call me any weekday morning or e-mail me at <singh@pcexec.com> if you have any questions, need further information, or would like to set up an interview. Thank you for your consideration.

Sincerely,

Sarah Singh

Sarah Singh

Enclosure: Résumé

FIGURE A–12. Application Letter (College Graduate Applying for a Job)

522 Beethoven Drive
Roanoke, VA 24017
November 15, 2003

Ms. Cecilia Smathers
Vice President, Dealer Sales
Hamilton Office Machines, Inc.
6194 Main Street
Hampton, VA 23661

Dear Ms. Smathers:

During the recent NOMAD convention in Washington, one of your sales
representatives, Karen Jarrett, informed me of a possible opening for a
district manager in your Dealer Sales Division. My extensive background in the office systems industry makes me highly qualified for the
position.

I was with Technology, Inc., Dealer Division from its formation in 1990
until its closing last year. During that period, I was involved in all areas
of dealer sales, both within Technology, Inc., and through personal contact with a number of independent dealers. From 1996 to 2002, I served
as Assistant to the Dealer Sales Manager as a Special Representative.
My education and work experience are indicated in the enclosed
résumé.

I would like to discuss my qualifications in an interview at your convenience. Please write to me, telephone me at (804) 449-6743 any weekday, or e-mail me at <gm302.476@sys.com>.

Sincerely,

Gregory Mindukakis

Gregory Mindukakis

Enclosure: Résumé

FIGURE A–13. Application Letter (Applicant with Years of Experience)

Each is adapted according to the emphasis, tone, and style to fit its particular audience.

- In Figure A–10 on page 42, a college student responds to a specific job advertisement in a local newspaper for an internship.
- In Figure A–11 on page 43, a college student seeks an internship in a retailing business.
- In Figure A–12 on page 44, a recent college graduate applies for a job in an advertising company.
- In Figure A–13 on page 45, a person with many years of work experience applies for a job as a district manager.

appositives

An *appositive* is a <u>noun</u> or noun <u>phrase</u> that follows and amplifies another noun or noun phrase. It has the same grammatical function as the noun it complements.

- Dennis Gabor, *the famous British scientist,* experimented with coherent light in the 1940s.
- The famous British scientist *Dennis Gabor* experimented with coherent light in the 1940s.

For detailed information on the use of <u>commas</u> with appositives, see <u>restrictive and nonrestrictive elements</u>.

   If you are in doubt about the <u>case</u> of an appositive, check it by substituting the appositive for the noun it modifies.

- My boss gave the two of us, Jim and *me,* the day off.
  [You would not say, "My boss gave *I* the day off."]

articles

Articles (*a, an, the*) function as <u>adjectives</u> because they modify the items they designate by either limiting them or making them more precise. The two kinds of articles are indefinite and definite.

   The indefinite articles, *a* and *an,* denote an unspecified item.

- *A* package was delivered yesterday.
  [This is not a specific package but an unspecified package.]

The choice between *a* and *an* depends on the sound rather than on the letter following the article, as described in the entry <u>a/an</u>.

The definite article, *the*, denotes a particular item.

- *The* package was delivered yesterday.
  [This is not just any package but *the* specific package.]

Do not omit all articles from your writing. Including articles costs nothing; eliminating them makes reading more difficult. (See also **telegraphic style**.) However, do not overdo it. An article can be superfluous.

ESL TIPS FOR USING ARTICLES

Whether to use a definite or an indefinite article is determined by what you can safely assume about your audience's knowledge. In each of these sentences, you can safely assume that the reader can clearly identify the noun. Therefore, use a definite article.

- *The* sun rises in the east. [The Earth has only one *sun*.]

- Did you know that yesterday was *the* coldest day of the year so far? [The modified noun refers to *yesterday*.]

- *The* man who left his briefcase in the conference room was in a hurry. [The relative phrase *who left his briefcase in the conference room* restricts and, therefore, identifies the meaning of *man*.]

In the following sentence, however, you cannot assume that the reader can clearly identify the noun.

- *A* package is on the way.
  [It is impossible to identify specifically what package is meant.]

A more important question for some nonnative speakers of English is when *not* to use articles. These generalizations will help. Do not use articles with:

singular proper nouns
- Utah, Main Street, Harvard University, Mount Hood

plural nonspecific countable nouns (when making generalizations)
- Helicopters are the new choice of transportation for the rich and famous.

singular uncountable nouns
- She loves chocolate.

plural countable nouns used as complements
- Those women are physicians.

- Fill with *a* half *a* pint of fluid.
  [Choose one article and eliminate the other.]

Do not capitalize articles in titles except when they are the first word ("*The Scientist* reviewed *Saving the Humboldt Penguin*"). See also **capitalization** and **English as a second language**.

as / because / since

As, because, and *since* are commonly used to mean "because." To express cause, *because* is the strongest and most specific connective; *because* is unequivocal in stating a causal relationship. (*Because* she did not have an engineering degree, she was not offered the job.)

Since is a weak substitute for *because* as a connective to express cause. However, *since* is an appropriate connective when the emphasis is on circumstance, condition, or time rather than on cause and effect. (*Since* it went public, the company has earned a profit every year.)

As is the least definite connective to indicate cause; its use for that purpose is best avoided. (See also **subordination**.) The word *as* can also contribute to **awkwardness** by appearing too many times in a sentence.

> *We*   *the moment*   *that*
- ~~As we~~ realized ~~as soon as~~ we began the project, the problem
  needed a solution.

Avoid colloquial, nonstandard, or wordy phrases sometimes used instead of *as, because,* or *since.* See also **affectation** and **as such**.

PHRASE	REPLACE WITH
being as/being that	*because/since*
inasmuch as	*because*
insofar as	*since*
on account of	*because*
on the grounds of/that	*because*

as much as / more than

The phrases *as much as* and *more than* are sometimes incorrectly combined, especially when intervening phrases delay the completion of the phrase.

- The engineers had as much, ~~if not more,~~ influence in planning the
  as　　　　　　　　　　　*, if not more*
  program ~~than~~ the accountants .
  　　　　　　^　　　　　　　　^

as such

The phrase *as such* is seldom useful and should be omitted.

- Templates, ~~as such,~~ are useful in developing online documentation.

as to whether (see whether)

as well as

Do not use *as well as* with *both*. The two expressions have similar meanings; use one or the other.

　　　　　　　　　　　　and
- Both General Motors ~~as well as~~ Ford are developing electric cars.
  　　　　　　　　　　　　^

　　　　　　　　　　　　　　　　　is
- ~~Both~~ General Motors as well as Ford ~~are~~ developing electric cars.
  　　　　　　　　　　　　　　　　　^

audience

Although the word *audience* can have a slightly different meaning for a writer than a speaker, it is crucial to both. The writer and the speaker must know as much as possible about the people they are trying to reach with their message, regardless of its method of delivery. For the specific requirements of each, see **readers, presentations, international correspondence**, and **global communication**.

augment / supplement

Augment means to increase or magnify in size, degree, or effect. (Many employees *augment* their incomes by freelancing.) *Supplement* means to add something to make up for a deficiency. (He will *supplement* his diet with vitamins.)

average / median / mean

The *average* is determined by adding two or more quantities and dividing the sum by the number of items totaled. For example, if one report is 10 pages, another is 30 pages, and a third is 20 pages, their *average* length is 20 pages. It is incorrect to say that "each report averages 20 pages" because each report is a specific length.

The three reports average
● ~~Each report averages~~ 20 pages.
       ^

The *median* is the middle number in a sequence of numbers. (The *median* of the series 1, 3, 4, 7, 8, is 4.) The word *mean* ("something midway between extremes") can apply to either the average or the median.

awhile / a while

Awhile is an adverb that means "for a short time." It is not preceded by the word *for* because the meaning of *for* is inherent in the meaning of *awhile*. *A while* is a noun phrase that means "a period of time."

● Wait ~~for~~ awhile before testing the sample.

      a while
● Wait for ~~awhile~~ before testing the sample.
         ^

awkwardness

Any writing that strikes the reader as awkward—that is, as forced or unnatural—impedes the reader's understanding. Awkwardness has many causes, but the following checklist and the entries indicated will help you smooth out most awkward passages.

Writer's Checklist: Eliminating Awkwardness

☑ Strive for **clarity** and **coherence** during your **revision**.

☑ Check for **organization** to tighten your writing.

☑ Keep sentences as direct and simple as possible (see **sentence construction**).

☑ Use **subordination** appropriately and avoid needless **repetition**.

Writer's Checklist: Eliminating Awkwardness (continued)

- ☑ Correct any <u>**logic errors**</u> within your sentences.
- ☑ Revise for <u>**conciseness**</u> and avoid <u>**expletives**</u> where possible.
- ☑ Use the active <u>**voice**</u> unless you have a justifiable reason to use the passive voice.
- ☑ Eliminate any jammed <u>**modifiers**</u>.
- ☑ Apply the tactics in <u>**garbled sentences**</u> for particularly awkward constructions.

B

bad / badly

Bad is the <u>adjective</u> form that follows such linking <u>verbs</u> as *feel* and *look*. (We don't want to look *bad* at the meeting.) *Badly* is an <u>adverb</u>. (The test model performed *badly* during the test.) To say "I feel *badly*" would mean, literally, that your sense of touch is impaired. See also <u>good/well</u>.

balance / remainder

One meaning of *balance* is "a state of equilibrium"; another meaning is "the amount of money in a bank account after deposits and withdrawals have been credited and debited." *Remainder*, in all applications, means "what is left over."

be sure to

The <u>phrase</u> *be sure and* is colloquial and unidiomatic when used for *be sure to*. See also <u>idioms</u>.

- When you sign the contract, be sure ~~and~~ ^to^ keep a copy.

because (see *as / because / since*)

being as / being that (see *as / because / since*)

beside / besides

Besides, meaning "in addition to" or "other than," should be carefully distinguished from *beside*, meaning "next to" or "apart from." (*Besides* the two of us from the Systems Department, three people from Production stood *beside* the president during the ceremony.)

between / among

Between is normally used to relate two items or persons. (Preferred stock offers a middle ground *between* bonds and common stock.) *Among* is used to relate more than two. (The subcontracting was distributed *among* the three firms.)

between you and me

The expression *between you and I* is incorrect. Because the <u>pronouns</u> are <u>objects</u> of the <u>preposition</u> *between,* the objective form of the personal pronoun (*me*) must be used. See also <u>case (grammar)</u>.

- Between you and ~~I~~, Joan should be promoted.
  ^{*me*}

bi- / semi-

When used with periods of time, *bi-* means "two" or "every two." *Bimonthly* means "once in two months"; *biweekly* means "once in two weeks." When used with periods of time, *semi-* means "half of" or "occurring twice within a period of time." *Semimonthly* means "twice a month"; *semiweekly* means "twice a week." Both *bi-* and *semi-* normally are joined with a following element without a space or a hyphen.

biannual / biennial

In conventional usage, *biannual* means "twice during the year," and *biennial* means "every other year." See also <u>bi-/semi-</u>.

B

biased language

Biased language refers to words and expressions that offend because they make inappropriate assumptions or stereotypes about gender, ethnicity, physical or mental disability, age, or sexual orientation. The easiest way to avoid bias is simply not to mention differences among people unless the differences are relevant to the discussion. Keep current with accepted usage, and, if you are unsure of the appropriateness of an expression or the tone of a passage, have several colleagues review the material and give you their honest assessment.

Sexist Language

Sexist language can be an outgrowth of sexism, the arbitrary stereotyping of men and women in their roles in life. Sexism, a form of biased language, can breed and reinforce inequality. To avoid sexism in your writing, treat men and women equally, and do not make assumptions about traditional or occupational roles. Accordingly, use nonsexist occupational descriptions in your writing.

INSTEAD OF	USE
chairman	chair, chairperson
foreman	supervisor
manpower	staff, personnel, workers
policeman/policewoman	police officer
salesman	salesperson

Use parallel terms to describe men and women.

INSTEAD OF	USE
man and wife	husband and wife
Ms. Jones and Bernard Weiss	Ms. Jones and Mr. Weiss; Mary Jones and Bernard Weiss
ladies and men	ladies and gentlemen; women and men

Sexism can creep into your writing by the unthinking use of male pronouns where a reference could apply equally to a man and a woman. One way to avoid such usage is to rewrite the sentence in the plural.

- *All employees*   *their managers*   *their travel vouchers.*
  ~~Every employee~~ will have ~~his manager~~ sign ~~his travel voucher.~~

Other possible solutions are to use *his or her* instead of *his* alone or to omit the pronoun completely if it is not essential to the meaning of the sentence. See also **pronoun reference**.

B

- Everyone must submit ~~his~~ *an* expense report by Monday.

He or she can become monotonous when repeated constantly, and a pronoun cannot always be omitted without changing the meaning of a sentence. Another solution is to omit troublesome pronouns by using the imperative <u>mood</u> whenever possible.

- *Submit all* ~~Everyone must submit his or her~~ expense ~~report~~ *reports* by Monday.

Other Types of Biased Language

Identifying people by racial, ethnic, or religious categories is simply not relevant in most workplace writing. Telling readers that an engineer is Native American or that a professor is African American almost never conveys useful information. It also reinforces stereotypes, implying that it is rare for a person of a certain background to have achieved such a position. It also is inappropriate stereotyping to link a profession or some characteristic to race or ethnicity: a Jewish lawyer, an African-American jazz musician, an Asian-American mathematics prodigy.

Consider how you refer to people with disabilities. If you refer to "a disabled employee," you imply that the part (*disabled*) is as significant as the whole (*employee*). Use "an employee with a disability" instead. Similarly, the preferred usage is "a person who uses a wheelchair" rather than "a wheelchair-bound person"; the latter expression inappropriately equates the wheelchair with the person. Generally, however, you should refer to a disability only if it is necessary.

Terms that refer to a person's age are also open to inappropriate stereotyping. Referring to older colleagues as "geezers" and to younger colleagues as "kids" is derogatory, at the least.

In most workplace writing, such issues are simply not relevant. Of course, there are contexts in which race, ethnicity, or religion should be identified. For example, if you are writing an Equal Employment Opportunity Commission report about your firm's hiring practices, the racial composition of the workforce is relevant. In such cases, you need to present the issues in ways that respect and do not demean the individuals or groups to which you refer. See also <u>ethics in writing</u>.

bibliographies

A *bibliography* is an alphabetical list of books, articles, Web sources, and other materials that is often used to record the works consulted in preparing a document. It provides a convenient alphabetical listing of these sources in a standardized form for readers interested in getting further information on the topic or in assessing the scope of the research.

While a list of references (APA style), footnotes/endnotes (*Chicago Manual* style or *CMS* system), and a works-cited list (MLA style) refer to works actually cited in the text, a bibliography includes works consulted for general background information. For information on using APA and MLA styles and *CMS* system, see **documenting sources**.

Entries in a bibliography are listed alphabetically by the author's last name. If an author is unknown, the entry is alphabetized by the first word in the title (other than *A, An,* or *The*). Entries also can be arranged by subject and then ordered alphabetically within those categories.

An annotated bibliography includes complete bibliographic information about a work (author, title, publisher) followed by a brief description or evaluation of what the work contains. The following annotation concisely summarizes and evaluates a book of historical interest:

> Rickard, T. A. *A Guide to Technical Writing.* San Francisco: Mining and Scientific Publishers, 1908.
>
> > This book is of particular historical interest because it is the first published book on technical writing for the professional. The author comments, "It has been said that in this age the man of science appears to be the only one who has anything to say, and he is the one that least knows how to say it. . . . Write simply and clearly, be accurate and careful; above all, put yourself in the other fellow's place. Remember the reader." Geared to the mining and metallurgical sciences, the 17 short, unnumbered chapters cover matters of language, usage, grammar, and mechanics slanted toward the needs of the technical writer. The book ends with a paper the author read before the American Association for the Advancement of Science, at Denver on August 28, 1901: "A Plea for Greater Simplicity in the Language of Science." "We must remember," the author suggests, "that the language in relation to ideas is a solvent, the purity and clearness of which affect what it bears in solution."

See **formal reports** for guidance on the placement of a bibliography in a report.

both . . . and

Statements using the *both . . . and* construction should always be balanced grammatically and logically. See also **parallel structure**.

- For success in engineering, it is necessary *both* to develop writing
  to master
  skills *and* ~~mastering~~ mathematics.
  ^

brackets

The primary use of brackets ([ ]) is to enclose a word or words inserted by the writer or editor into a quotation.

- The text stated, "Hypertext systems can be categorized as either modest [not modifiable] or robust [modifiable]."

Brackets are used to set off a parenthetical item within parentheses.

- We should be sure to give Emanuel Foose (and his brother Emilio [1812–1882]) credit for his role in founding the institute.

Brackets are also used in academic writing to insert the Latin word *sic,* which indicates that the writer has quoted material exactly as it appears in the original, even though it contains an error.

- Dr. Smith wrote that "the earth does not revolve around the son [*sic*] at a constant rate."

If you are following MLA style in your writing, use brackets around ellipsis dots to show that some words have been omitted from the original source. See also **ellipses** and **quotations**.

- "The vast majority of the Internet's inhabitants are [. . .] between the ages of eighteen and thirty-four" (5).

brainstorming

Brainstorming, a form of free association used to generate ideas, can be done individually or in groups. Brainstorming can stimulate creative thinking about a topic and reveal fresh perspectives and new connections. When preparing to write, jot down as many random ideas as you can think of about the topic. When a group brainstorms, a designated person writes down the words or phrases as the other group members suggest them. Do not stop to analyze ideas or hold back looking for only the "best" ideas; just write down everything that comes to mind. After compiling a list of initial ideas, ask *what, when, who, where, how,* and *why* for each idea, then list additional details that those questions bring to mind. When you run out of ideas, analyze each one you recorded, discarding those that are redundant. Then group the items in the most logical order, based on the purpose of the document and the needs of the **readers**; that will result in a tentative

B

<u>outline</u> of the document. Although the outline will be sketchy and in-complete, it will show where further brainstorming or research is needed and provide a framework for any new details that additional research yields.

Many writers find a technique called *clustering* (also called *mind mapping*), as shown in Figure B–1, helpful in recording and organizing ideas created during a brainstorming session. To cluster, begin with a blank sheet of paper or a flip chart. Think of a keyword or phrase that best characterizes your topic and put it in a circle at the center of the paper. Then put the subtopics most closely related to the main topic in circles or boxes around the main topic, connecting each to the center circle, like adding spokes to a wheel hub. Repeat the exercise for each subtopic. Each subtopic should stimulate additional subtopics, which you add in circles connected to the parent subtopic, as you did with the main topic at the center of the page. Continue the process until you ex-haust all possible ideas. The resulting "map" will show clusters of terms grouped around the central concept.

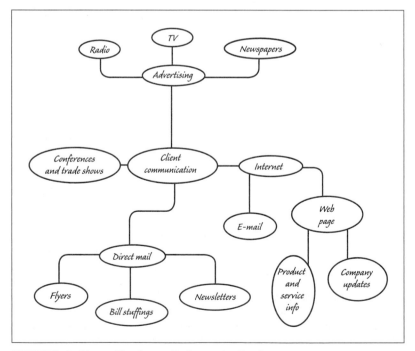

FIGURE B–1. Cluster Map from a Brainstorming Session

brochures

Brochures are printed publications that promote the products and services offered by a business or that promote the image of a business or organization by providing information important to a target audience. They can also provide speecific technical information. The goal of a brochure is to inform or to persuade or both.

Types of Brochures

There are two major types of brochures: *sales brochures* and *informational brochures*. *Sales brochures* are created specifically to sell a company's products and services. A technically oriented sales brochure can be created to promote the performance data of various models of a product or to list the many benefits of different types of equipment available for a specific job. For example, the brochure of a microwave-oven manufacturer might describe the various models and their accessories and how the ovens are to be installed over stovetops. *Informational brochures* are created to inform and educate the **reader** as well as to promote goodwill and raise the profile of an organization. For example, a technically oriented informational brochure for a pesticide company might show the reader how to identify various types of pests like silverfish and termites, detail the damage they can do, and explain how to protect your house from them.

Designing the Brochure

Before you begin to write, you must decide on the specific **purpose** of the brochure—to sell a product? to provide technical information about a product? You must also identify your target audience—general reader? expert? potential client? Understanding your purpose and audience is crucial to creating content and design that will be both rhetorically appropriate and persuasive to your target audience.

The main goal of the cover panel of a sales or promotional brochure is to gain the attention of your target audience. The cover panel, which should clearly identify the company being promoted, usually features a carefully selected visual image geared toward the interest of your audience. Accompanying the image, there may be a minimal amount of text—for example, a statement about the company's mission and success, or a brief promotional quotation from a satisfied customer. Figure B–2 on page 60 shows the cover of a brochure for the U.S. Department of Education that hopes to attract visitors to its Web site with free teaching and learning resources.

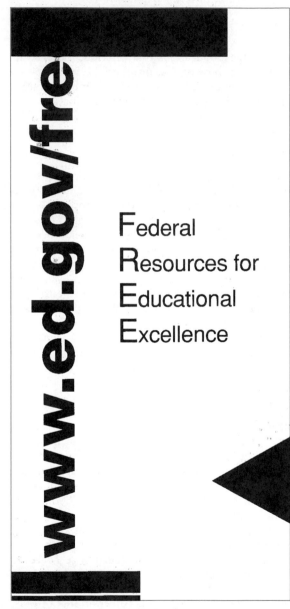

FIGURE B–2. Brochure Cover Panel

The first inside panel of the sales or promotional brochure should again identify the company and attract the reader with headlines and brief, readable content, such as that used in advertising. Figure B-3 on page 62 shows the first inside panel of the Department of Education brochure for free teaching and learning resources. Notice that the inside panel opens with a <u>table of contents</u> for the topics covered at the agency's Web site. It begins with a question ("What is FREE?") that positions the resource as a useful, convenient service for those seeking information. The text that follows describes the popularity of the site and its usefulness to those who have visited. This panel not only uses a bulleted <u>list</u> and white space effectively but also uses text that spans to the next panel, thus drawing readers to further information. See also <u>layout and design</u>.

In subsequent panels of your brochure, describe your product, service, or options or accessories with your readers' needs in mind, clearly stating the benefits and solutions to problems that your product offers. Installation information could be included here, as well as relevant and accurate supporting facts and visuals, and you might further establish credibility with a company history, a product history, and a list of current clients or testimonials. Use subheadings and bulleted lists to break up the text and make the brochure easy to read. In the final panel, be clear about the action you want the reader to take, such as calling for an appointment, e-mailing for technical support, or mailing an enclosed reply card.

As you plan the design for your brochure, create a rough planning sketch. To find the best way to present your content, experiment with margins, spacing, and the arrangement and amount of text on each panel (you may want to make changes to content or length, based on concerns of layout). Allow adequate white space for readability. Experiment with various fonts and formatting. Consider, for example, enlarging the first letter of the first word in a paragraph, but do not overdo the use of unusual fonts or alternative styles (such as running type vertically). Choose a style that unifies your brochure, and use it consistently throughout. Be judicious about color choices. For some brochures, black and white, which is cheaper, is more effective; in other circumstances (such as brochures for travel) color, although more expensive, is a must. Depending on your budget and the scale of your project, consider using a professional printer. See also <u>visuals</u>.

B

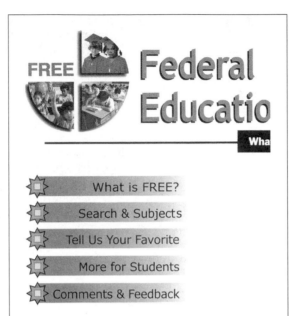

FREE Federal Educatio Wha

What is FREE?

Search & Subjects

Tell Us Your Favorite

More for Students

Comments & Feedback

Hundreds of teaching and learning resources from across the federal government are now available at one of the most popular federal websites for education.

Find out how partnerships of teachers and federal agencies are developing online learning modules and communities. Also, use the "Looking for Partners" page to describe learning resources *you* would like to create with a federal agency.

http://www.

FIGURE B–3. Brochure First Inside Panel (title spans to second panel)

Writer's Checklist: Creating Sales or Informational Brochures   **B**

- ☑ Determine your audience and what content and <u>visuals</u> will most appeal to your <u>readers</u>. See also <u>persuasion</u>.
- ☑ Outline content and select visuals, color schemes, and the number of panels appropriately. Create a rough sketch that maps out the content and visuals of each panel.
- ☑ Be sure that your first panel provides basic information about your company and grabs your readers' attention with a unique title or headline.
- ☑ In the body of the brochure, offer your products, services, processes, and respective equipment in terms of the potential benefits to readers or as solutions to problems. Clearly describe different models and accessories and use facts to back up your claims. See also <u>clarity</u>.
- ☑ Choose a <u>layout and design</u> that best presents your content to your audience. Is the design unified and consistent? Is it easy to read?
- ☑ Evaluate your content and design. Is the content persuasive? If it's technical, is it clearly written? Is a concise <u>glossary</u> necessary? Does the design complement the content and make the brochure more persuasive? Have you adequately considered the needs of your readers? When your brochure is distributed, will it fulfill its purposes?

buzzwords

Buzzwords are words that suddenly become popular and, because of an intense period of overuse, lose their freshness and preciseness. They may become popular through their association with science, technology, or even sports. We include them in our vocabulary because they seem to give force and vitality to our language. Ordinarily, buzzwords sound pretentious in everyday writing. (See also <u>word choice</u>.)

interface [as a verb]	deliverables	dialog [as a verb]
impact [as a verb]	newbie	cutting edge
skill sets	scalability	robust

Obviously, some buzzwords are appropriate in the right context. It is when they are used outside that context that imprecision becomes a problem.

- An *interface* was established between the computer and the satellite hardware.
  [*Interface* is appropriately used as a noun.]

B

cooperate
- We must ~~interface~~ with the Human Resources Department.
  ^

  [*Interface* is inappropriately used as a verb; *cooperate* is more precise.]

 **WEB LINK BUZZWORDS**

Former newspaper editor John Walston offers BuzzWhack, a lighthearted site dedicated to the meaningful use of buzzwords. See <*www.bedford stmartins.com/alred*> and select *Links for Technical Writing.*

C

can / may

In writing, *can* refers to capability and *may* refers to possibility or permission.

- I *can* have the project finished by the end of the year. [capability]
- I *may* be in Boston on Thursday. [possibility]
- *May* I proceed with the project? [permission]

cannot

Cannot is one word.

- We ~~can not~~ ^{cannot} meet the deadline specified in the contract.

canvas / canvass

Canvas refers to "heavy, coarse, closely woven cotton or hemp fabric." (The maintenance crew spread the *canvas* over the equipment.) *Canvass* means "to solicit votes or opinions." (The executive committee decided to *canvass* the employees.)

capital / capitol

Capital refers either to financial assets or to the city that hosts the government of a state or a nation. *Capitol* refers to the building in which the state or national legislature meets. *Capitol* is often written with a small *c* when it refers to a state building, but it is always capitalized when it refers to the home of the U.S. Congress in Washington, D.C.

C

capitalization

The use of capital, or uppercase, letters is determined by custom. Capital letters are used to call attention to certain words, such as proper <u>nouns</u> and the first word of a sentence. Exercise care in using capital letters because they can affect the meaning of words (march/March, china/China, turkey/Turkey). The proper use of capital letters can help eliminate ambiguity.

Proper Nouns

Proper nouns name specific persons, places, things, concepts, or qualities and are capitalized (Physics 101, Microsoft, Jennifer Wilde, Argentina).

Common Nouns

Common nouns name general classes or categories of people, places, things, concepts, or qualities rather than specific ones and are not capitalized (a physics class, a company, a person, a country).

First Words

The first letter of the first word in a sentence is always capitalized. (Of all the plans submitted, ours seems best.) The first word after a <u>colon</u> may be capitalized if the statement following is a complete sentence or if it is a formal resolution or question.

- Today's meeting will deal with only one issue: What is the firm's role in environmental protection?

If a subordinate element follows the colon or if the thought is closely related, use a lowercase letter following the colon.

- We had to keep working for one reason: the approaching deadline.

The first word of a complete sentence in <u>quotation marks</u> is capitalized.

- Dr. Vesely stated, "Decisions should not be made until all the relevant information is assembled."

The first word in the salutation (Dear Mr. Smith:) and complimentary close (Sincerely yours,) of a letter is capitalized. See also <u>correspondence</u>.

Specific Groups

Capitalize the names of ethnic groups, religions, and nationalities (Native American, Italian, Jewish). Do not capitalize the names of social and economic groups (middle class, working class, unemployed).

Specific Places

Capitalize the names of all political divisions (Ward Six, Chicago, Cook County, Illinois). Capitalize the names of geographical divisions (Europe, Asia, North America, the Middle East). Do not capitalize geographic features unless they are part of a proper name.

- The *mountains* in some areas, such as the *Great Smoky Mountains,* make television transmission difficult.

The words *north, south, east,* and *west* are capitalized when they refer to sections of the country. They are not capitalized when they refer to directions.

- I may travel *north* when I relocate, but my family will remain in the *South.*

Capitalize the names of stars, constellations, and planets (Saturn, Andromeda, Milky Way, Orion). Do not capitalize *earth, sun,* and *moon* except when they are referred to formally as astronomical bodies.

- My workday was so long that I saw the *sun* rise over the lake and the *moon* appear as darkness settled over the *earth.*
- The various effects of the *Sun* on *Earth* and the *Moon* were discussed at the symposium.

Specific Institutions, Events, and Concepts

Capitalize the names of institutions, organizations, and associations (Association of Professional Communication Consultants). An organi-

C

zation usually capitalizes the names of its internal divisions and departments (Faculty, Human Resources). Types of organizations are not capitalized unless they are part of an official name (a business communication association; the Association for Business Communication). Capitalize historical events (the Great Depression). Capitalize words that designate holidays, specific periods of time, months, or days of the week (Labor Day, the Renaissance, January, Monday). Do not capitalize seasons of the year (spring, autumn, winter, summer).

Capitalize the scientific names of classes, families, and orders but not the names of species or English derivatives of scientific names (Mammalia/mammal, Carnivora/carnivorous).

Titles of Works

Capitalize the initial letters of the first and last words of the title of a book, article, play, or film, as well as all major words in the title. Do not capitalize articles (*a, an, the*), coordinating conjunctions (*and, but*), or short prepositions (*at, in, on, of*) unless they begin or end the title (*The Lives of a Cell*). Capitalize prepositions that contain more than four letters (*between, because, until, after*). The same rules apply to the subject line of a memo or an e-mail.

Personal, Professional, and Job Titles

Titles preceding proper names are capitalized (Ms. March, Professor Galbraith). Appositives following proper names normally are not capitalized (Charles Schumer, *senator* from New York; Senator Schumer). However, the word *President* usually is capitalized when it refers to the chief executive of a national government.

Job titles used with personal names are capitalized (Ho-shik Kim, *Division Manager*). Job titles used without personal names are not capitalized.

- The *division manager* will meet with us on Wednesday.

Use capital letters to designate family relationships only when they occur before a name or substitute for a name (my uncle; Uncle Fred).

Abbreviations

Capitalize **abbreviations** if the words they stand for would be capitalized, such as UCLA (University of California at Los Angeles).

Letters

Capitalize letters that serve as names or indicate shapes (vitamin B, T-square, U-turn, I-beam).

Miscellaneous Capitalizations

The first word of a complete sentence enclosed in dashes, brackets, or parentheses is not capitalized when it appears as part of another sentence.

- We must make an extra effort in safety this year (accidents last year were up 10 percent).

Certain units, such as parts and chapters of books and rooms in buildings, when specifically identified by number, are capitalized (Chapter 5, Ch. 5; Room 72, Rm. 72). Minor divisions within such units are not capitalized unless they begin a sentence (page 11, verse 14, seat 12).

case (grammar)

Grammatically, *case* indicates the functional relationship of a <u>noun</u> or a <u>pronoun</u> to the other words in a sentence. Nouns change form only in the possessive case; pronouns may show change for the subjective, the objective, or the possessive case. The case of a noun or a pronoun is always determined by its function in a phrase, clause, or sentence. If it is the subject of a phrase, clause, or sentence, it is in the subjective case; if it is an <u>object</u> in a phrase, clause, or sentence, it is in the objective case; if it reflects possession or ownership and modifies a noun, it is in the possessive case. The subjective case can indicate the person or thing acting ("*He* sued the vendor"), the person or thing acted upon ("*He* was sued by the vendor"), or the topic of description ("*He* is the vendor"). The objective case can indicate the thing acted on ("The vendor sued *him*") or the person or thing acting but in the objective position ("The vendor was sued by *him*"). (See also <u>voice</u>.) The possessive case indicates the person or thing owning or possessing something ("It was *his* company").

The different forms of a noun or a pronoun indicate whether it is functioning as a subject (subjective case), as a <u>complement</u> (usually objective case), or as a <u>modifier</u> (possessive case). Figure C–1 on page 70 is a table of pronouns in the subjective, objective, and possessive cases.

C

SINGULAR	SUBJECTIVE	OBJECTIVE	POSSESSIVE
First person	I	me	my, mine
Second person	you	you	your, yours
Third person	he, she, it	him, her, it	his, her, hers, its

PLURAL	SUBJECTIVE	OBJECTIVE	POSSESSIVE
First person	we	us	our, ours
Second person	you	you	your, yours
Third person	they	them	their, theirs

FIGURE C–1. Pronoun Case

Subjective Case

A pronoun is in the *subjective case* (also called the *nominative case*) when it represents the person or thing acting or is the receiver of the action but in the subject position.

- *I* wrote a proposal.
- *I* was praised for my proposal writing.

A linking <u>verb</u> links a pronoun to its antecedent to show that they identify the same thing. Because they represent the same thing, the pronoun is in the subjective case even when it follows the verb, which makes it a subjective complement.

- *He* is the head of the Quality Control Group. [subject]
- The head of the Quality Control Group is *he*. [subjective complement]

The subjective case is used after the words *than* and *as* because of the understood (although unstated) portion of the clauses in which those words appear.

- George is as good a designer as *I* [am].
- Our subsidiary can do the job better than *we* [can].

Objective Case

A pronoun is in the *objective case* (also called the *accusative case*) when it indicates the person or thing receiving the action expressed by a verb in the active voice.

- They informed *me* by letter that they had received my résumé.

Pronouns that follow action verbs (which excludes all forms of the verb *be*) must be in the objective case. Do not be confused by an additional name.

- The company promoted *me* in June.

- The company promoted John and *me* in June.

A pronoun is in the objective case when it is the object of a gerund or a **preposition** or the subject of an infinitive.

- Between *you* and *me*, his facts are questionable.
  [objects of a preposition]

- Many of *us* attended the conference. [object of a preposition]

- Training *him* was the best thing I could have done.
  [object of a gerund]

- We asked *them* to return the deposit. [subject of an infinitive]

English does not differentiate between direct objects and indirect objects; both require the objective form of the pronoun. See also **complements**.

- The interviewer seemed to like *me*. [direct object]

- They wrote *me* a letter. [indirect object]

Possessive Case

A noun or pronoun is in the *possessive case* when it represents a person, place, or thing that possesses something. To make a singular noun possessive, add *'s* (the *manufacturer's* robotic inventory system). With plural nouns that end in *s*, show the possessive by placing an apostrophe after the *s* that forms the plural (a *managers'* meeting). For other guidelines, see **possessive case**.

Appositives

An **appositive** is a noun or noun phrase that follows and amplifies another noun or noun phrase. Because it has the same grammatical function as the noun it complements, an appositive should be in the same case as the noun with which it is in apposition.

- Two auditors, Jim Knight and *I,* were asked to review the books.
  [subjective case]

- The group leader selected two members to represent the department—Mohan Pathak and *me*. [objective case]

Determining the Case of Pronouns

C

One test to determine the proper case of a pronoun is to try it with some transitive verb such as *resembled* or *hit*. If the pronoun would logically precede the verb, use the subjective case; if it would logically follow the verb, use the objective case.

- *She* [*he, they*] resembled her father. [subjective case]

- Angela resembled *him* [*her, them*]. [objective case]

In the following example, try omitting the noun to determine the case of the pronoun.

SENTENCE	(*We/Us*) pilots fly our own airplanes.
INCORRECT	*Us* fly our own airplanes. [This incorrect usage is obviously wrong.]
CORRECT	*We* fly our own airplanes. [This correct usage sounds right.]

To determine the case of a pronoun that follows *as* or *than*, mentally add the words that are omitted but understood.

- The other operator is not paid as well as *she* [is paid]. [You would not write, "*Her* is paid."]

- His partner was better informed than *he* [was informed]. [You would not write, "*Him* was informed."]

If pronouns in compound constructions cause problems, try using them singly to determine the proper case.

SENTENCE	(*We/Us*) and the Johnsons are going to the Grand Canyon.
CORRECT	*We* are going to the Grand Canyon. [You would not write, "*Us* are going to the Grand Canyon."]

Using *Who/Whom*

Who and *whom* cause much trouble in determining case. *Who* is the subjective case form, and *whom* is the objective case form. When in doubt about which form to use, try substituting a personal pronoun to see which one fits. If *he, she,* or *they* fits, use *who*.

- *Who* is the representative from the 45th District? [*He* is the representative from the 45th District.]

If *him, her,* or *them* fits, use *whom.*

- It depended on *whom?* [It depended on *them.*]

It is becoming common to use *who* for the objective case when it begins a clause or sentence, although some readers still object to such a construction, especially in formal contexts.

- *Who* should I call to report a fire?

The best advice is to know your <u>readers</u>.

case (usage)

The word *case* is often merely filler. Be critical of the word and eliminate it if it contributes nothing. See also <u>conciseness/wordiness</u>.

- An exception was made ~~in the case of~~ *for* those closely connected with the project.

cause-and-effect method of development

The cause-and-effect <u>method of development</u> is a common strategy to explain why something happened or why you think something will happen. The goal of the cause-and-effect method of development is to make as plausible as possible the relationship between a situation and either its cause or its effect. The conclusions you draw about the relationships should be based on evidence you have gathered. Because not all evidence will be of equal value, keep the following guidelines in mind as you gather evidence.

Evaluating Evidence

Your facts and arguments should be relevant to your topic. Be careful not to draw a conclusion that your evidence does not lead to or support. You may have researched some statistics, for example, showing that an increasing number of Americans are licensed to fly small airplanes. You cannot use that information as evidence for a decrease in new car sales in the United States—the evidence does not lead to that conclusion.

Your evidence should be adequate. Incomplete evidence can lead to false conclusions.

C

- Driver training classes do not help prevent auto accidents. Two people I know who completed driver training classes were involved in accidents.

A thorough investigation of the usefulness of driver training classes in keeping down the accident rate would require more than one or two examples. It also would require a comparison of the driving records of those who had completed driver training with the driving records of those who had not.

Your evidence should be representative. If you conduct a survey to obtain your evidence, do not solicit responses only from individuals or groups whose views are identical to yours; be sure you obtain a representative sampling.

Your evidence should also be plausible. Two events that occur close to each other in time or place may or may not be causally related. You must demonstrate the relationship with pertinent facts and arguments. See also <u>logic errors</u>.

Linking Causes to Effects

To show a true relationship between a cause and an effect, you must demonstrate that the existence of the one *requires* the existence of the other. It is often difficult to establish beyond any doubt that one event was the cause of another event. More often, a result will have more than one cause. As you research a subject, your task is to determine which cause or causes are most plausible.

When several probable causes are equally valid, report your findings accordingly, as in the following excerpt from an article on the use of an energy-saving device called a furnace-vent damper. The damper is a metal plate that fits inside the flue or vent pipe of a natural-gas or fuel-oil furnace. When the furnace is on, the damper opens to allow gases to escape up the flue. When the furnace shuts off, the damper closes, thus preventing warm air from escaping up the flue stack. The dampers are potentially dangerous, however. If a damper fails to open at the proper time, poisonous furnace gases could back up into the house and asphyxiate the occupants. Tests run on several dampers showed a number of probable causes for their malfunctioning.

- One damper was sold without proper installation instructions, and another was wired incorrectly. Two of the units had slow-opening dampers (15 seconds) that prevented the [furnace] burner from firing. And one damper jammed when exposed to a simulated fuel temperature of more than 700 degrees.

—Don DeBat, "Save Energy But Save Your Life, Too," *Family Safety*

The investigator located more than one cause of damper malfunctions and reported on them. Without such a thorough account, recommendations to prevent malfunctions would be based on incomplete evidence.

---

center on

Use the phrase *center on* in writing, not *center around*. (The experiments *center on* the new discovery.) Often the idea intended by *center on* is better expressed by other words.

- The hearings on computer security ~~centered on~~ *dealt with* access codes.

---

chair / chairperson

The words *chair, chairperson, chairman,* and *chairwoman* are used to refer to a presiding officer. The titles *chair* and *chairperson,* however, avoid any sexual bias that might be implied by the other titles. (Mary Roberts preceded John Stephens as *chair* of the committee.) See also <u>biased language</u>.

---

chronological method of development

The chronological <u>method of development</u> arranges the events under discussion in sequential order, beginning with the first event and continuing chronologically to the last. <u>Trip reports</u>, <u>laboratory reports</u>, work schedules, some <u>minutes of meetings</u>, and certain <u>trouble reports</u> are among the types of writing in which information is organized chronologically. Chronological order is typically used in <u>narration</u>.

In the memo shown in Figure C–2 on page 76, a retail-store manager describes the steps taken over a one-year period to reduce shoplifting at his store. After providing important background information, the writer presents the steps taken in chronological order.

C

The Rack
Interoffice Memorandum

To: Joanna Sanchez, Vice President for Marketing
From: Larry Brown, Manager, Downtown Branch   *LB*
Date: September 9, 2003
Subject: Reducing Shoplifting at the Downtown Store

Over the past year, my staff and I have taken a number of measures to try to reduce the amount of shoplifting in the downtown store. As you know, we have spent much time, effort, and money on the problem, which we hope will be alleviated during the Christmas shopping season. Let me recap the specific steps we have taken.

Task Force
We formed a task force of salespeople, buyers, managers, and executive staff to recommend ways of curtailing shoplifting and methods of implementing our recommendations. We met four times during January and twice in March to reach our final recommendations.

Mark IV Surveillance System
In April, we installed a Mark IV System, which uses closed-circuit TV cameras at each exit. The cameras, which are linked with our security office, are capable of taping signals from all exits simultaneously. The task force felt the Mark IV System might be useful in detecting a pattern of specific individuals entering and leaving the store. This system, which became operational on April 20, has been very helpful in reducing the number of thefts.

Employee Training
During May and June, we held employee workshops on detecting shoplifters. Security, Inc., a consulting firm, led the workshops and provided not only lectures and tips on spotting shoplifters but also demonstrations of common techniques shoplifters use to divert store personnel. All those who attended thought the workshops were quite helpful.

Other Steps Taken
Because the task force determined that certain items were particularly vulnerable to shoplifters, we decided in July to restructure some of the display areas. Our purpose was to make those areas less isolated from the view of clerks and other store personnel. The remodeling, most of which was relatively minor, was completed over the summer months.

For the fall and holiday sales, we have hired extra uniformed security guards. The guards, also from Security, Inc., should be able to deter first-time shoplifters, although we know that this step will not eliminate the problem altogether.

We believe the steps we have taken will substantially reduce our losses from theft. Of course, after we have reviewed the figures at the end of the year, the task force will meet again in January to assess the success of the methods we have used. If you need more details, please let me know.

FIGURE C–2. Chronological Method of Development

cite / sight / site

Cite means "acknowledge" or "quote an authority." (The speaker *cited* several famous economists.) *Sight* is the ability to see. (He feared that he might lose his *sight*.) *Site* is a plot of land (a construction *site*) or the place where something is located (a Web *site*).

clarity

Clarity is essential to effective communication with your <u>readers</u>. You cannot achieve your <u>purpose</u> or a goal like <u>persuasion</u> without clarity. Many factors contribute to clarity, just as many other elements can defeat it.

A logical <u>method of development</u> and an <u>outline</u> will help you avoid presenting your reader with a jumble of isolated thoughts. A method of development and outline that puts your thoughts in a logical, meaningful sequence brings <u>coherence</u> as well as <u>unity</u> to your writing. Clear <u>transition</u> contributes to clarity by providing the smooth flow that enables the reader to connect your thoughts with one another without conscious effort.

Proper <u>emphasis</u> and <u>subordination</u> are mandatory if you want to achieve clarity. If you do not use those two complementary techniques wisely, all your clauses and sentences will appear to be of equal importance. Your reader will only be able to guess which are most important, which are least important, and which fall somewhere in the middle. The <u>pace</u> at which you present your ideas is also important to clarity; if the pace is not carefully adjusted to both the topic and the reader, your writing will appear cluttered and unclear.

<u>Point of view</u> establishes through whose eyes or from what vantage point the reader views the subject. A consistent point of view is essential to clarity; if you switch from the first person to the third person in midsentence, you are certain to confuse your reader.

Precise <u>word choice</u> contributes to clarity and helps eliminate <u>ambiguity</u> and <u>awkwardness</u>. <u>Vague words</u>, <u>clichés</u>, poor use of <u>idiom</u>, and inappropriate <u>usage</u> detract from clarity.

That <u>conciseness</u> is a requirement of clearly written communication should be evident to anyone who has ever attempted to decipher an insurance policy or a legal contract. Although words are our chief means of communication, too many of them can impede clear communication, just as too many cars on a highway can impede traffic. For the sake of clarity, remove unnecessary words from your writing.

clauses

C

A *clause* is a group of words that contains a subject and a predicate and that functions as a sentence or as part of a sentence. (See <u>sentence construction</u>.) Every subject-predicate word group in a sentence is a clause, and every sentence must contain at least one independent clause; otherwise, it is a <u>sentence fragment</u>.

A clause that could stand alone as a simple sentence is an *independent clause.*

- *The scaffolding fell* when the rope broke.

A clause that could not stand alone if the rest of the sentence were deleted is a *dependent* (or *subordinate*) *clause.*

- I was at the St. Louis branch *when the decision was made.*

Dependent clauses are useful in making the relationship between thoughts clearer and more succinct than if the ideas were presented in a series of simple sentences or compound sentences.

| FRAGMENTED | The recycling facility is located between Millville and Darrtown. Both villages use it. [two thoughts of approximately equal importance] |
| SUBORDINATED | The recycling facility, *which is located between Millville and Darrtown,* is used by both villages. [one thought subordinated to the other] |

Subordinate clauses are especially effective for expressing thoughts that describe or explain another statement. Too much <u>subordination</u>, however, can be confusing.

- He selected instructors whose classes ~~had a slant that was~~ *were* specifically directed ~~toward students who intended to go into engineering.~~ *at engineering students.*

A clause can be connected with the rest of its sentence by a coordinating <u>conjunction</u>, a subordinating conjunction, a relative <u>pronoun</u>, or a conjunctive <u>adverb</u>.

- It was 500 miles to the facility, *so* we made arrangements to fly. [coordinating conjunction]

- Mission control will have to be alert *because* at launch the space laboratory will contain a highly flammable fuel. [subordinating conjunction]

- It was Robert M. Fano *who* designed and developed the earliest multiple-access computer system at MIT. [relative pronoun]

- It was dark when we arrived; *nevertheless,* we began the tour of the facility. [conjunctive adverb]

clichés

Clichés are expressions that have been used for so long that they are no longer fresh but come to mind easily because they are so familiar. Clichés are often wordy as well as vague and can be confusing, especially to nonnative speakers of English. For each of the following clichés, a better, more direct word or phrase is given.

INSTEAD OF	USE
all over the map	scattered, unfocused
outside the box	unconventional, creative
last but not least	last, finally

Some writers use clichés in a misguided attempt to appear casual or spontaneous, just as other writers try to impress readers with trendy words (see also **affectation**). Although clichés may come to mind easily while you are **writing a draft**, eliminate them during **revision**. See also **buzzwords**, **conciseness/wordiness**, and **vague words**.

clipped words

When the beginning or the end of a word is cut off to create a shorter word, the result is a *clipped word* (lab, demo, memo, fax, Net). The word *specification,* for example, is often shortened to *spec*. Although acceptable in conversation, most clipped forms should not appear in writing unless they are commonly accepted as part of the special vocabulary of an occupational group. Some clipped forms, such as *phone, memo,* and *fax,* have nearly replaced their original, longer forms.

Apostrophes normally are not used with clipped forms of words (*fridge,* not *'fridge*). Because they are not strictly **abbreviations**, clipped forms are not followed by **periods** (*lab,* not *lab.*).

Do not use clipped forms of spelling (*thru, nite, lite,* and so on) in technical writing.

C

closings (*see* conclusions)

coherence

Writing is coherent when the relationships among ideas are clear to readers. Each idea should relate clearly to the other ideas, with one idea flowing smoothly to the next. The major components of coherent writing are a logical sequence of related ideas and clear transitions between these ideas. See also organization.

Presenting ideas in a logical sequence is the most important requirement in achieving coherence. The key to achieving a logical sequence is the use of a good outline (see outlining). An outline forces you to establish a beginning (introduction), a middle (body), and an end (conclusion). That structure contributes greatly to coherence by enabling you to experiment with sequences and to lay out the most direct route to your purpose without digressing.

Thoughtful transitions are also essential; without them, your writing cannot achieve the smooth flow from sentence to sentence and from paragraph to paragraph that is required for coherence.

During revision, check your draft carefully. If your writing is not coherent, you are not communicating effectively with your readers. See also clarity and unity.

collaborative writing

Collaborative writing occurs when two or more writers work together to produce a single document for which they share responsibility and decision-making authority. Collaborative writing teams are formed when (1) the size of a project or the time constraints imposed on it require collaboration, (2) the project involves multiple areas of expertise, or (3) the project requires the melding of divergent views into a single perspective that is acceptable to the whole team or to another group.

The collaborating writers strive to achieve a compatible working relationship by dividing the work in a way that uses each member's expertise and experience to their collective advantage. To do so, the team should designate one person as its coordinator. This person does not normally have decision-making authority—he or she merely coordinates the team members' activities and organizes the final project. If the team often works together, the coordinator's duties can be determined by mutual agreement or assigned on a rotating basis.

> **DIGITAL TIP USING COLLABORATIVE SOFTWARE**
>
> If you must collaborate over distance, collaborative software can offer innovative ways to bridge the gap. Free instant messaging software such as that from Yahoo!, AOL, and Jabber allows you to share files and conduct live text chats (or even audio and video chats). Some free chat tools like Microsoft's Messenger and Netmeeting even let you share a virtual whiteboard or your application windows with your collaborators, allowing you to jointly compose as if you were all in the same room. If you have access to more sophisticated groupware like Lotus Notes or Groove, you'll also be able to create and manage group schedules and organize and file materials in secure, group-accessible databases. For more on this topic, see <*www .bedfordstmartins.com/alred*> and select *Digital Tips.*

Tasks of the Collaborative Writing Team

The collaborative writing team normally performs four tasks. It plans the document, researches and writes the drafts, reviews the drafts of other team members, and revises drafts on the basis of these reviews.

Planning. The team collectively identifies the readers, purpose, and scope of the project. The team conceptualizes the document to be produced, creates a broad outline of the document, divides the document into segments, and assigns each segment to individual team members, often on the basis of expertise. For a review of the writing process, see also "Five Steps to Successful Writing."

In the planning stage, the team projects a schedule and sets any writing style standards that the team is expected to follow. The schedule includes due dates for drafts, reviews of the drafts, revisions, and the final document. Milestone deadlines must be met, even if the drafts are not as polished as the individual writers would like: One missed deadline can delay the entire project.

Research and Writing. Planning is followed by research and writing, a period of intense independent activity by members of the team. Each member researches his or her assigned segment of the document, fleshes out the broad outline in greater detail, and produces a draft from the detailed outline. (See outlining.) Then, by the deadline established for the drafts, writers submit copies of the drafts to their teammates for review.

Reviewing. During the review stage, team members assume the role of the reader to address any potential problems. Each member critically

C

yet diplomatically reviews the work of the other team members. Reviewers evaluate colleagues' drafts, from the **organization** to the **clarity** of each **paragraph** and sentence. They offer advice to help the writer improve his or her segment of the document. Team members can easily solicit feedback by sharing files on a network system, by e-mailing documents back and forth, or by exchanging disks. Redlining or highlighting allows the reviewer to show the suggested changes without deleting the original text. The author can then easily accept or reject the proposed changes. See also **proofreaders' marks**.

Revising.   In this stage, individual writers evaluate their colleagues' reviews and accept or reject their suggestions. Once each member revises his or her draft, all drafts can be consolidated into a final master copy maintained by the team coordinator. Integral to this process is that team members evaluate their colleagues' suggestions objectively and accept criticism constructively. See also **revision**.

Conflict

Team members may not agree on every subject, and differing perspectives can easily lead to conflict, ranging from mild differences over minor points to major showdowns. However, creative differences resolved respectfully can energize the team and, in fact, strengthen a finished document by compelling writers to reexamine assumptions and issues in unanticipated ways.

Writer's Checklist: Writing Collaboratively

☑ Designate one person as the team coordinator.

☑ Collectively identify the audience, purpose, and scope of the project.

☑ Create a working outline of the document.

☑ Assign segments or tasks to each team member.

☑ Establish a schedule: due dates for drafts, revisions, and final documents.

☑ Agree on a standard reference guide for style and format.

☑ Research and write drafts of document segments.

☑ Exchange segments for team member reviews.

☑ Revise segments as needed.

☑ Meet your established deadlines.

As you collaborate, be ready to tolerate some disharmony, but temper it with mutual respect.

colons

The *colon* (:) is a mark of introduction that alerts readers to the close connection between the first statement and what follows.

A colon is used to connect a list or series to a word, clause, or phrase that identifies or renames another expression (thus, in apposition).

- Two topics will be discussed: the new accounting system and the new bookkeeping procedures.

Do not, however, place a colon between a <u>verb</u> and its <u>objects</u>.

- Three fluids that clean pipettes are water, alcohol, and acetone.

Do not use a colon between a <u>preposition</u> and its object.

- I would like to be transferred to Phoenix, Boston, or Miami.

One common exception is made when a preposition or verb is followed by a stacked <u>list</u>; however, it may be possible to make a complete sentence.

- *The following corporations* :
  Corporations that manufacture computers include:

Apple	Toshiba	Micron
IBM	Dell	Gateway

A colon is used to link one statement to another statement that develops, explains, amplifies, or illustrates the first.

- Any organization is confronted with two separate, though related, information problems: It must maintain an effective internal communication system and an effective external communication system.

A colon is used to link an <u>appositive</u> phrase to its related statement if more emphasis is needed and if the phrase comes at the end of the sentence.

- There is only one thing that will satisfy Mr. Sturgess: our finished report.

Colons are used to link numbers that signify different nouns.

- 9:30 a.m. [9 hours, 30 minutes]
- Matthew 14:1 [chapter 14, verse 1]

C

In proportions, colons indicate ratios of amounts to each other (7:3 = 14:*x*). A colon follows the salutation in business <u>correspondence</u>, even when the salutation refers to a person by first name.

- Dear Ms. Jeffers: *or* Dear George:

A colon always goes outside <u>quotation marks</u>.

- This was the real meaning of his "suggestion": The division must show a profit by the end of the year.

When quoting material that ends in a colon, drop the colon and replace it with an <u>ellipsis</u>. (See also <u>quotations</u>.)

- "Any large corporation is confronted with two separate, though re-

  lated, information problems . . . . "

The first word after a colon may be capitalized if the statement following the colon is a complete sentence or a formal resolution or question. (See also <u>capitalization</u>.)

- The conference passed a single resolution: Voting will be open to associate members next year.

If the element following the colon is subordinate, use a lowercase letter to begin that element.

- There is only one way to stay within our present budget: to reduce expenditures for research and development.

---

comma splice

A *comma splice* is a grammatical error in which two independent <u>clauses</u> are joined by only a <u>comma</u>.

> **INCORRECT**   It was 500 miles to the facility, we arranged to fly.

A comma splice can be corrected in several ways.

1. Substitute a <u>semicolon</u>, a semicolon and a conjunctive <u>adverb</u>, or a comma and a coordinating <u>conjunction</u>.

   It was 500 miles to the facility*;* we arranged to fly.

   It was 500 miles to the facility*; therefore,* we arranged to fly.

   It was 500 miles to the facility, *so* we arranged to fly.

2. Create two sentences.

   It was 500 miles to the facility. *We* arranged to fly.

3. Subordinate one clause to the other.

    Because it was 500 miles to the facility, we arranged to fly.

commas

Like all **punctuation**, the *comma* (,) helps readers understand the writer's meaning and prevents **ambiguity**. Notice how the comma helps make the meaning clear in the second example.

    AMBIGUOUS   To be successful managers with MBAs must continue to learn.
                [At first glance, the sentence seems to be about "successful managers with MBAs."]

    CLEAR   To be successful, managers with MBAs must continue to learn.
                [The comma makes clear where the main part of the sentence begins.]

Do not follow the old myth that you should insert a comma wherever you would pause if you were speaking. Although you would pause wherever you encounter a comma, you should not insert a comma wherever you might pause. Effective use of commas depends instead on an understanding of **sentence construction**.

Linking Independent Clauses

Use a comma before a coordinating **conjunction** (*and, but, or, nor,* and sometimes *so, yet,* and *for*) that links independent **clauses**.

- The new microwave disinfection system was delivered, *but* the installation will require an additional week.

C

However, if two independent clauses are short and closely related—and there is no danger of confusing the reader—the comma may be omitted. Both of the following examples are correct.

- The cable snapped and the power failed.
- The cable snapped, and the power failed.

Enclosing Elements

Commas are used to enclose nonessential information in nonrestrictive clauses, **phrases**, and parenthetical elements. (For other means of punctuating parenthetical elements, see **dashes** and **parentheses** as well as **restrictive and nonrestrictive elements**.)

- Our new factory, *which began operations last month,* should add 25 percent to total output. [nonrestrictive clause]
- The accountant, *working quickly and efficiently,* finished early. [nonrestrictive phrase]
- We can, *of course,* expect their lawyer to call us. [parenthetical element]

Yes and *no* are set off by commas in such uses as the following:

- I agree with you, *yes.*
- *No,* I do not think we can finish as soon as we would like.

A **direct address** should be enclosed in commas.

- You will note, *Mark,* that the surface of the brake shoe complies with the specifications.

An appositive phrase (which re-identifies another expression in the sentence) is enclosed in commas.

- Our company, *Envirex Medical Systems,* won several awards last year.

Interrupting parenthetical and transitional words or phrases are usually set off with commas. See also **transition**.

- The report, *therefore,* needs to be revised.
- The estimated cost, *it turns out,* was incorrect.

Commas are omitted when the word or phrase does not interrupt the continuity of thought.

- I *therefore* suggest that we begin construction.

Introducing Elements

Clauses and Phrases.   It is generally a good idea to put a comma after an introductory clause or phrase. Identifying where the introductory element ends helps indicate where the main part of the sentence begins.

Always place a comma after a long introductory clause.

- *Because we have not yet contained the new strain of influenza,* we recommend vaccination for high-risk patients.

A long modifying phrase that precedes the main clause should always be followed by a comma.

- *During the first series of field-performance tests at our Colorado proving ground,* the new engine failed to meet our expectations.

When an introductory phrase is short and closely related to the main clause, the comma may be omitted.

- *In two seconds* a 20° temperature rise occurs in the test tube.

A comma should always follow an absolute phrase, which modifies the whole sentence.

- *The tests completed,* we organized the data for the final report.

Words and Quotations.   Certain types of introductory words are followed by a comma. One example is a transitional word or phrase (*however, in addition*) that connects the preceding clause or sentence with the thought that follows.

- *Furthermore,* steel can withstand a humidity of 99 percent, provided that there is no chloride or sulfur dioxide in the atmosphere.
- *For example,* this change will make us more competitive in the global marketplace.

When <u>adverbs</u> closely modify the <u>verb</u> or the entire sentence, they should not be followed by a comma.

- *Perhaps* we can still solve the environmental problem.
- *Certainly* we should try.

A proper noun used in an introductory direct address or <u>interjection</u> (such as *oh, well, why, indeed, yes,* and *no*) is followed by a comma.

- *Nancy,* enclosed is the article you asked me to review.
- *Indeed,* I will ensure that your request is forwarded.

Use a comma to separate a direct quotation from its introduction.

- Morton and Lucia White said, "People live in cities but dream of the countryside."

Do not use a comma when giving an indirect quotation. See also <u>quotations</u>.

- Morton and Lucia White said that people dream of the countryside, even though they live in cities.

Separating Items in a Series

Although the comma before the last item in a series is sometimes omitted, it is generally clearer to include it. The <u>ambiguity</u> that may result from omitting the comma is illustrated in the following sentence.

AMBIGUOUS	Random House, Bantam, Doubleday and Dell were individual publishing companies. [Does "Doubleday and Dell" refer to one company or two?]
CLEAR	Random House, Bantam, Doubleday, and Dell were individual publishing companies.

Phrases and clauses in coordinate series, like words, are punctuated with commas.

- Plants absorb noxious gases, act as receptors of dirt particles, and cleanse the air of other impurities.

When <u>adjectives</u> modifying the same noun can be reversed and make sense, or when they can be separated by *and* or *or,* they should be separated by commas.

- The drawing was of a *modern, sleek, swept-wing* airplane.

When an adjective modifies a phrase, no comma is required.

- She was investigating the *damaged inventory-control system.*
  [The adjective *damaged* modifies the phrase *inventory-control system.*]

Never separate a final adjective from its noun.

- He is a conscientious, honest, reliable, worker.

Clarifying and Contrasting

If you find you need a comma to prevent misreading when a word is repeated, rewrite the sentence.

AWKWARD   The assets we had, had surprised us.

IMPROVED   We had been surprised at our assets.

Use a comma to separate two contrasting thoughts or ideas.

- The project was finished on time, but not within the budget.

Use a comma after an independent clause that is only loosely related to the dependent clause that follows it or that could be misread without the comma.

- I should be able to finish the plan by July, even though I lost time because of illness.

See also restrictive and nonrestrictive elements.

Showing Omissions

A comma sometimes replaces a verb in certain elliptical constructions.

- Some were punctual; *others, late.* [The comma replaces *were.*]

It is better, however, to avoid such constructions in workplace writing.

Using with Other Punctuation

Conjunctive adverbs (*however, nevertheless, consequently, for example, on the other hand*) that join independent clauses are preceded by a semicolon and followed by a comma. Such adverbs function both as modifiers and as connectives.

- The idea is good; *however,* our budget is not sufficient.

Use a semicolon to separate phrases or clauses in a series when one or more of the phrases or clauses contain commas.

- Our new products include amitriptyline, which has sold very well; dipyridamole, which has not sold well; and cholestyramine, which was just released.

When an introductory phrase or clause ends with a <u>parenthesis</u>, the comma separating the introductory phrase or clause from the rest of the sentence always appears outside the parenthesis.

C

- Although we left late (at 7:30 p.m.), we arrived in time for the keynote address.

Commas always go inside <u>quotation marks</u>.

- The operator placed the discharge bypass switch at "normal," which triggered a second discharge.

Except with <u>abbreviations</u>, a comma should not be used with a <u>period</u>, <u>question mark</u>, <u>exclamation mark</u>, or <u>dash</u>.

- "Have you finished the project?" I asked.

Using with Numbers and Names

Commas are conventionally used to separate distinct items. Use commas between the elements of an address written on the same line (but not between the state and the zip code).

- Kristen James, 4119 Mill Road, Dayton, Ohio 45401

A date can be written with or without a comma following the year if the date is in the month-day-year format.

- October 26, 2002, was the date the project began.
- October 26, 2002 was the date the project began.

If the date is in the day-month-year format, as is typical in <u>international correspondence</u>, do not set off the date with commas.

- The project begun on 26 October 2002 was completed ahead of schedule.

Use commas to separate the elements of Arabic numbers (1,528,200 feet). However, because many countries use the comma as the decimal marker, consider using periods or spaces rather than commas in international documents.

- 1.528.200 meters *or* 1 528 200 meters

A comma may be substituted for the colon in the salutation of a personal letter or <u>e-mail</u>. Do not, however, use a comma in a business letter, even if you use the person's first name.

- Dear Marie, [personal letter or e-mail]
- Dear Marie: [business letter or e-mail]

Use commas to separate the elements of geographical names.

- Toronto, Ontario, Canada

Use a comma to separate names that are reversed or that are followed by an abbreviation.

- Smith, Alvin
- Jane Rogers, Ph.D.

Avoiding Unnecessary Commas

A number of common writing errors involve placing commas where they do not belong. As stated earlier, such errors often occur because writers assume that a pause in a sentence should be indicated by a comma. Be careful not to place a comma between a subject and verb or between a verb and its <u>object</u>.

- The conditions at the test site in the Arctic,/ made accurate readings difficult.
- She has often said,/ that one company's failure is another's opportunity.

Do not use a comma between the elements of a compound subject or a compound predicate consisting of only two elements.

- The director of the design department,/ and the supervisor of the quality-control section were opposed to the new schedules.
- The design director listed five major objections,/ and asked that the new schedule be reconsidered.

Placing a comma after a coordinating conjunction such as *and* or *but* is a common error.

- The chairperson formally adjourned the meeting, *but,/* the members of the committee continued to argue.

Do not place a comma before the first item or after the last item of a series.

- The new products we are considering include,/ calculators, scanners, and cameras.
- It was a fast, simple, inexpensive,/ process.

Do not use a comma to separate a prepositional phrase from the rest of the sentence unnecessarily.

- We discussed the final report,/ on the new project.

committee

C

Committee is a collective <u>noun</u> that usually takes a singular <u>verb</u> or <u>pronoun</u>. (The *committee is* to meet at 3:30 p.m.) When the individuals in a group are thought of separately, rather than as a unit, the collective noun may be considered plural. (The *committee* returned to *their* offices.) A better way to emphasize the individuals is to use a phrase with the plural form. (*The committee members* returned to *their* offices.)

compare / contrast

When you *compare* things, you point out similarities or both similarities and differences. (He *compared* the two brands before making his choice.) When you *contrast* things, you point out only the differences. (Their speaking styles *contrasted* sharply.) In either case, you compare or contrast only things that are part of a common category.

When *compare* is used to establish a general similarity, it is followed by *to*. (*Compared to* the computer, the typewriter is a primitive device.) When *compare* is used to indicate a close examination of similarities or differences, it is followed by *with*. (We *compared* the features of the new copier *with* those of the old one.)

Contrast is normally followed by *with*. (The new policy *contrasts* sharply *with* the earlier one.) When the <u>noun</u> form of *contrast* is used, one speaks of the *contrast between* two things or of one thing being in *contrast to* the other.

- There is a sharp *contrast between* the old and new policies.
- The new policy is in sharp *contrast to* the earlier one.

comparison

When you are making a comparison, be sure that both or all of the elements being compared are clearly evident to your <u>reader</u>.

- The Nicom 3 software is better *than the Nicom 2 software* .

The things being compared must be of the same kind.

- Imitation alligator hide is almost as tough as ~~a~~ real alligator *hide* .

Be sure to point out the parallels or differences between the things being compared. Do not assume your reader will know what you mean.

- Washington is farther from Boston than ^*it is from* Philadelphia.

A double comparison in the same sentence requires that the first comparison be completed before the second one is stated.

- The discovery of electricity was one of the great ~~if not the greatest~~ scientific discoveries in history ^*, if not the greatest* .

Do not attempt to compare things that are not comparable.

- Farmers say that ~~storage space is reduced by 40 percent compared with~~ *requires 40 percent less storage space than loose hay requires* baled hay ^ .

  [*Storage space* is not comparable to *baled hay*.]

comparison method of development

As a <u>method of development,</u> comparison points out similarities and differences between the elements of your subject. The comparison method of development can be especially effective because it can explain a difficult or unfamiliar subject by relating it to a simpler or more familiar one.

You must first determine the basis for the <u>comparison</u>. For example, if you were comparing bids from contractors for a remodeling project at your company, you most likely would compare such factors as price, previous experience, personnel qualifications, availability at a time convenient for you, and completion date.

Once you have determined the basis (or bases) for comparison, you can determine the most effective way to structure your comparison: whole by whole or part by part. In the *whole-by-whole method,* all the relevant characteristics of one item are discussed before those of the next item are considered. In the *part-by-part method,* the relevant features of each item are compared one by one. The discussion of typical woodworking glues in Figure C–3 on page 94 is organized according to the whole-by-whole method, so it describes each type of glue and its characteristics before going on to the next one. The purpose of this whole-by-whole method of comparison is to weigh the advantages and disadvantages of each glue for certain kinds of woodworking. The comparison could be expanded, of course, by comparing additional types of glue.

C

White glue is the most useful all-purpose adhesive for light construction, but it cannot be used on projects that will be exposed to moisture, high temperature, or great stress. Wood that is being joined with white glue must remain in a clamp until the glue dries, which takes about 30 minutes.

Aliphatic resin glue has a stronger and more moisture-resistant bond than white glue. It must be used at temperatures above 50°F. The wood should be clamped for about 30 minutes. . . .

Plastic resin glue is the strongest of the common wood adhesives. It is highly moisture resistant, though not completely waterproof. Sold in powdered form, this glue must be mixed with water and used at temperatures above 70°F. It is slow setting, and the joint should be clamped for four to six hours. . . .

Contact cement is a very strong adhesive that bonds so quickly it must be used with great care. It is ideal for mounting sheets of plastic laminate on wood. It is also useful for attaching strips of veneer to the edges of plywood. Because this adhesive bonds immediately when two pieces are pressed together, clamping is not necessary, but the parts to be joined must be carefully aligned before being placed together. Most brands are flammable, and the fumes can be harmful if inhaled. To meet current safety standards, this type of glue must be used in a well-ventilated area, away from flames or heat.

FIGURE C–3. Whole-by-Whole Method of Comparison

If your purpose were to consider, one at a time, the various characteristics of all the glues, the information might be arranged according to the part-by-part method of development, as in Figure C–4. The part-by-part method could accommodate further comparison. Comparisons might be made according to temperature ranges, special warnings, common use, and so on.

Woodworking adhesives are rated primarily according to their bonding strength, moisture resistance, and setting times.

Bonding strengths are categorized as very strong, moderately strong, or adequate for use with little stress. Contact cement and plastic resin glue bond very strongly, while aliphatic resin glue bonds moderately strongly. White glue provides a bond least resistant to stress.

The *moisture resistance* of woodworking glues is rated as high, moderate, and low. Plastic resin glues are highly moisture resistant, aliphatic resin glues are moderately moisture resistant, and white glue is least moisture resistant.

Setting times for the glues vary from an immediate bond to a four-to-six-hour bond. Contact cement bonds immediately and requires no clamping. Because the bond is immediate, surfaces being joined must be carefully aligned before being placed together. White glue and aliphatic resin glue set in 30 minutes; both require clamping to secure the bond. Plastic resin, the strongest wood glue, sets in four to six hours and also requires clamping.

FIGURE C–4. Part-by-Part Method of Comparison

complaint letters

The <u>tone</u> of a complaint letter or e-mail is important; the most effective ones do not sound complaining. If you write a letter that reflects only your annoyance and anger, you may not be taken seriously. Assume that the recipient will be conscientious in correcting the problem. However, anticipate reader reactions or rebuttals.

● I reviewed carefully the "safe operating guidelines" in the user manual before I installed the device.

Without such explanations, readers may be tempted to dismiss your complaint.

Although the circumstances and severity of the problem may vary, effective complaint letters generally follow this pattern:

1. Identify the problem or faulty item(s) and include relevant invoice numbers, part names, and dates. It is often a good idea to include a copy of the receipt, bill, or contract.
2. Explain logically, clearly, and specifically what went wrong, especially for a problem with a service. (Avoid guessing why you *think* some problem occurred.)
3. State what you expect the reader to do to solve the problem.

Begin by checking to see if the company's Web site provides instructions for submitting a complaint. Otherwise, for large organizations, you may address your complaint to Customer Service. In smaller organizations, you might write to a vice president in charge of sales or service, or directly to the owner. As a last resort, you may find that sending copies of a complaint letter to more than one person in the company will get faster results. Figure C–5 on page 96 shows a typical complaint letter. See also <u>adjustment letters</u> and <u>refusal letters</u>.

complement / compliment

Complement means "anything that completes a whole" (see also <u>complements</u>). It is used as either a <u>noun</u> or a <u>verb</u>.

● A *complement* of four employees would bring our staff up to its normal strength. [noun]

● The two programs *complement* one another perfectly. [verb]

C

BAKER MEMORIAL HOSPITAL

Diagnostic Services Department
501 Main Street
Springfield, OH 45321
(513) 683-8100
Fax (513) 683-8000

September 23, 2003

Manager, Customer Relations
Computer Solutions, Inc.
521 West 23rd Street
New York, NY 10011

Subject: HV3 Monitors

On July 9, I ordered nine HV3 monitors for your model MX-15 diagnostic scanner. The monitors were ordered from your Web site.

On August 2, I received from your Newark, New Jersey, parts warehouse seven HL monitors. I immediately returned these monitors with a note indicating the mistake that had been made. However, not only have I failed to receive the HV3 monitors I ordered, but I have also been billed repeatedly.

I have enclosed a copy of my confirmation e-mail, the shipping form, and the most recent bill. If you cannot send me the monitors I ordered by November 1, please cancel my order.

Sincerely,

Paul Denlinger

Paul Denlinger
Manager
pld@baker.org

Enclosures

www.baker.org

FIGURE C–5. Complaint Letter

Compliment means "praise." It too is used as either a noun or a verb.

- The manager's *compliment* boosted staff morale. [noun]
- The manager *complimented* the staff on its efficient job. [verb]

complements (grammar)

A *complement* is a word, <u>phrase</u>, or <u>clause</u> used in the predicate of a sentence to complete the meaning of the sentence. See also <u>sentence construction</u>.

- Pilots fly *airplanes*. [word]
- To live is *to risk death*. [phrase]
- John knew *that he would be late*. [clause]

Four kinds of complements are generally recognized: direct object, indirect object, objective complement, and subjective complement.

A *direct object* is a <u>noun</u> or noun equivalent that receives the action of a transitive <u>verb</u>; it answers the question *What?* or *Whom?* after the verb.

- I designed *a Web page*. [noun]
- I like *to work*. [verbal]
- I like *her*. [pronoun]
- I like *what I saw*. [noun clause]

An *indirect object* is a noun or noun equivalent that occurs with a direct <u>object</u> after certain kinds of transitive verbs such as *give, wish, cause,* and *tell*. It answers the question *To whom or to what?* or *For whom or for what?*

- We should buy the *Milwaukee office* a *color copier*.
  [*Color copier* is the direct object and *Milwaukee office* is the indirect object.]

An *objective complement* completes the meaning of a sentence by revealing something about the object of its transitive verb. An objective complement may be either a noun or an <u>adjective</u>.

- They call him *a genius*. [noun phrase]
- We painted the building *white*. [adjective]

A *subjective complement,* which follows a linking verb rather than a transitive verb, describes the subject. A subjective complement may be either a noun or an adjective.

- Her sister is *a consultant*. [noun phrase]
- His brother is *ill*. [adjective]

compose / constitute / comprise

Compose and *constitute* both mean "make up the whole." The parts *compose* or *constitute* the whole.

- The 13 offices *compose* the division.
- The contract outlined activities that *constitute* cause for dismissal.

Comprise means "include," "contain," or "consist of." The whole *comprises* the parts. (The division *comprises* 13 offices.)

compound words

A *compound word* is made from two or more words that function as a single concept. A compound may be hyphenated, written as one word, or written as separate words. If you are not certain whether a compound word should use a <u>hyphen</u>, check a dictionary.

editor-in-chief	high-energy	low-level
courthouse	nevertheless	online
home page	post office	Web site

Be careful to distinguish between compound words (*greenhouse*) and words that simply appear together but do not constitute compound words (*green house*).

For plurals of compound words, generally add *s* to the last letter (*bookcases* and *Web sites*). However, when the first word of the compound is more important to its meaning than the last, the first word takes the *s* (*editors-in-chief*). Possessives are formed by adding '*s* to the end of the compound word (the *post office's* hours; the *pipeline's* diameter). See also <u>possessive case</u>.

concept / conception

A *concept* is a thought or an idea. A *conception* is the sum of a person's ideas, or concepts, on a subject. (From the *concept* of compression, Rudolf Diesel developed the *conception* of the internal combustion engine.)

conciseness / wordiness

Conciseness means that extraneous words, <u>phrases</u>, <u>clauses</u>, and sentences have been removed from writing without sacrificing <u>clarity</u> or appropriate detail. Conciseness is not a synonym for brevity; a long report may be concise, while its <u>abstract</u> may be brief and concise. Conciseness is always desirable, but brevity may or may not be desirable in a given passage, depending on the writer's objective. Although concise sentences are not guaranteed to be effective, wordy sentences always sacrifice some of their readability and coherence.

Causes of Wordiness

<u>Modifiers</u> that repeat an idea implicit or present in the word being modified contribute to wordiness by being redundant. See also <u>reason is because</u>.

basic essentials	*completely* finished
final outcome	*present* status

Coordinated synonyms that merely repeat each other contribute to wordiness.

each and every	*basic and fundamental*
finally and for good	*first and foremost*

Excess qualification also contributes to wordiness.

perfectly clear	*completely* accurate

<u>Expletives</u>, relative <u>pronouns</u>, and relative <u>adjectives</u>, although they have legitimate purposes, often result in wordiness.

WORDY   *There are* [expletive] many Web designers *who* [relative pronoun] are planning to attend the conference, *which* [relative adjective] is scheduled for May 13–15.

CONCISE   Many Web designers plan to attend the conference scheduled for May 13–15.

Circumlocution (a long, indirect way of expressing things) is a leading cause of wordiness. See also <u>gobbledygook</u>.

WORDY   The payment to which a subcontractor is entitled should be made promptly so that in the event of a subsequent contractual dispute we, as general contractors, may not be held in default of our contract by virtue of nonpayment.

C

CONCISE	Pay subcontractors promptly. Then if a contractual dispute occurs, we cannot be held in default of our contract because of nonpayment.

When conciseness is overdone, writing can become choppy and ambiguous. (See also **telegraphic style**.) Too much conciseness can produce a style that is not only too brief but also too blunt, especially in **correspondence**.

Writer's Checklist: Achieving Conciseness

Wordiness is understandable when you are **writing a draft**, but it should not survive **revision**.

☑ Use **subordination** to achieve conciseness.

- The financial report was carefully ~~documented, and it covered five pages.~~
  - *five-page* ^ *documented.* ^

☑ Use simple words and phrases.

WORDY	It is the policy of the company to provide Web access to enable employees to conduct the online communication necessary to discharge their responsibilities; such should not be utilized for personal communications or nonbusiness activities.
CONCISE	Employee Web access should be used only for appropriate company business.

☑ Eliminate redundancy.

WORDY	Postinstallation testing, which is offered to all our customers at no further cost to them whatsoever, is available with each Line Scan System One purchased from this company.
CONCISE	Free postinstallation testing is offered with each Line Scan System One.

☑ Change the passive **voice** to the active and the indicative **mood** to the imperative.

WORDY	Bar codes normally are used when an order is intended to be displayed on a computer, and inventory numbers normally are used when an order is to be placed with the manufacturer.
CONCISE	Use bar codes to display the order on a computer, and use inventory numbers to place the order with the manufacturer.

Writer's Checklist: Achieving Conciseness (continued)

☑   Eliminate or replace wordy introductory phrases or pretentious words and phrases (*it may be said that, it appears that, in the case of, needless to say*). See <u>affectation</u>.

REPLACE	WITH
in order to, with a view to	to
due to the fact that, for the reason that, owing to the fact that, the reason for	because
by means of, by using, in connection with, through the use of	by, with
at this time, at this point in time, at present, at the present	now

☑   Do not overuse modifiers, such as *very, more, most, best, quite, great, really, especially*. Instead provide specific and useful details. See <u>intensifiers</u> and <u>very</u>.

conclusions

The *conclusion* of a document ties the main ideas together and can do so emphatically and persuasively by making a final significant point. This final point may, for example, recommend a course of action, make a prediction or a judgment, or merely summarize main points.

The way you conclude depends on both the <u>purpose</u> of your writing and your <u>readers'</u> needs. For example, a committee <u>report</u> about possible locations for a new production facility might end with a recommendation; a lengthy sales proposal might conclude persuasively with a summary of the proposal's salient points and the company's main strong points. The following examples are typical concluding strategies.

RECOMMENDATION   These results indicate that you need to alter your testing procedure to eliminate the impurities we found in specimens A through E.

PREDICTION   Although I have exceeded my original estimate for equipment, I have reduced my original labor estimate; therefore, I will easily stay within the original bid.

JUDGMENT   Although our estimate calls for a substantially higher budget than in the three previous years, we believe that it is reasonable given our planned expansion.

C

SUMMARY	As this letter indicates, we would attract more recent graduates with the following strategies:
	1. Establish a Web site where students can register and submit online résumés.
	2. Increase our advertising in local student newspapers and our attendance at college career fairs.
	3. Expand our local co-op program.

The concluding statement may merely present ideas for consideration, call for action, or deliberately provoke thought.

IDEAS FOR CONSIDERATION	The new prices become effective the first of the year. Price adjustments are routine for the company, but some of your customers will not consider them acceptable. Please bear in mind the needs of both your customers and the company as you implement these new prices.
CALL FOR ACTION	Send us a check for $250 now if you wish to keep your account active. If you have not responded to our previous letters because of some special hardship, I will be glad to work out a solution with you personally.
THOUGHT-PROVOKING STATEMENT	Can we continue to accept the losses incurred by inefficiency? Or should we consider steps to control it now?

Be especially careful not to introduce a new topic when you conclude. A conclusion should always relate to and reinforce the ideas presented earlier in your writing. Moreover, the conclusions must be consistent with what the <u>introduction</u> promised the report would examine (its purpose) and how it would do so (its method). Figure C–6 is a conclusion from a proposal to reduce health-care costs by increasing employee fitness through health-club subsidies; it makes recommendations that pull the various parts of the proposal together.

For guidance about the location of the conclusion section in a report, see <u>formal reports</u>. For letter and other short closings, see <u>correspondence</u> and entries on specific types of documents throughout this book.

CONCLUSION AND RECOMMENDATION

As shown earlier, building and equipping fitness centers at all five company locations would require an initial investment of nearly $2 million. Such facilities would also occupy valuable office space. Therefore, this option would be costly.

Enrolling employees in the corporate program at AeroFitness would allow them to attend on a trial basis. Those interested in continuing could join the club and pay half of the $400 annual membership cost, less a 30-percent discount. The other half of the membership ($140) would be paid for by First Investment. Employees who leave the company would be given the option to purchase First Investment's share of the membership.

I recommend that First Investment, Inc., participate in the corporate membership program at AeroFitness Clubs by subsidizing employee memberships. First Investment benefits from such a program in several ways: We demonstrate our commitment to a fit workforce, we augment our already generous package, and we boost employee morale. Most importantly, implementing this program will help First Investment, Inc., reduce its health-care costs both by building a healthier workforce and by qualifying for insurance premium discounts.

FIGURE C-6. Conclusion

conjunctions

A *conjunction* connects words, **phrases**, or <u>clauses</u> and can also indicate the relationship between the elements it connects.

A *coordinating conjunction* joins two sentence elements that have identical functions. The coordinating conjunctions are *and, but, or, for, nor, yet,* and *so.*

- Nature *and* technology are only two conditions that affect petroleum operations around the world. [joins two <u>nouns</u>]

- To hear *and* to listen are two different things. [joins two phrases]

- I would like to include the survey, *but* that would make the report too long. [joins two clauses]

Correlative conjunctions are used in pairs. The correlative conjunctions are *either . . . or, neither . . . nor, not only . . . but also, both . . . and,* and *whether . . . or.*

- The auditor will arrive *either* on Wednesday *or* on Thursday.

A *subordinating conjunction* connects sentence elements of different weights, normally independent and dependent clauses. The most frequently used subordinating conjunctions are *so, although, after, because, if, where, than, since, as, unless, before, that, though, when,* and *whereas.*

- I left the office *after* finishing the report.

A *conjunctive adverb* has the force of a conjunction because it joins two independent clauses. The most common conjunctive <u>adverbs</u> are *however, moreover, therefore, further, then, consequently, besides, accordingly, also,* and *thus.*

- The engine performed well in the laboratory; *however,* it failed under road conditions.

Coordinating conjunctions in the titles of books, articles, plays, and movies should not be capitalized unless they are the first or last word in the title. See also <u>capitalization</u>.

- The *Journal of Business and Technical Communication* is in our library.

Occasionally, a conjunction may begin a sentence; in fact, conjunctions can be strong transitional words and at times can provide <u>emphasis</u>. See also <u>transition</u>.

- I realize that the project is more difficult than expected and that you have encountered personnel problems. *But* we must meet our deadline.

connotation / denotation

The *denotations* of a word are its literal meanings, as defined in a dictionary. The *connotations* of a word are its meanings and associations beyond its literal, dictionary definitions. For example, the denotations of *Hollywood* are "a district of Los Angeles" and "the U.S. movie industry as a whole"; its connotations are "romance, glittering success, and superficiality." Use words with both the most accurate denotations and the most appropriate connotations. See also <u>defining terms</u> and <u>word choice</u>.

consensus

Because *consensus* normally means "harmony of opinion," the phrases *consensus of opinion* and *general consensus* are redundant. (See also <u>conciseness/wordiness</u>.) The word *consensus* can be used to refer only to a group, never to one or two people.

- The ~~general~~ consensus ~~of opinion~~ of the members was that the committee should change its name.

continual / continuous

Continual means "happening over and over" or "frequently repeated." (Writing well requires *continual* practice.) *Continuous* means "occurring without interruption" or "unbroken." (The *continuous* roar of the machinery was deafening.)

contractions

A *contraction* is a shortened spelling of a word or phrase with an **apostrophe** substituting for the missing letter or letters (*cannot/can't; have not/haven't; will not/won't; it is/it's*). Contractions are often used in speech and informal writing; they are generally not appropriate in **reports, proposals,** and formal **correspondence**. See also **technical writing style**.

copyright

The Copyright Act protects all original works from the moment of their creation, regardless of whether they are published or even contain a notice of copyright (©). However, all works created by U.S. government agencies are in the public domain — that is, they are not copyrighted — and can be used without prior approval. For information on the Copyright Act, visit the Library of Congress copyright Web site at <http://lcweb.loc.gov/copyright>.

Writer's Checklist: Using Copyrighted Materials (Printed and Online)

- ☑ In general, give credit to any source from which material is taken, unless it is boilerplate or common knowledge, as described in **plagiarism**. See also **documenting sources**.

- ☑ With the exception of works created by U.S. government agencies, any work first published after March 1, 1989, receives copyright protection regardless of whether it bears a notice of copyright. Works published before March 1, 1989, without a notice of copyright are in the public domain.

Writer's Checklist: Using Copyrighted Materials (Printed and Online) (continued)

C

☑ A small amount of material from a copyrighted source may be used without permission or payment as long as the use satisfies the fair-use criteria, as described at the Library of Congress Web site at <http://lcweb.loc.gov/copyright>.

☑ Copyright law applies to electronic works just as it does to their print counterparts. Web pages that are copyrighted include the symbol © and provide information on terms of use.

☑ If you plan to reproduce or further distribute copyrighted works posted on the Internet, you must obtain permission from the copyright holder, unless the fair-use provision of the copyright law applies to your intended use. To obtain permission, read the site's terms-of-use information and e-mail your request to the appropriate party. As with printed works, document the source of all materials (text, graphics, tables) obtained.

correspondence

DIRECTORY

The process of writing letters or **memos** (or **e-mails** that function as either) involves many of the same steps that go into writing most other documents, as described in "Five Steps to Successful Writing." One important consideration in correspondence is the impression you convey to **readers**. To convey a professional image—of yourself and your company or organization—take particular care with the **tone** and **style** of your writing. See also **technical writing style**.

Writing Style and Accuracy

Letter-writing style varies from informal (or casual), as in a letter to a close business associate, to formal (or restrained) as in a letter to someone you do not know.

INFORMAL	It worked! The new process is better than we had dreamed.
RESTRAINED	You will be pleased to know that the new process is more effective than we had expected.

You will probably find yourself using the restrained style more frequently than the casual one. Remember that an overdone attempt to sound casual or friendly can sound insincere. However, do not adopt so formal a style that your letters read like legal contracts. See also **affectation**.

AFFECTED	In response to your query, we no longer possess an original copy of the brochure requested. Please be advised that a photographic copy is enclosed herewith. Address further correspondence to this office for assistance as required.
IMPROVED	Because we are currently out of original copies of our brochure, I am sending you a photocopy. If I can help further, please let me know.

The improved version is more concise. However, do not be so concise that you become blunt. Responding to a written request that is vague with "Your request was unclear" or "I don't understand" could easily offend your reader. What you need to do is ask for more information and establish goodwill to encourage your reader to provide the information.

- I will need more information before I can answer your request. Specifically, can you give me the title and the date of the report you are looking for?

Although this version is a bit longer, it is more tactful and will elicit a faster response. See also **technical writing style** and **telegraphic style**.

Check your letters for accuracy. Incorrect punctuation or grammar and unconventional usage can undermine your credibility. Likewise, facts, figures, and dates that are incorrect or misleading may cost time, money, and goodwill. Remember that when you sign a letter, you are accepting responsibility for it. Therefore, allow yourself time to review correspondence carefully before sending it.

Audience: Tone and Goodwill

C

Although correspondence must convey a professional image, it is always more personal than reports or other forms of technical writing. To achieve a conversational style, imagine your reader sitting across the desk from you and write to the reader as if you were talking face-to-face.

Take into account your reader's needs and feelings. Ask yourself, "How might I feel if I received this letter?" and then tailor your message accordingly. Remember, an impersonal and unfriendly letter to a customer or client can tarnish the image of you and your business, but a thoughtful and sincere letter can enhance it. Suppose, for example, you received a refund request from a customer who forgot to enclose the receipt with the request. In a response to that customer, you might write the following:

- We must receive the sales receipt with your letter before we can process a refund.

If you consider how you might keep the customer's goodwill, you might word the request this way:

- Please mail or fax the sales receipt with your letter so that we can process your refund.

You can put the reader's needs and interests foremost in the letter by writing from the reader's perspective. Often, doing so means using the words *you* and *your* rather than *we, our, I,* and *mine*—a technique called the "you" viewpoint. Consider the following revision, which is written with the "you" viewpoint:

- So you can receive your refund promptly, please mail or fax the sales receipt with your letter.

This revision stresses the reader's benefit and interest. By emphasizing the reader's needs, the writer will be more likely to accomplish the objective: to get the reader to act. See also **purpose**.

Keep in mind that, if overdone, goodwill and the "you" viewpoint can produce writing that is fawning, insincere, and overreaching. Messages that are full of excessive praise and inflated language may be ignored—or even resented by the reader.

EXCESSIVE	You are the sort of intelligent, forward-thinking person whose outstanding good judgment is obvious from your selection of the K-50 Copier—the best anywhere!
REASONABLE	Congratulations on selecting the K-50 Copier: We believe you will find that the K-50 is the finest copier on the market.

Writer's Checklist: Using Tone to Build Goodwill

The following guidelines will help you achieve a tone that builds goodwill with the recipients of your correspondence.

☑ Be respectful, not demanding.

DEMANDING   Submit your answer in one week.

RESPECTFUL   I would appreciate your answer within one week.

☑ Be modest, not arrogant.

ARROGANT   My report is thorough, and I'm sure that you won't be able to continue without it.

MODEST   I have tried to be as thorough as possible in my report, and I hope you find it useful.

☑ Be polite, not sarcastic.

SARCASTIC   I just now received the shipment we ordered six months ago. I'm sending it back—we can't use it now. Thanks a lot!

POLITE   I am returning the shipment we ordered on March 12. Unfortunately, it arrived too late for us to be able to use it.

☑ Be positive and tactful, not negative and condescending.

NEGATIVE   Your complaint about our prices is way off target. Our prices are definitely not any higher than those of our competitors.

TACTFUL   Thank you for your suggestion concerning our prices. We believe, however, that our prices are competitive with, and in some cases are below, those of our competitors.

Good-News and Bad-News Patterns

Although the relative directness of correspondence may vary, it is generally more effective to present good news directly and bad news indirectly, especially if the stakes are high.* This principle is based on the fact that readers form their impressions and attitudes very early in letters and that you as a writer may want to subordinate the bad news to reasons that make the bad news understandable. Further, if you are writing international correspondence, be aware that far more cultures are generally indirect in business messages than are direct. See also <u>international correspondence</u>.

---

*Alred, Gerald J. "'We Regret to Inform You': Toward a New Theory of Negative Messages," in *Studies in Technical Communication,* ed. Brenda R. Sims (Denton: University of North Texas and NCTE, 1993), 17–36.

C

Consider the thoughtlessness and direct rejection in Figure C–7. Although the letter is concise and uses the pronouns *you* and *your*, the writer does not consider how the recipient is likely to feel as she reads the letter. Its pattern is (1) the bad news, (2) an explanation, and (3) the close.

A better general pattern for bad-news letters is (1) an opening that provides context (often called a "buffer"), (2) an explanation, (3) the bad news, and (4) goodwill. (See also **refusal letters**.) The opening introduces the subject and establishes a professional tone. The body provides an explanation by reviewing the facts that make the bad news understandable. Although bad news is never pleasant, information that

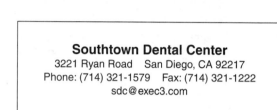

Southtown Dental Center
3221 Ryan Road   San Diego, CA 92217
Phone: (714) 321-1579   Fax: (714) 321-1222
sdc@exec3.com

November 11, 2003

Ms. Barbara L. Mauer
157 Beach Drive
San Diego, CA 92113

Dear Ms. Mauer:

Your application for the position of records administrator at Southtown Dental Center has been rejected. We have found someone more qualified than you.

Sincerely,

Mary Hernandez

Mary Hernandez
Office Manager

FIGURE C–7. A Poor Bad-News Letter

either puts the bad news in perspective or makes it seem reasonable maintains goodwill between the writer and the reader. The closing should reinforce a positive relationship through goodwill or helpful information. Consider, for example, the rejection letter shown in Figure C–8. It carries the same disappointing news as does the letter in Figure C–7, but the writer is careful to thank the reader for her time and effort, explain why she was not chosen for the job, and offer her encouragement.

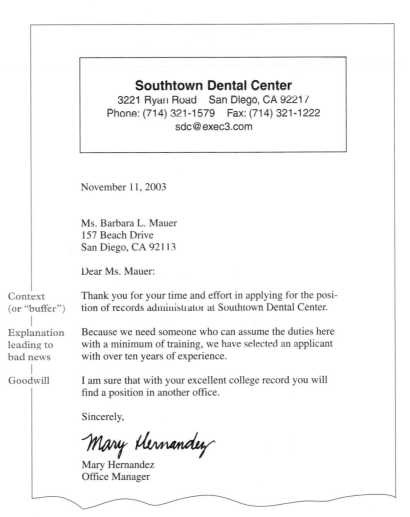

FIGURE C–8. A Courteous Bad-News Letter

Presenting good news is, of course, easier. Present good news at the beginning of the letter. By presenting the good news first, you increase the likelihood that the reader will pay careful attention to details, and you achieve goodwill from the start. The pattern for good-news letters should be (1) the good news, (2) an explanation of facts, and (3) goodwill. Figure C–9 is a good example of a good-news letter.

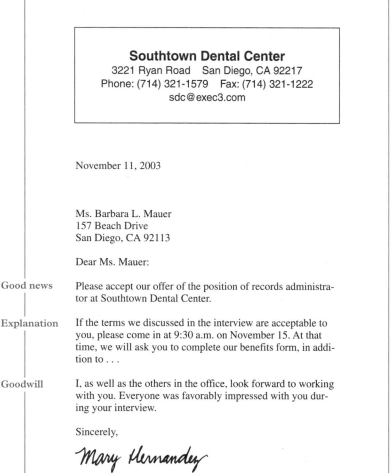

Southtown Dental Center
3221 Ryan Road   San Diego, CA 92217
Phone: (714) 321-1579   Fax: (714) 321-1222
sdc@exec3.com

November 11, 2003

Ms. Barbara L. Mauer
157 Beach Drive
San Diego, CA 92113

Dear Ms. Mauer:

Good news — Please accept our offer of the position of records administrator at Southtown Dental Center.

Explanation — If the terms we discussed in the interview are acceptable to you, please come in at 9:30 a.m. on November 15. At that time, we will ask you to complete our benefits form, in addition to . . .

Goodwill — I, as well as the others in the office, look forward to working with you. Everyone was favorably impressed with you during your interview.

Sincerely,

Mary Hernandez

Mary Hernandez
Office Manager

FIGURE C–9. A Good-News Letter

Openings and Closings

To focus the relevance of any correspondence for the reader, identify your subject in the opening.

- Yesterday, I received your letter and the pager, number AJ 50172. I sent the pager to our quality-control department for tests.

   Carol Moore, our lead technician, reports that preliminary tests indicate . . .

Your closing should let the reader know what he or she should do next and reinforce goodwill.

- Thanks again for the report, and let me know if you want me to send you a copy of the tests.

Because a closing is in a position of emphasis, be especially careful to avoid <u>clichés</u>. Of course, some very commonly used closings are so appropriate, even though they are routine, that they are hard to replace. (If you have further questions, please let me know.)

   Do not use such a closing just because it is easy. Make your closing work for you. It may be helpful to provide prompts to which the reader can respond.

- If you would like further information, such as a copy of the questionnaire we used, please e-mail me at delgado@prn.com.

See also <u>introductions</u> and <u>conclusions</u>.

Format and Design

Although word-processing software provides templates for correspondence, it may not provide specific dimensions and spacing. To achieve a professional appearance, center the letter on the page vertically and horizontally. Although one-inch margins are the default standard in many word-processing programs, it is more important to establish a picture frame of blank space surrounding the page of text. When you use organizational letterhead stationery, consider the bottom of the letterhead as the top edge of the paper. The right margin should be approximately as wide as the left margin. To give a fuller appearance to very short letters, increase both margins to about an inch and a half. Use your computer's full-page or print-preview feature to check for proportion.

   The two most common formats for business letters are the *full-block style* shown in Figure C–10 on page 114 and the *modified-block style* shown in Figure C–11 on page 115. In the *full-block style,* which should be used only with letterhead, the entire letter is aligned at the

C

Letterhead

520 Niagara Street
Braintree, MA 02184

Phone: (781) 787-1175
Fax: (781) 787-1213
E-mail: 92000.121@CompuServe.com

Date

May 15, 2003

Inside
address

Mr. George W. Nagel
Director of Operations
Boston Transit Authority
57 West City Avenue
Boston, MA 02210

Salutation

Dear Mr. Nagel:

Body

Enclosed is our final report evaluating the safety measures for the Boston Intercity Transit System.

We believe that the report covers the issues you raised and that it is self-explanatory. However, if you have any further questions, we would be happy to meet with you at your convenience.

We would also like to express our appreciation to Mr. L. K. Sullivan of your committee for his generous help during our trips to Boston.

Compli-
mentary
close

Sincerely,

Signature

Carolyn Brown

Typed name
Title

Carolyn Brown, Ph.D.
Director of Research

Additional
information

CB/ls
bt515.doc
Enclosure: Final Safety Report
cc: ITS Safety Committee Members

FIGURE C–10. Full-Block-Style Letter (with Letterhead)

Center

C

Heading from center to right	3814 Oak Lane Dedham, MA 02180 December 8, 2003
Inside address	Dr. Carolyn Brown Director of Research Evans and Associates Transportation Engineers 520 Niagara Street Braintree, MA 02184
Salutation	Dear Dr. Brown:
Body	Thank you very much for allowing me to tour your testing facilities. The information I gained from the tour will be of great help to me in preparing the report for my class at Marshall Institute. The tour has also given me some insight into the work I may eventually do as a laboratory technician. I especially appreciated the time and effort Vikram Singh spent in showing me your facilities. His comments and advice were most helpful.
Complimen- tary close aligned with heading	Again, thank you. Sincerely,
Signature	*Leslie Warden*
Typed name	Leslie Warden

Center

FIGURE C–11. Modified-Block-Style Letter (Without Letterhead) left margin. In the *modified- block style*, the return address, date, and complimentary closing begin at the center of the page and the other elements are aligned at the left margin. All other letter styles are variations of the full-block and modified-block styles.

If your employer requires a particular format, use it. Otherwise, follow the guidelines provided here, and review the examples shown in Figures C–10 and C–11.

Heading.   Place your full address and the date in the heading. Because your name appears at the end of the letter, it need not be included in the heading. Spell out words like *street, avenue, first,* and *west* rather than abbreviating them. You may either spell out the name of the state in full or use the standard Postal Service abbreviation. The date usually goes directly beneath the last line of the return address. Do not abbreviate the name of the month. If you are using company letterhead that gives the address, enter only the date, three lines below the last line of printed copy or two inches from the top of the page.

Inside Address.   Include the recipient's full name, title, and address in the inside address, two to six lines below the date, depending on the length of the letter. The inside address should be aligned with the left margin, and the left margin should be at least one inch wide.

Salutation.   Place the salutation, or greeting, two lines below the inside address and align it with the left margin. In most business letters, the salutation contains the recipient's personal title (such as *Mr., Ms., Dr.*) and last name, followed by a colon. If you are on a first-name basis with the recipient, use only the first name in the salutation.

Address women as *Ms.,* unless they have expressed a preference for *Miss* or *Mrs.* However, other titles (such as *Professor, Senator, Major*) take precedence over *Ms.* If you do not know whether the recipient is a man or a woman, use a title appropriate to the context of the letter. The following are examples of the kinds of titles you may find suitable: Dear Customer, Dear Colleague, Dear IT Professional.

If you are writing to a large company and you do not know the name or title of the recipient, you may address the letter to an appropriate department or identify the subject in a subject line and use no salutation.

- National Business Systems
  501 West National Avenue
  Minneapolis, MN 55407

  Attention: Customer Relations Department

  I am returning three pagers that failed to operate. . . .

- National Business Systems
  501 West National Avenue
  Minneapolis, MN 55407

Subject: Defective Parts for SL-100 Pagers

I am returning three pagers that failed to operate. . . .

When a person's first name could be either feminine or masculine, one solution is to use both the first and last names in the salutation (Dear Pat Smith:). Avoid "To Whom It May Concern" because it is impersonal and dated.

For multiple recipients, the following salutations are appropriate:

- Dear Professor Allen and Dr. Rivera: [two recipients]

- Dear Ms. Becham, Ms. Moore, and Mr. Stein: [three recipients]

- Dear Colleagues: [Members, or other suitable collective term]

Body.   The body of the letter should begin two lines below the salutation (or any element that precedes the body, such as a subject or attention line). Single-space within paragraphs, and double-space between paragraphs. To provide a fuller appearance to a very short letter, you can increase the side margins or increase the font size. You can also insert extra space above the inside address, the typed (signature) name, and the initials of the person keying the letter—but do not exceed twice the recommended space for each of these elements.

Complimentary Closing.   Type the complimentary closing two spaces below the body. Use a standard expression like *Sincerely, Sincerely yours,* or *Yours truly.* (If the recipient is a friend as well as a business associate, you can use a less formal closing, such as *Best wishes* or *Best regards* or, simply, *Best.*) Capitalize only the initial letter of the first word, and follow the expression with a comma. Place your full name four lines below, aligned with the closing. On the next line include your business title, if it is appropriate to do so. Sign the letter in the space between the complimentary closing and your name.

Second Page.   If a letter requires a second page, always carry at least two lines of the body text over to that page. Use plain (nonletterhead) paper of quality equivalent to that of the letterhead stationery for the second page. It should have a header with the recipient's name, the page number, and the date. The heading can go in the upper left-hand corner or across the page, as shown in Figure C–12 on page 118.

Additional Information.   Business letters sometimes require additional information that is placed at the left margin, two spaces below

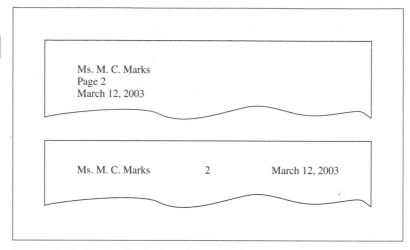

FIGURE C–12. Headers for the Second Page of a Letter

the typed name and title of the writer in a long letter, four spaces below in a short letter.

Reference initials identify the person keying the letter if that person is not the writer. Show the letter writer's initials in capital letters, followed by a slash mark, and then the initials of the person keying the letter in lowercase letters, as shown in Figure C–10. When the writer is also the person keying the letter, no initials are needed.

File-name notation is sometimes included to indicate where the document is stored electronically, as shown in Figure C–10. This notation (*bt515.doc*) allows writers and recipients to refer to the precise document.

Enclosure notations indicate that the writer is sending material along with the letter (an invoice, an article, and so on). Note that you must mention the enclosure in the body of the letter. Enclosure notations may take several forms:

- Enclosure: Final Safety Report
- Enclosures (2)
- Enc. or Encs.

Copy notation (cc:) tells the reader that a copy of the letter is being sent to the named recipients (see Figure C–10). Use a blind-copy notation (bcc:) when you do not want the addressee to know that a copy is being sent to someone else. A blind-copy notation appears only on the copy, not on the original (bcc: Dr. Brenda Shelton).

For additional details on letter format and design, you may wish to consult a guide such as *The Gregg Reference Manual* (ninth edition) by William A. Sabin.

Writer's Checklist: Writing Correspondence

☑ Establish your purpose, analyze your audience, and determine your **scope**.

☑ Prepare an outline, even if it is only a list of points to be covered in the order you want to cover them (see **outlining**).

☑ Write the first draft (see **writing a draft**).

☑ Allow for a cooling period prior to **revision** or seek a colleague's advice, especially for correspondence that addresses a problem.

☑ Revise the draft, checking for key problems in clarity and **coherence**.

☑ Use the appropriate or prescribed format.

☑ Check for accuracy: Make sure that all facts, figures, and dates are correct.

☑ Check for appropriate **punctuation** and use effective **proofreading** techniques.

☑ Allow yourself time to review a letter before you send it.

☑ Remember that when you sign a letter, initial a memo, or send an e-mail, you are accepting responsibility for it.

For information on specific types of correspondence, see the Topical Key.

cover letters

A *cover letter* (sometimes called a *transmittal*) accompanies a **report**, **proposal**, or other material you send to someone by mail or by **e-mail**. It identifies an item that is being sent, the person to whom it is being sent, and the reason that it is being sent: It provides a permanent record for both the writer and the **reader**. For cover letters to **résumés**, see **application letters**.

Your opening should explain what is being sent and why. Then you might highlight or briefly summarize the information enclosed or attached. A cover letter for a proposal, for example, might point out sections in the proposal of particular interest to the reader and go on to present a key point or two explaining why the writer's firm is the best one to do the job. This paragraph could also mention the conditions under which the proposal was prepared, such as limitations of time or budget. The closing paragraph might contain acknowledgments, offer additional assistance, or express the hope that the material will fulfill its purpose.

The example in Figure C–13 on page 120 is brief and to the point. The example in Figure C–14 on page 121 is a bit more detailed; it touches on the manner in which the information was gathered. See also **correspondence**.

C

ECOLOGY SYSTEMS AND SERVICES

39 Beacon Street
Boston, Massachusetts 02106
(617) 351-1223
Fax: (617) 351-2121

May 23, 2003

Mario Espinoza, Chief Engineer
Louisiana Chemical Products
3452 River View Road
Baton Rouge, LA 70893

Dear Mr. Espinoza:

Enclosed is the final report on our installation of pollution-control
equipment at Eastern Chemical Company, which we send with
Eastern's permission. Please call me collect (ext. 1206) or send
me an e-mail message at the address below if I can answer any
questions.

Sincerely,

Susan Wong

Susan Wong, Ph.D.
Technical Services Manager
swong.ecology@omni.com

SW/ls
Enclosure: Report

FIGURE C–13. Brief Cover Letter (for a Report)

C

WATERFORD PAPER PRODUCTS
P.O. Box 413
WATERFORD, WI 53474

Phone: (414) 738-2191
Fax: (414) 738-9122

January 16, 2003

Mr. Roger Hammersmith
Ecology Systems, Inc.
1015 Clarke Street
Chicago, IL 60615

Dear Mr. Hammersmith:

Enclosed is the report estimating our power consumption for the
year as requested by John Brenan, Vice President, on September 4.

The report is a result of several meetings with the Manager of Plant
Operations and her staff and an extensive survey of all our employees. The survey was delayed by the transfer of key staff in Building
"A." We believe, however, that the report will provide the information you need in order to furnish us with a cost estimate for the installation of your Mark II Energy Saving System.

We would like to thank Diana Biel of ESI for her assistance in
preparing the survey. If you need any more information, please let
me know.

Sincerely,

James G. Evans

James G. Evans
New Projects Office
jge.waterford@omni.com

Enclosure

FIGURE C–14. Long Cover Letter (for a Report)

credible / creditable

Something is *credible* if it is believable. (The statistics in this report are *credible*.) Something is *creditable* if it is worthy of praise or credit. (The lead engineer did a *creditable* job.)

criteria / criterion

Criterion is a singular <u>noun</u> meaning "an established standard for judging or testing." *Criteria* and *criterions* are both acceptable plural forms of *criterion*, but *criteria* is generally preferred.

critique

A *critique* is a written or oral evaluation of something. Avoid using *critique* as a <u>verb</u> meaning "criticize."

- Please critique his job description.
  prepare a   *of*

D

dangling modifiers

Phrases that do not clearly and logically refer to the correct <u>noun</u> or <u>pronoun</u> are called *dangling modifiers.* Dangling modifiers usually appear at the beginning of a sentence as an introductory phrase.

DANGLING	*While eating lunch in the cafeteria,* the computer malfunctioned. [*Who* was eating lunch in the cafeteria?]
CORRECT	While *I* was eating lunch in the cafeteria, the computer malfunctioned.

Dangling modifiers can appear at the end of the sentence as well.

DANGLING	The program gains efficiency by *eliminating the superfluous instructions.* [*Who* eliminates the superfluous instructions?]
CORRECT	The program gains efficiency *when you* eliminate the superfluous instructions.

To test whether a <u>phrase</u> is a dangling modifier, turn it into a <u>clause</u> with a subject and a <u>verb</u>. If the expanded phrase and the independent clause do not have the same subject, the phrase is dangling.

DANGLING	*After finishing the research,* the proposal was easy to write. [The implied subject of the phrase is *I,* but the subject of the independent clause is *the proposal.*]
CORRECT	After finishing the research, *I found that* the proposal was easy to write. [Now the subject of the independent clause agrees with the implied subject of the introductory phrase.]
CORRECT	After *I* finished the research, the proposal was easy to write. [Here the phrase is a dependent clause with an explicit subject.]

For a discussion of misplaced modifiers, see <u>modifiers</u>.

dashes

The *dash* (—) can perform all the duties of punctuation: linking, separating, and enclosing. It is an emphatic mark that is easily overused. Use the dash cautiously to indicate more informality, <u>emphasis</u>, or abruptness than the other punctuation marks would show.

A dash can emphasize a sharp turn in thought.

- The project will end January 15 — unless the company provides additional funds.

A dash can indicate an emphatic pause.

- The job will be done — after we are under contract.

Sometimes, to emphasize contrast, a dash is used with *but.*

- We may have produced work more quickly — but the result was not as good.

A dash can be used before a final summarizing statement or before repetition that has the effect of an afterthought.

- It was hot near the ovens — steaming hot.

Such a statement may also complete the meaning of the clause preceding the dash.

- We try to speak as we write — or so we believe.

A dash can be used to set off an explanatory or appositive series.

- Three of the applicants — John Evans, Rosalita Fontiana, and Kyong-shik Choi — seem well qualified for the job.

Dashes set off parenthetical elements more sharply and emphatically than <u>commas</u>. Unlike dashes, <u>parentheses</u> tend to reduce the importance of what they enclose. Compare the following sentences:

- Only one person — the president — can authorize such activity.
- Only one person, the president, can authorize such activity.
- Only one person (the president) can authorize such activity.

The first word after a dash is capitalized only if it is a proper <u>noun</u>.

data / datum

In much informal writing, *data* is considered a collective singular <u>noun</u>. In formal scientific and scholarly writing, however, *data* is generally used as a plural, with *datum* as the singular form. Base your decision on whether your readers should consider the data as a single collection or as a group of individual facts. Whatever you decide, be sure that your <u>pronouns</u> and <u>verbs</u> agree in number with the selected <u>usage</u>.

- The *data are* voluminous. *They indicate* a link between cigarette smoking and lung cancer. [formal]

- The *data is* now ready for evaluation. *It is* in the mail. [less formal]

See also <u>agreement</u> and <u>English, varieties of</u>.

dates

In the United States, dates are generally indicated by the month-day-year format, with a comma separating the figures (October 26, 2013). The day-month-year system used in other parts of the world and by the U.S. military does not require commas (26 October 2013). A date can be written with or without a comma following the year if the date is in the month-day-year format.

- October 26, 2013, is the payoff date.
- October 26, 2013 is the payoff date.

Use the strictly numerical form for dates (10/26/13) sparingly and never in business letters or formal documents because the date is not always immediately clear. In fact, the numerical form may be confusing in <u>international correspondence</u>. Writing out the name of the month makes the entire date immediately clear to all readers.

Month and Day

When writing days of the month without the year, use the cardinal <u>number</u> rather than the ordinal number.

> INCORRECT   We will meet on March 4th. [ordinal]

> CORRECT   We will meet on March 4. [cardinal]

Of course, in speaking or <u>presentations</u>, use the ordinal number ("March fourth").

D

Centuries

Confusion often occurs because the names given to centuries do not correspond to the numeral designations. The twentieth century, for example, is the 1900s: 1900–1999. When the century is written as a <u>noun</u>, do not use a hyphen.

- The sixteenth century produced great literature.

When the century is written as an <u>adjective</u>, however, use a hyphen.

- Twenty-first-century technology relies on clear communications.

---

decreasing-order-of-importance method of development (*see* order-of-importance method of development)

---

defective / deficient

If something is *defective*, it is faulty. (The wiring was *defective*.) If something is *deficient*, it is lacking or is incomplete in an essential component. (The company was found to be *deficient* in meeting its legal obligations.)

---

defining terms

Good writing ensures that readers understand key terms and concepts used. Terms can be defined either formally or informally, depending on your <u>purpose</u> and your <u>readers</u>.

A *formal definition* is a form of classification. You define a term by placing it in a category and then identifying the features that distinguish it from other members of the same category.

TERM	CATEGORY	DISTINGUISHING FEATURES
An *annual* is	a plant	that completes its life cycle, from seed to natural death, in one growing season.

An *informal definition* explains a term by giving a more familiar word or phrase as a <u>synonym</u>.

- Plants have a *symbiotic,* or *mutually beneficial,* relationship with certain kinds of bacteria.

State definitions positively; focus on what the term *is* rather than on what it is not.

D

NEGATIVE	In a legal transaction, *real property* is not personal property.
POSITIVE	*Real property* is legal terminology for the right or interest a person has in land and the permanent structures on that land.

For a discussion of when negative definitions are appropriate, see <u>definition methods of development</u>.

Avoid circular definitions, which merely restate the term to be defined and therefore fail to clarify it.

CIRCULAR	*Spontaneous combustion* is fire that begins spontaneously.
REVISED	*Spontaneous combustion* is the self-ignition of a flammable material through a chemical reaction.

In addition, avoid "is when" and "is where" definitions. Such definitions fail to include the category and are too indirect.

a medical procedure in which a tissue sample is removed for testing.
- A *biopsy* is ~~when a tissue sample is removed for testing.~~
  ^

In technical writing, as illustrated in this example, definitions often contain the purpose of what is defined.

definite / definitive

Definite and *definitive* both apply to what is precisely defined, but *definitive* more often refers to what is complete and authoritative. (When the committee took a *definite* stand, the president made it the *definitive* company policy.)

definition method of development

DIRECTORY

D

To *define* something is to identify precisely its fundamental qualities. In technical writing, clear and accurate definitions are critical. Sometimes simple, dictionary-like definitions suffice (see <u>defining terms</u>), but at other times definitions need to be expanded with additional details, examples, comparisons, or other explanatory devices. The most common explanatory devices are (1) extended definition, (2) definition by analogy, (3) definition by cause, (4) definition by components, (5) definition by exploration of origin, and (6) negative definition.

Extended Definition

When more than a phrase or a sentence or two is needed to explain an idea, use an extended definition, which explores a number of qualities of the item being defined. How an extended definition is developed depends on your readers' needs and on the complexity of the subject. Readers familiar with a topic might be able to handle a long, fairly complex definition, whereas readers less familiar with a topic might require simpler language and more basic information.

The easiest way to give an extended definition is with specific examples. Listing examples gives readers easy-to-picture details that help them see and thus understand the term being defined.

- Form, which is the shape of landscape features, can best be represented by both small-scale features, such as *trees* and *shrubs,* and by large-scale elements, such as *mountains* and *mountain ranges.*

Definition by Analogy

Another useful way to define a difficult concept, especially when you are writing for nonspecialists, is to use an <u>analogy</u> to link the unfamiliar concept with a simpler or more familiar one. An analogy can help the reader understand an unfamiliar term by showing its similarities with a more familiar term. In the following description of radio waves in terms of their length (long) and frequency (low), notice how the writer develops an analogy to show why a low frequency is advantageous.

- The low frequency makes it relatively easy to produce a wave having virtually all its power concentrated at one frequency. Think, for example, of a group of people lost in a forest. If they hear sounds of a search party in the distance, they all will begin to shout for help in different directions. Not a very efficient process, is it? But suppose all the energy that went into the production of this noise could be concentrated into a single shout or whistle. Clearly the chances that the group will be found would be much greater.

Definition by Cause

Some terms are best defined by an explanation of their causes. In the following example from a professional journal, a nurse describes an apparatus used to monitor blood pressure in severely ill patients. Called an *indwelling catheter,* the device displays blood-pressure readings on an oscilloscope and on a numbered scale. Users of the device, the writer explains, must understand what a *dampened wave form* is.

- The dampened wave form, the smoothing out or flattening of the pressure wave form on the oscilloscope, is usually caused by an obstruction that prevents blood pressure from being freely transmitted to the monitor. The obstruction may be a small clot or bit of fibrin at the catheter tip. More likely, the catheter tip has become positioned against the artery wall and is preventing the blood from flowing freely.

Definition by Components

Sometimes a formal definition of a concept can be made simpler by breaking the concept into its component parts. In the following example, the formal definition of *fire* is given in the first paragraph, and the component parts are given in the second.

FORMAL DEFINITION	Fire is the visible heat energy released from the rapid oxidation of a fuel. A substance is "on fire" when the release of heat energy from the oxidation process reaches visible light levels.
COMPONENT PARTS	The classic fire triangle illustrates the elements necessary to create fire: *oxygen, heat,* and *burnable material (fuel).* Air provides sufficient oxygen for combustion; the intensity of the heat needed to start a fire depends on the characteristics of the burnable material. A burnable substance is one that will sustain combustion after an initial application of heat to start the combustion.

Definition by Exploration of Origin

Under certain circumstances, the meaning of a term can be clarified and made easier to remember by an exploration of its origin. Medical terms, because of their sometimes unfamiliar Greek and Latin roots, benefit especially from an explanation of this type. Tracing the derivation of a word also can be useful when you want to explain why a word has favorable or unfavorable associations, particularly if your goal is to influence your reader's attitude toward an idea or an activity. (See also **persuasion**.)

D

- Efforts to influence legislation generally fall under the head of *lobbying,* a term that once referred to people who prowl the lobbies of houses of government, buttonholing lawmakers and trying to get them to take certain positions. Lobbying today is all of this, and much more, too. It is a respected—and necessary—activity. It tells the legislator which way the winds of public opinion are blowing, and it helps inform [legislators] of the implications of certain bills, debates, and resolutions [that they must face].
  —Bill Vogt, *How to Build a Better Outdoors*

Negative Definition

In some cases, it is useful to point out what something is not to clarify what it is. A negative definition is effective only when the reader is familiar with the item with which the defined item is contrasted. If you say *x* is not *y*, your readers must understand the meaning of *y* for the explanation to make sense. In a crane operator's manual, for instance, a negative definition is used to show that, for safety reasons, a hydraulic crane cannot be operated in the same manner as a lattice boom crane.

- A hydraulic crane is *not* like a lattice boom crane in one very important way. In most cases, the safe lifting capacity of a lattice boom crane is based on the *weight needed to tip the machine.* Therefore, operators of friction machines sometimes depend on signs that the machine might tip to warn them of impending danger. This practice is very dangerous with a hydraulic crane. . . .
  —*Operator's Manual* (Model W-180), Harnishfeger Corporation

description

The key to effective description is the accurate presentation of details, whether for simple or complex descriptions. In Figure D–1, notice that the description in the purchase order includes five specific details in addition to the part number.

Complex descriptions, of course, involve more details. In describing a mechanical device, for example, describe the whole device and its function before giving a detailed description of how each part works. The description should conclude with an explanation of how each part contributes to the functioning of the whole.

PURCHASE ORDER

PART NO.	DESCRIPTION	QUANTITY
IW 8421	Infectious-waste bags, 12″ × 14″, heavy-gauge polyethylene, red double closures with self-sealing adhesive strips	5 boxes containing 200 bags per box

D

FIGURE D-1. Description

In descriptions intended for <u>readers</u> who are unfamiliar with the topic, details are crucial. For these readers, show or demonstrate (as opposed to "tell") primarily through the use of images and details. Notice the use of color, shapes, and images in the following description of a company's headquarters. The writer assumes that the reader knows such terms as *colonial design* and *champaign elm.*

- Their corporate headquarters, which reminded me of a rural college campus, are located just north of the city in a 90-acre wooded area. The complex consists of five three-story buildings of colonial design. The buildings are spaced about 50 feet apart and are built in a U shape, with the main building at the bottom of the U. The driveway that runs in front of the buildings is lined with overarching champaign elms. Employee and visitor parking lots are neatly hidden behind each building.

You can also use <u>figures of speech</u> to explain unfamiliar concepts in terms of familiar ones, such as "U shape" in the previous example or "armlike block" in Figure D–2 on page 132.

<u>Visuals</u> can be powerful aids in descriptive writing, especially when they show details too intricate to explain completely in words. (See also <u>layout and design</u>.) The example in Figure D–2 uses an illustration to help describe a mechanical assembly for an assembler/repairperson. Note that the description concentrates on the number of pieces — their sizes, shapes, and dimensions — and on their relationship to one another to perform their function. It also specifies the materials of which the hardware is made. Because the description is illustrated (with identifying labels) and is intended for technicians who have been trained on the equipment, it does not require the use of "bridging devices" to explain the unfamiliar in relation to the familiar. Even so, an important term is defined (*chad*), and crucial alignment dimensions are specified (± .003″).

D

The Die Block Assembly consists of two machined block sections, eight Code Pins, and a Feed Pin. The larger section, called the Die Block, is fashioned of a hard, noncorrosive beryllium-copper alloy. It houses the eight Code Pins and the smaller Feed Pin in nine finely machined guide holes.

Number of features and relative size differences noted

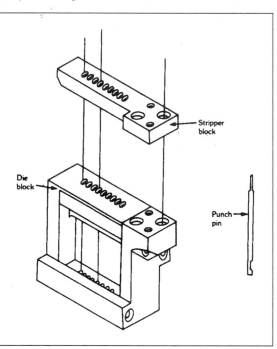

The guide holes in the upper part of the Die Block are made smaller to conform to the thinner tips of the Feed Pins. Extending over the top of the Die Block and secured to it at one end is a smaller, armlike block called the Stripper Block. The Stripper Block is made from hardened tool steel, and it also has been drilled through with nine finely machined guide holes. It is carefully fitted to the Die Block at the factory so that its holes will be precisely above those in the Die Block and so that the space left between the blocks will measure .015" (± .003"). The residue from the oper-ation, called chad, is pushed out through the top of the Stripper Block and guided out of the assembly by means of a plastic Residue Collector and Residue Collector Extender.

Figure of speech

Definition

FIGURE D–2. Complex Description

design (*see* layout and design)

despite / in spite of

Although there is no literal difference between *despite* and *in spite of*, *despite* suggests an effort to avoid blame.

- *Despite* our best efforts, the plan failed.
  [We are not to blame for the failure.]

- *In spite of* our best efforts, the plan failed.
  [We did everything possible, but failure overcame us.]

Despite and *in spite of* (both meaning "notwithstanding") should not be blended into *despite of.*

> *In spite of*
- ~~Despite of~~ our best efforts, the plan failed.
  ^

diacritical marks

Diacritical marks are **symbols** added to letters to indicate their specific sounds or pronunciation values. Diacritical marks include the phonetic symbols used in dictionaries as well as the marks used with foreign words. When you write **international correspondence,** be sure to use appropriate diacritical marks, especially for names (Müller, Conceição). Some dictionaries place a reference list of common marks and sounds at the bottom of each page. Figure D–3 on page 134 lists common diacritical marks with their equivalent sound values. See also **foreign words in English**.

diagnosis / prognosis

Because they sound somewhat alike, these words are often confused with each other. *Diagnosis* means "an analysis of the nature of something" or "the conclusions reached by such analysis." (The CEO *diagnosed* the problem as too few dollars for research.) *Prognosis* means "a forecast or prediction." (He offered his *prognosis* that the problem would be solved next year.)

NAME	SYMBOL	EXAMPLE	MEANING
macron	¯	cāke	indicates the standard pronunciation ("long" sound) of the letter
breve	˘	brăcket	indicates the "short" sound of the letter, in contrast to the standard pronunciation
dieresis	¨	coöperate	indicates that the second of two consecutive vowels is to be pronounced separately; this mark is also used as an "umlaut" to indicate blended vowel sounds, especially in German (universität = universitaet)
acute accent	´	cliché	indicates a primary vocal stress on the indicated letter
grave accent	`	crèche	indicates a deep sound articulated toward the back of the mouth
circumflex	^	crêpe	indicates a very soft sound
tilde	~	cañon	indicates the palatal nasal *ny* sound; a Spanish diacritical mark
cedilla	¸	garçon	placed beneath the letter *c* in French, Portuguese, and Spanish; indicates an *s* sound

FIGURE D–3. Diacritical Marks

diction

The term *diction* is often misunderstood because it means both "the choice of words used in writing and speech" *and* "the degree of distinctness of enunciation of speech." In discussions of writing, diction applies to <u>word choice</u>.

dictionaries

Dictionaries, available in both print and electronic form, give more than just information about words' meanings. They often provide words' etymologies (origins and history), forms, pronunciations, <u>spellings</u>, uses as <u>idioms</u>, and functions as <u>parts of speech</u>. (See Figure D–4 on page 135.) For certain words, a dictionary lists <u>synonyms</u> and may also provide illustrations, if appropriate, such as tables, maps, photographs, and drawings.

Types of Dictionaries

Desk and Multimedia Dictionaries. Desk and multimedia dictionaries are often abridged versions of larger dictionaries. There is no single "best" dictionary, but there are several guidelines for selecting a

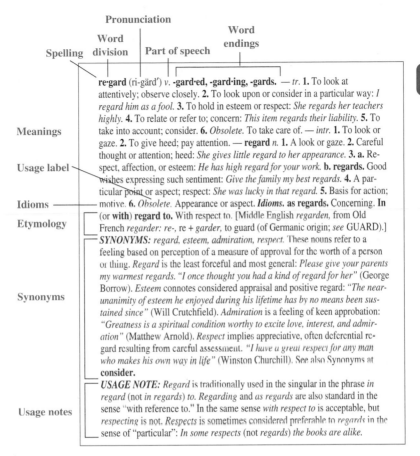

Pronunciation

Word division

Word endings

Spelling

Part of speech

Meanings

Usage label

Idioms

Etymology

Synonyms

Usage notes

re·gard (ri-gärd′) *v.* **-gard·ed, -gard·ing, -gards.** — *tr.* **1.** To look at attentively; observe closely. **2.** To look upon or consider in a particular way: *I regard him as a fool.* **3.** To hold in esteem or respect: *She regards her teachers highly.* **4.** To relate or refer to; concern: *This item regards their liability.* **5.** To take into account; consider. **6.** *Obsolete.* To take care of. — *intr.* **1.** To look or gaze. **2.** To give heed; pay attention. — **regard** *n.* **1.** A look or gaze. **2.** Careful thought or attention; heed: *She gives little regard to her appearance.* **3. a.** Respect, affection, or esteem: *He has high regard for your work.* **b. regards.** Good wishes expressing such sentiment: *Give the family my best regards.* **4.** A particular point or aspect; respect: *She was lucky in that regard.* **5.** Basis for action; motive. **6.** *Obsolete.* Appearance or aspect. *Idioms.* **as regards.** Concerning. **In (or with) regard to.** With respect to. [Middle English *regarden*, from Old French *regarder*: re-, re + *garder*, to guard (of Germanic origin; *see* GUARD).]

— *SYNONYMS: regard, esteem, admiration, respect.* These nouns refer to a feeling based on perception of a measure of approval for the worth of a person or thing. *Regard* is the least forceful and most general: *Please give your parents my warmest regards.* *"I once thought you had a kind of regard for her"* (George Borrow). *Esteem* connotes considered appraisal and positive regard: *"The near-unanimity of esteem he enjoyed during his lifetime has by no means been sustained since"* (Will Crutchfield). *Admiration* is a feeling of keen approbation: *"Greatness is a spiritual condition worthy to excite love, interest, and admiration"* (Matthew Arnold). *Respect* implies appreciative, often deferential regard resulting from careful assessment. *"I have a great respect for any man who makes his own way in life"* (Winston Churchill). See also Synonyms at **consider.**

— *USAGE NOTE: Regard* is traditionally used in the singular in the phrase *in regard* (not *in regards*) *to.* *Regarding* and *as regards* are also standard in the sense "with reference to." In the same sense *with respect to* is acceptable, but *respecting* is not. *Respects* is sometimes considered preferable to *regards* in the sense of "particular": *In some respects* (not *regards*) *the books are alike.*

FIGURE D-4. Dictionary Entry

good one. Choose the most recent edition. The older the dictionary, the less likely it is to have the up-to-date information you need. Select a dictionary with upward of 125,000 entries. Dictionaries available on CD-ROM include a human-voice pronunciation of each word. Many general and specialized dictionaries are available at a variety of Internet sites. Excellent sites are also available that translate words from one language to another; one such site is "Research It!" at <www.itools .com/research-it/>.

The following are considered good desk dictionaries.

The American Heritage College Dictionary, book and CD editions, 1997

Encarta World English Dictionary, book and CD editions, 1999

Merriam-Webster's Collegiate Dictionary, 10th indexed ed., 1998

Merriam-Webster's Collegiate Dictionary and Thesaurus, deluxe audio edition, 2000

Random House Webster's College Dictionary, 2nd rev. and updated ed., 2000

D

Unabridged Dictionaries.   Unabridged dictionaries provide complete and authoritative linguistic information. The printed versions are impractical for desk use because of their size and the CD versions are expensive; libraries make available both versions of these important reference sources.

* *The Oxford English Dictionary,* 2nd ed., is the standard historical dictionary of the English language. Its 20 volumes contain over 600,000 words and give the chronological developments of over 240,000 words, providing numerous examples of uses and sources. It is also available on CD-ROM for Windows.

* *The Random House Dictionary of the English Language,* 2nd ed., contains about 315,000 entries and uses copious examples. It gives a word's most current meaning first and includes biographical and geographical names.

* *Webster's Third New International Dictionary of the English Language, Unabridged,* contains over 450,000 entries. Word meanings are listed in historical order, with the current meaning given last. This dictionary does not list biographical and geographical names, nor does it include usage information.

ESL Dictionaries.   English-as-a-second-language (ESL) dictionaries are more helpful to the nonnative speaker than are regular English dictionaries or bilingual dictionaries. The pronunciation symbols in ESL dictionaries are based on the international phonetic alphabet rather than on English phonetic systems, and useful grammatical information is included in both the entries and special grammar sections. In addition, the definitions usually are easier to understand than those in regular English dictionaries; for example, a regular English dictionary defines *opaque* as "impervious to the passage of light," while an ESL dictionary defines the word as "not allowing light to pass through." The definitions in ESL dictionaries also are usually more thorough than those in bilingual dictionaries. For example, a bilingual dictionary might indicate that *obstacle* and *blockade* are synonymous but not indicate that although both words can refer to physical objects ("The soldiers went around the *blockade* [or *obstacle*] in the road"), only *obstacle* can be used for abstract meanings ("Lack of money can be an *obstacle* [not a *blockade*] to a college education").

The following dictionaries and references provide helpful information for nonnative speakers of English.

Longman Dictionary of American English: A Dictionary for Learners of English

Longman Advanced American Dictionary

Oxford American Wordpower Dictionary and CD version, Ruth Urbom (Oxford)

Longman Dictionary of Contemporary English, with CD [British English]

Longman Dictionary of English Idioms, Laurence Urdang

D

Subject Dictionaries.   For the meanings of words too specialized for a general dictionary, a subject dictionary is useful. Subject dictionaries define terms used in a particular field, such as business, geography, architecture, or consumer affairs. Definitions in subject dictionaries are generally more detailed and comprehensive than those found in general dictionaries. One example is *Dictionary of Personal Computing and the Internet,* edited by Simon Collin.

Although subject dictionaries are specialized and offer detailed definitions of field-specific terms, they are written in language that is straightforward enough to be understood by nonspecialists. Many general and specialized dictionaries are available on CD-ROM as well as on the Web.

differ from / differ with

Differ from suggests that two things are not alike. (Our earlier proposal *differs from* the current one.) *Differ with* indicates disagreement between persons. (The architect *differed with* the contractor on the proposed site.)

different from / different than

In formal writing, the preposition *from* is used with *different.* (The Quantum PC is *different from* the Macintosh computer.) *Different than* is used when it is followed by a clause. (The job cost was *different than* we had estimated it.)

direct address

D

Direct address refers to a sentence or phrase in which the person being spoken or written to is explicitly named. It is often used in speech and in e-mail messages. Notice that the person's name in a direct address is set off by commas.

- *John,* call me as soon as you arrive at the airport.
- Call me, *John,* as soon as you arrive at the airport.

discreet / discrete

Discreet means "having or showing prudent or careful behavior." (Because the matter was personal, he asked Bob to be *discreet.*) *Discrete* means something is "separate, distinct, or individual." (Several *discrete* strands of copper wire form the cable.)

disinterested / uninterested

Disinterested means "impartial, objective, unbiased."

- Like good judges, scientists should be passionately interested in the problems they tackle but completely *disinterested* when they seek to solve those problems.

Uninterested means simply "not interested." (Despite Asha's enthusiasm, her manager remained *uninterested* in the project.)

division-and-classification method of development

An effective way to organize information about a complex subject is to divide it into manageable parts and then discuss each part separately. You might use this approach, called *division,* to describe a physical object, such as the parts of a fax machine; to examine an organization, such as a company; or to explain the components that make up the Internet global network. The emphasis in division as a method of development is on breaking down a complex whole into a number of like units—it is easier to consider smaller units and to examine the relationship of each to the other than to attempt to discuss the whole.

If you were a financial planner describing the types of mutual funds available to your investors, you could divide the variety available into three broad categories: money-market funds, bond funds, and stock funds. Such division would be accurate, but it would be only a first-level grouping of a complex whole. The three broad categories could, in turn, be subdivided into additional groups based on investment strategy, as follows:

Money-market funds
• taxable money market
• tax-exempt money market

Bond funds
• taxable bonds
• tax-exempt bonds
• balanced (mix of stocks and bonds)

Stock funds
• balanced
• equity-income
• domestic growth
• growth and income
• international growth
• small-capitalization
• aggressive growth
• specialized

Specialized stock funds could be further subdivided as follows:

Specialized funds
• communications
• energy
• environmental services
• financial services
• gold
• health services
• technology
• utilities
• worldwide capital goods

The process of *classification* is similar to the process of division. While division is the separation of a whole into its component parts, classification is the grouping of a number of units (such as people, objects, or ideas) into related categories. Consider the following list:

triangular file	steel tape ruler	needle-nose pliers
vise	pipe wrench	keyhole saw
mallet	tin snips	C-clamps
rasp	hacksaw	plane
glass cutter	ball-peen hammer	steel square
spring clamp	claw hammer	utility knife
crescent wrench	folding extension ruler	slip-joint pliers
crosscut saw	tack hammer	utility scissors

To group the items in the list, you would first determine what they have in common. The most obvious characteristic they share is that they all belong in a carpenter's tool chest. With that observation as a starting point, you can begin to group the tools into related categories. Pipe wrenches belong with slip-joint pliers because both tools grip objects. The rasp and the plane belong with the triangular file because all three tools smooth rough surfaces. By applying this kind of thinking to all the items in the list, you can group the tools according to function (Figure D–5).

To divide or classify a subject, you must first divide the subject into its largest number of comparable units. The basis for division depends, of course, on your subject and your purpose. For explaining the functions of carpentry tools, the classifications in Figure D–5 (smoothing, hammering, measuring, gripping, and cutting) are excellent. For recommending which tools a new homeowner should buy first, however, those classifications are not helpful—each group contains tools that a new homeowner might want to purchase right away. To give homeowners advice on purchasing tools, you probably would clarify the types of repairs they most likely will have to do. That classification could serve as a guide to tool purchase.

Once you have established the basis for the division, apply it consistently, putting each item in only one category. For example, it might seem logical to classify needle-nose pliers as both a tool that cuts and a tool that grips because most needle-nose pliers have a small section for cutting wires. However, the primary function of needle-nose pliers is to grip. So listing them only under "tools that grip" would be most consistent with the basis used for listing the other tools. See also instructions and process explanation.

documentation

The word *documentation* in academic work refers to the giving of formal credit to sources used or quoted in research papers, trade journal articles, and reports—that is, the form used in bibliographies, footnotes,

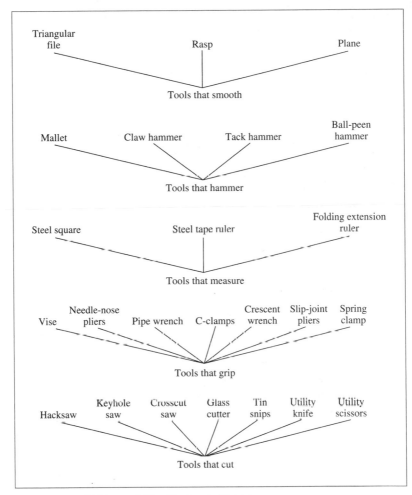

FIGURE D–5. Division and Classification (of Tools)

and references. (See <u>copyright</u>, <u>documenting sources</u>, and <u>plagiarism</u>.) In the workplace, *documentation* also may refer to information that is recorded, or documented, on paper or in electronic files. The technical manuals that manufacturers provide to customers are examples of such documentation. Even flowcharts and engineering drawings are considered documentation.

documenting sources

D

Documenting sources achieves three important purposes:

- It allows readers to locate and consult the sources used and to find further information on the subject.

- It enables writers to support their assertions and arguments in such documents as proposals, reports, and trade journal articles.

- It helps writers to give proper credit to others and thus avoid **plagiarism** by identifying the sources of facts, ideas, **quotations**, and paraphrases. See also **paraphrasing**.

This entry shows citation models and sample pages for three principal documentation systems.

- The American Psychological Association (APA) system of citation is often used in the social sciences. It is referred to as an author/date method of documentation because parenthetical in-text citations and a references list (at the end of the paper) in APA style emphasize the author(s) and date of publication so that the currency of the research is clear.

- The system set forth in *The Chicago Manual of Style (CMS)* is often used in business, communications, economics, and history. The *CMS* system of footnotes (at the bottom of the page) or endnotes (at the end of the document) and a bibliography plainly identifies the sources of information.

- The Modern Language Association (MLA) system is used in the humanities. MLA style uses parenthetical citations and a list of works cited and places greater importance on the pages on which cited information can be found than on the publication date.

APA IN-TEXT CITATION	(Author's last name, Date of publication)
CMS **FOOTNOTE/ ENDNOTE**	#. Author's first name Last name, *Title* (Place of publication: Publisher, Date of publication), Pages.
MLA IN-TEXT CITATION	(Author's last name, Pages)

BOOK, ONE AUTHOR

APA REFERENCES ENTRY	Author's last name, First initial. (Year of publication). *Title*. Place of publication: Publisher.
CMS **BIBLIOGRAPHY ENTRY**	Author's last name, First name. *Title*. Place of publication: Publisher, Date of publication.
MLA WORKS-CITED ENTRY	Author's last name, First name. *Title*. Place of publication: Publisher, Date of publication.

JOURNAL ARTICLE

APA REFERENCES ENTRY	Author's last name, First initial. (Date of publication). Title of article. *Journal, Volume number, Pages.*
CMS **BIBLIOGRAPHY ENTRY**	Author's last name, First name. "Title of Article," *Journal* Volume number (Year): Pages.
MLA WORKS-CITED ENTRY	Author's last name, First name. "Title of Article." *Journal* Volume number (Year): Pages.

These systems are described in full detail in the following publications:

American Psychological Association. *Publication Manual of the American Psychological Association*. 5th ed. Washington, D.C.: American Psychological Association, 2001. See also <www .apastyle.org>.

The Chicago Manual of Style. 14th ed. Chicago: University of Chicago Press, 1993. See also <www.press.uCMS.edu /Misc/CMS/cmosfaq.html>.

MLA Handbook for Writers of Research Papers. 5th ed. New York: Modern Language Association of America, 1999. See also <www.mla.org>.

For additional bibliographic advice and documentation models, consult these style manuals or those listed at the end of this entry. See also **bibliographies**, **copyright**, **Internet research**, and **library research**.

APA Documentation

APA In-Text Citations.   APA parenthetical documentation within the text of a paper gives a brief citation — in parentheses — of the author, year of publication, and relevant page numbers if it helps locate a passage in a lengthy document.

When APA parenthetical citations are added midsentence, place them after the closing quotation marks and continue with the rest of the sentence. If the APA parenthetical citation follows a block quotation, place it after the final punctuation mark.

- World War II was the occasion of radar's first application in warfare, and Great Britain led the way in radar research (Butrica, 2000).

- According to Butrica (2000), the use of radar as an offensive and defensive warfare agent made World War II "the first electronic war" (p. 2).

- The "first electronic war" (Butrica, 2000, p. 2) was fought as much in the research laboratory as on the battlefield.

When a work has two authors, cite both names joined by an ampersand: (Hey & Walters, 1997). For the first citation of a work with up to and including five authors, include all names. For subsequent citations and works with five or more authors, include only the name of the first author followed by *et al.* When two or more works by different authors are cited in the same parentheses, list the citations alphabetically and use semicolons to separate them: (Hey & Walters, 1997; Ostro, 1993).

APA Documentation Models

BOOKS

Single Author

Hassab, J. C. (1997). *Systems management: People, computers, machines, materials.* New York: CRC Press.

Multiple Authors

Testerman, J. O., Kuegler, T. J., Jr., & Dowling, P. J., Jr. (1998). *Web advertising and marketing* (2nd ed.). Rocklin, CA: Prima.

Corporate Author

Ernst and Young. (1999). *Ernst and Young's retirement planning guide.* New York: Wiley.

Edition Other Than First

Estes, J. C., & Kelley, D. R. (1998). *McGraw-Hill's interest amortization tables* (3rd ed.). New York: McGraw-Hill.

Multivolume Work

Standard and Poor. (1998). *Standard and Poor's register of corporations, directors and executives* (Vols. 1–3). New York: McGraw-Hill.

Work in an Edited Collection

Thorne, K. S. (1997). Do the laws of physics permit wormholes for interstellar travel and machines for time travel? In Y. Terzian & E. Bilson (Eds.), *Carl Sagan's universe* (pp. 121–134). Cambridge, England: Cambridge University Press.

Encyclopedia or Dictionary Entry

Gibbard, B. G. (1997). Particle detector. In *World Book encyclopedia* (Vol. 15, pp. 186–187). Chicago: World Book.

ARTICLES IN PERIODICALS (*See also* Electronic Sources)
Magazine Article

Coley, D. (1997, June). Compliance for the right reasons. *Business Geographics, 12,* 30–32.

Journal Article

Rossouw, G. J. (1997). Business ethics in South Africa. *Journal of Business Ethics, 16,* 1539–1547.

Newspaper Article

Mathews, A. W. (1997, October 1). The Internet generation taps into Morse code. *Wall Street Journal,* pp. B1, B7.

Article with Unknown Author

American City adds Nashville, Memphis. (1997, September 26). *The Business Journal,* 16.

ELECTRONIC SOURCES
An Entire Web Site

The APA recommends that at minimum, a reference to a Web source should provide a document title or description, a date (of the publication or retrieval of the document), an address (URL) that links directly to the document or section, and an author, whenever possible. On the rare occasion that you need to cite multiple pages of a Web site (or the entire site), provide a URL that links to the site's homepage.

Society for Technical Communication. (2002). Retrieved April 20, 2002, from http://www.stc.org

A Document on a Web Site, with an Author

Locker, K. O. (1995). *The history of the association for business communication.* Retrieved April 20, 2001, from the Association for Business Communication Web site: http://www.theabc.org/history.html

A Document on a Web Site, with a Corporate Author

General Motors. (2001). *Company profile.* Retrieved April 20, 2001, from http://www.gm.com/company/corp_info/profiles/

A Document on a Web Site, with an Unknown Author

Forgotten inventors. (2001). PBS Online. Retrieved April 19, 2001, from http://www.pbs.org/wgbh/amex/telephone/sfeature/index.html

Article or Other Work from a Database

Goldbort, R. C. (2001, March). Scientific writing as an art and as a science. *Journal of Environmental Health, 63*(7). Retrieved April 19, 2001, from Expanded Academic ASAP database.

Article in an Online Periodical

Tiernen, R. (2001, April 18). Waiting for wireless. *SmartMoney.com.* Retrieved April 19, 2001, from http://www.smartmoney.com /techmarket/index.cfm?story=20010418

E-mail

Personal communications (including e-mail and messages from discussion groups and electronic bulletin boards) are not cited in an APA reference list. They can be cited in the text as follows: "According to J. D. Kahl (personal communication, October 2, 2001), Web pages need to reflect. . . ."

Publication on CD-ROM

Money 99. (1997). [CD-ROM]. Redmond, WA. Microsoft.

MULTIMEDIA SOURCES (Print and Electronic)
Map or Chart

Asia. (2001). [Map]. Maps.com. Retrieved April 20, 2001, from http:// www.maps.com/explore/atlas/political/asia.html

Wisconsin. (2000). [Map]. Chicago: Rand.

Film or Video

Lawrence, D. (Director), & Christopher, J. (Editor). (2001). *Emergency film group video* [Video]. Retrieved April 20, 2001, from http://www .efilmgroup.com/video1.rm

Massingham, G. (Director), & Christopher, J. (Editor). (1998). *Introduction to hazardous chemicals* [Motion picture]. Edgartown, MA: Emergency Film Group.

Radio or Television Program

Norris, R. (Host). (2001, April 3). Energy supplies. *All things considered* [Radio broadcast]. Boston: WGBH. Retrieved April 20, 2001, from http://www.npr.org/programs/atc

Novak, R. (Host). (2001, April 16). Do Americans really want a tax cut? *CNN: Crossfire* [Television broadcast]. Washington, DC: CNN.

OTHER SOURCES

Published Interview

Gates, B. (2000, April 17). The view from the very top [Interview]. *Newsweek, 135,* 36–39.

Personal Interview and Letters

Personal communications (including telephone conversations) are not cited in a reference list. They can be cited in the text as follows: "According to J. D. Kahl (personal communication, October 2, 2001), Web pages need to reflect. . . ."

Brochure or Pamphlet

Library of Congress. U.S. Copyright Office. (1999). *Copyright registration for online works* [Brochure]. Washington, DC: U.S. Government Printing Office.

Government Document

U.S. Department of Energy. (1998). *The energy situation in the next decade* (Technical Publication No. 11346-53). Washington, DC: U.S. Government Printing Office.

Report

Bertot, J. C., McClure, C. R., & Zweizig, D. L. (1996). *The 1996 national survey of public libraries and the Internet: Progress and issues: Final report.* Washington, DC: U.S. Government Printing Office.

Unpublished Data

Morrell, K. (2002). [Genetics counseling statistics for northern New Jersey]. Unpublished raw data.

APA Sample Pages

Shortened
title and page
number.

This report examines the nature and disposition of the 3,458 ethics

cases handled companywide by CGF's ethics officers and managers

during 2002. The purpose of such reports is to provide the Ethics

and Business Conduct Committee with the information necessary for

assessing the effectiveness of the first year of CGF's Ethics Program

(Davis, Roland, & Tegge, 2001). According to Matthias Jonas

(1999), recommendations are given for consideration "in planning

for the second year of the Ethics Program" (p. 152).

One-inch
margins. Text
double-
spaced.

The Office of Ethics and Business Conduct was created to ad-

minister the Ethics Program. The director of the Office of Ethics and

Business Conduct, along with seven ethics officers throughout CGF,

was given the responsibility for the following objectives, as de-

scribed by Rossouw (1997):

Long quote
indented five
to seven
spaces,
double-
spaced, with-
out quotation
marks.

Communicate the values, standards, and goals of CGF's Pro-

gram to employees. Provide companywide channels for em-

ployee education and guidance in resolving ethics concerns.

Implement companywide programs in ethics awareness and

recognition. Employee accessibility to ethics information and

guidance is the immediate goal of the Office of Business

Conduct in its first year. (p. 1543)

The purpose of the Ethics Program, established by the Committee,

is to "promote ethical business conduct through open communica-

tion and compliance with company ethics standards" (Jonas, 2001,

In-text cita-
tion gives
name, date,
and page
number.

p. 89). This report examines the nature and disposition of the 3,458

ethics cases handled companywide by CGF's ethics officers and

managers during 2002. The purpose of such

FIGURE D–6. APA Sample Page

D

APA

Heading centered.

References

List alphabetized by authors' last names and double-spaced.

Davis, W. C, Roland, M., & Tegge, D. (2001). *Working in the system: Five new management principles.* New York: St. Martin's Press.

Hassab, J. C. (1997). *Systems management: People, computers, machines, materials.* New York: CRC Press.

Jonas, M. (1999). The Internet and ethical communication: Toward a new paradigm. *Journal of Ethics and Communication, 27,* 147–177.

Jonas, M. (2001). Ethics in organizational communication: A review of the literature. *Journal of Ethics and Communication, 29,* 79–99.

Library of Congress. U.S. Copyright Office. (1999). *Copyright registration for online works.* [Brochure]. Washington, DC: U.S. Government Printing Office.

Hanging-indent style used for entries.

The one-minute manager. (1998). [CD-ROM]. Boston: Bedford/St. Martin's.

Rossouw, G. J. (1997). Business ethics in South Africa. *Journal of Business Ethics, 16,* 1539–1547.

FIGURE D–7. APA Sample List of References

CMS Documentation

CMS *Footnotes and Endnotes.* CMS footnote and endnote citations give superscript numerals within the text that correspond to numbered footnotes (appearing at the bottom of the page where referenced) or endnotes (listed on a separate page at the end of the paper).

Place superscript numbers at the end of the quotation or sentence after the punctuation marks. Indent the first line of the footnote or endnote entry five spaces ($\frac{1}{2}$ inch). Use the number corresponding to the text, but do not make it superscript. Include the author's name (first name first), title, publication information (enclosed in parentheses), and page number(s), separated by commas.

- The year 1776 can be considered a watershed; political events in that year spawned the industrial revolution and, in turn, the burgeoning modern economy.[4]

- 4. Mark Skousen, *The Making of Modern Economics: The Lives and the Ideas of the Great Thinkers* (Armonk, N.Y.: M. E. Sharpe, 2001), 15.

In subsequent footnotes or endnotes from the same source, use just the author's last name and the page number: 4. Skousen, 381. If you are using two or more works by the same author, give a shortened version of the title: 4. Skousen, *Modern Economics,* 381.

CMS *Documentation Models*

BOOKS
Single Author
1. Joseph C. Hassab, *Systems Management: People, Computers, Machines, Materials* (New York: CRC, 1997), 23.

Hassab, Joseph C. *Systems Management: People, Computers, Machines, Materials.* New York: CRC, 1997.

Two or Three Authors
2. Joshua O. Testerman, Thomas J. Kuegler Jr., and Paul J. Dowling Jr., *Web Advertising and Marketing,* 2d ed. (Roseville, Calif.: Prima, 1998), 56.

Testerman, Joshua O., Thomas J. Kuegler Jr., and Paul J. Dowling Jr. *Web Advertising and Marketing.* 2d ed. Roseville, Calif.: Prima, 1998.

Four or More Authors
3. Michael A. Lewis et al., *Ecotoxicology and Risk Assessment for Wetlands* (Pensacola, Fla.: Society of Environmental Toxicology and Chemistry, 1999), 277.

Lewis, Michael A., et al. *Ecotoxicology and Risk Assessment for Wetlands.* Pensacola, Fla.: Society of Environmental Toxicology and Chemistry, 1999.

Corporate Author

4. Ernst and Young, *Ernst and Young's Retirement Planning Guide* (New York: Wiley, 1999).

Ernst and Young. *Ernst and Young's Retirement Planning Guide.* New York: Wiley, 1999.

Edition Other Than First

5. Jack C. Estes and Dennis R. Kelley, *McGraw-Hill's Interest Amortization Tables,* 3d ed. (New York: McGraw-Hill, 1998).

Estes, Jack C., and Dennis R. Kelley. *McGraw-Hill's Interest Amortization Tables.* 3d ed. New York: McGraw-Hill, 1998.

Multivolume Work

6. Standard and Poor, *Standard and Poor's Register of Corporations, Directors and Executives,* 3 vols. (New York: McGraw-Hill, 1996).

Standard and Poor *Standard and Poor's Register of Corporations, Directors and Executives.* 3 vols. New York: McGraw-Hill, 1996.

Work in an Edited Collection

7. Judith M. Gueron, "Welfare and Poverty: Strategies to Increase Work," in *Reducing Poverty in America: Views and Approaches,* ed. Michael R. Darby (Thousand Oaks, Calif.: Sage, 1996), 240.

Gueron, Judith M. "Welfare and Poverty: Strategies to Increase Work." In *Reducing Poverty in America: Views and Approaches,* edited by Michael R. Darby, 237–55. Thousand Oaks, Calif.: Sage, 1996.

Encyclopedia or Dictionary Entry

8. *World Book Encyclopedia,* 1999 ed., s.v. "particle detector."

Well-known reference books, such as encyclopedias and dictionaries, are not included in the bibliography.

ARTICLES IN PERIODICALS
Magazine Article

9. Don Coley, "Compliance for the Right Reasons," *Business Geographics,* June 1997, 31.

Coley, Don. "Compliance for the Right Reasons." *Business Geographics,* June 1997, 30–32.

Journal Article
10. George J. Rossouw, "Business Ethics in South Africa," *Journal of Business Ethics* 16 (1997): 1543.

Rossouw, George J. "Business Ethics in South Africa." *Journal of Business Ethics* 16 (1997): 1539–47.

Newspaper Article
11. Anna Wilde Mathews, "The Internet Generation Taps into Morse Code," *Wall Street Journal,* 1 October 1997, sec. B.

Mathews, Anna Wilde. "The Internet Generation Taps into Morse Code." *Wall Street Journal,* 1 October 1997, sec. B.

Article with Unknown Author
12. "American City Adds Nashville, Memphis," *Business Journal,* 26 September 1997: 16.

"American City Adds Nashville, Memphis." *Business Journal,* 26 September 1997: 16.

ELECTRONIC SOURCES
The Chicago Manual of Style, 14th ed., does not include guidelines for documenting electronic sources; however, the University of Chicago Press recommends the system in *Online! A Reference Guide to Using Internet Sources,* 2000 ed. (New York: St. Martin's, 2000). (See <www.press.uchicago.edu/Misc/Chicago/cmosfaq.html>.) The following models are based on the guidelines set forth in *Online!*

An Entire Web Site
13. Society for Technical Communication, *Society for Technical Communication,* January 2002, <http://www.stc.org> (20 April 2002).

Society for Technical Communication. *Society for Technical Communication.* January 2002. <http://www.stc.org> (20 April 2002).

A Short Work from a Web Site, with an Author
14. Kitty O. Locker, "The History of the Association for Business Communication," *Association for Business Communication,* 25 October 1995, <http://www.theabc.org/history.htm> (20 April 2001).

Locker, Kitty O. "The History of the Association for Business Communication." *Association for Business Communication.* 25 October 1995. <http://www.theabc.org/history.htm> (20 April 2001).

A Short Work from a Web Site, with a Corporate Author
15. General Motors, "Company Profile," *General Motors,* 2001, <http://www.gm.com/company/corp_info/profiles> (19 April 2001).

General Motors. "Company Profile." *General Motors.* 2001. <http://www
.gm.com/company/corp_info/profiles> (19 April 2001).

A Short Work from a Web Site, with an Unknown Author
16. "Technology Timeline: 1752–1990," *The American Experience, PBS Online,* 2000, <http://www.pbs.org/wgbh/amex/telephone/timeline
/timeline_text.html> (10 December 2001).

"Technology Timeline: 1752–1990." *The American Experience. PBS Online.*
2000. <http://www.pbs.org/wgbh/amex/telephone/timeline
/timeline_text.html> (10 December 2001).

Article or Other Work from a Database
17. Robert C. Goldbort, "Scientific Writing as an Art and as a Science," *Journal of Environmental Health* 63, no. 7 (2001): 22. *Expanded Academic ASAP,* InfoTrac (19 April 2001).

Goldbort, Robert C. "Scientific Writing as an Art and as a Science." *Journal of Environmental Health* 63, no. 7 (2001): 22. *Expanded Academic ASAP.* InfoTrac (19 April 2001).

Online Book
18. Virginia Shea, *Netiquette,* 2000, <http://www.albion.com
/netiquette/book/index.html> (10 December 2001), 47.

Shea, Virginia. *Netiquette.* 2000. <http://www.albion.com/netiquette
/book/index.html> (10 December 2001).

E-mail Message
19. Jonathan D. Kahl, "Re: Web page," 1 October 2001, personal e-mail (2 October 2001).

Kahl, Jonathan D. "Re: Web page." 1 October 2001. Personal e-mail (2 October 2001).

MULTIMEDIA SOURCES (Print and Electronic)
Map or Chart
20. *Wisconsin,* map (Chicago: Rand, 2000).

Wisconsin. Map. Chicago: Rand, 2000.

Film or Video
> 21. *Introduction to Hazardous Chemicals,* dir. Gordon Massingham, ed. Jane Christopher, 30 min., Emergency Film Group, Edgartown, Mass., 1998, videocassette.

> *Introduction to Hazardous Chemicals.* Directed by Gordon Massingham, edited by Jane Christopher. 30 min. Emergency Film Group, Edgartown, Mass., 1998. Videocassette.

Television Interview
> 22. Mark Malloch Brown, interview by Charlie Rose. *The Charlie Rose Show,* Public Broadcasting System, 6 December 2001.

> Brown, Mark Malloch. Interview by Charlie Rose. *The Charlie Rose Show.* Public Broadcasting System, 6 December 2001.

OTHER SOURCES
Published Interview
> 23. Bill Gates, "The View from the Very Top," interview, *Newsweek,* 17 April 2000, 46.

> Gates, Bill. "The View from the Very Top." Interview. *Newsweek* 17 April 2000, 46.

Personal Interview
> 24. Mahmood Sariolgholam, interview by author, tape recording, Berkeley, Calif., 29 November 2000.

> Sariolgholam, Mahmood. Interview by author. Tape recording. Berkeley, Calif., 29 November 2000.

Personal Letter
> 25. Monica Pascatore, letter to author, 10 April 2002.

Personal communications are not included in the bibliography.

Brochure or Pamphlet
> 26. Library of Congress, U.S. Copyright Office, *Copyright Registration for Online Works* (Washington, D.C.: GPO, 1999), 10.

> Library of Congress. U.S. Copyright Office. *Copyright Registration for Online Works.* Washington, D.C.: GPO, 1999.

Government Document
> 27. Department of Energy, *The Energy Situation in the Next Decade* (Washington, D.C.: GPO, 1998), 11346.

U.S. Department of Energy. *The Energy Situation in the Next Decade.* Washington, D.C.: GPO, 1998.

Report

28. John Carlo Bertot, Charles R. McClure, and Douglas L. Zweizig, *The 1996 National Survey of Public Libraries and the Internet: Progress and Issues: Final Report* (Washington, D.C.: GPO, 1996).

Bertot, John Carlo, Charles R. McClure, and Douglas L. Zweizig. *The 1996 National Survey of Public Libraries and the Internet: Progress and Issues: Final Report.* Washington, D.C.: GPO, 1996.

CMS *Sample Pages*

D

CMS

Marks 14   Author's last name (op-tional) and page number.

This report examines the nature and disposition of the 3,458 ethics cases handled companywide by CGF's ethics officers and managers during 2002. The purpose of such reports is to provide the Ethics and Business Conduct Committee with the information necessary for as-

One-inch margins. Text double-spaced.

sessing the effectiveness of the first year of CGF's Ethics Program.[1] According to Matthias Jonas, recommendations are given for consideration "in planning for the second year of the Ethics Program."[2]

The Office of Ethics and Business Conduct was created to administer the Ethics Program. The director of the Office of Ethics and Business Conduct was given the responsibility for the following objectives, as described by Rossouw:

Long quote indented, set in smaller font size, without quo-tation marks.

> Communicate the values, standards, and goals of CGF's Program to employees. Provide companywide channels for employee education and guidance in resolving ethics concerns. Implement companywide programs in ethics awareness and recognition. Employee accessibility to ethics information and guidance is the immediate goal of the Office of Business Conduct in its first year.[3]

The purpose of the Ethics Program, according to Jonas, is to "promote ethical business conduct through open communication and compliance with company ethics standards."[4] This report examines

Raised num-bers in text correspond to footnotes or endnotes. First line of each entry indented five spaces.

1. W. C. Davis, Roland Marks, and Diane Tegge, *Working in the System: Five New Management Principles* (New York: St. Martin's, 2001), 142.

2. Matthias Jonas, "The Internet and Ethical Communication: Toward a New Paradigm," *Journal of Ethics and Communications* 27 (Fall 1999): 152.

3. George J. Rossouw, "Business Ethics in South Africa." *Journal of Business Ethics* 16 (Spring 1997): 1543.

4. Matthias Jonas, "Ethics in Organizational Communication: A Review of the Literature," *Journal of Ethics and Communication* 29 (Summer 2001): 89.

FIGURE D-8. *CMS* **Sample Page (with Footnotes)**

Marks 21

Heading
centered.

Bibliography

Association for Business Communication. *Association for Business*
Communication. August 1999. <http://www.theabc.org> (20

List alpha-
betized by
authors' last
names and
double-
spaced.

April 2001).

Davis, W. C., Roland Marks, and Diane Tegge. *Working in the System:*
Five New Management Principles. New York: St. Martin's, 2001.

Hassab, Joseph C. *Systems Management: People, Computers, Machines,*
Materials. New York: CRC, 1997.

Jonas, Matthias. "Ethics in Organizational Communication: A Review of
the Literature." *Journal of Ethics and Communication* 29
(Summer 2001): 77–79.

Jonas, Matthias. "The Internet and Ethical Communication: Toward a
New Paradigm." *Journal of Ethics and Communication* 27 (Fall

Hanging-
indent style
used for
entries.

1999): 147–77.

Library of Congress. U.S. Copyright Office. *Copyright Registration for*
Online Works. Washington, D.C.: GPO, 1999.

The One-Minute Manager. CD-ROM. Boston: Bedford/St. Martin's,
1998.

Rossouw, George J. "Business Ethics in South Africa." *Journal of*
Business Ethics 16 (1997): 1539–47.

Sariolgholam, Mahmood. Interview by author. Tape recording. Berkeley,
Calif.: 29 November 2000.

FIGURE D–9. *CMS* Sample Bibliography

MLA Documentation

MLA In-Text Citations.   MLA parenthetical citation within the text of a paper gives a brief citation — in parentheses — of the author and relevant page numbers.

- As Peterson writes, preparing a videotape of measurement methods is cost effective and can expedite training (151).

- The results of these studies have led even the most conservative managers to adopt technologies that will "catapult the industry forward" (Peterson 183–91).

If the parenthetical citation refers to a long indented quotation, place it outside the punctuation of the last sentence.

- . . . a close collaboration with the physics and technology staff is essential. (Minsky 42)

If no author is named or if you are using more than one work by the same author, give a shortened version of the title in the parenthetical citation, unless you mention it in a signal phrase in the text. A proper citation for Thomas J. Peterson's book *The Pursuit of Wow: Every Person's Guide to Topsy-Turvy Times* would appear as (Peterson, <u>Pursuit</u> (93).

MLA Documentation Models

BOOKS
Single Author

Hassab, Joseph C. <u>Systems Management: People, Computers, Machines, Materials</u>. New York: CRC, 1997.

Multiple Authors

Testerman, Joshua O., Thomas J. Kuegler, Jr., and Paul J. Dowling, Jr. <u>Web Advertising and Marketing</u>. 2nd ed. Roseville, CA: Prima, 1998.

Corporate Author

Ernst and Young. <u>Ernst and Young's Retirement Planning Guide</u>. New York: Wiley, 1999.

Edition Other Than First

Estes, Jack C., and Dennis R. Kelley. <u>McGraw-Hill's Interest Amortization Tables</u>. 3rd ed. New York: McGraw, 1998.

Multivolume Work

Standard and Poor. <u>Standard and Poor's Register of Corporations, Directors and Executives</u>. 3 vols. New York: McGraw, 1998.

Work in an Edited Collection

Gueron, Judith M. "Welfare and Poverty: Strategies to Increase Work." Reducing Poverty in America: Views and Approaches. Ed. Michael R. Darby. Thousand Oaks: Sage, 1996. 237–55.

Encyclopedia or Dictionary Entry

Gibbard, Bruce G. "Particle Detector." World Book Encyclopedia. 1999 ed.

ARTICLES IN PERIODICALS (See also Electronic Sources)
Magazine Article

Coley, Don. "Compliance for the Right Reasons." Business Geographics June 1997: 30–32.

Journal Article

Rossouw, G. J. "Business Ethics in South Africa." Journal of Business Ethics 16 (1997): 1539–47.

Newspaper Article

Mathews, Anna Wilde. "The Internet Generation Taps into Morse Code." Wall Street Journal 1 Oct. 1997, natl. ed.: B1+.

Article with Unknown Author

"American City Adds Nashville, Memphis." Business Journal 26 Sept. 1997: 16.

ELECTRONIC SOURCES
An Entire Web Site

Society for Technical Communication. Jan. 2002. Society for Technical Communication. 20 Apr. 2002 <http://www.stc.org/>.

A Short Work from a Web Site, with an Author

Locker, Kitty O. "The History of the Association for Business Communication." Association for Business Communication. 25 Oct. 1995. Assn. for Business Communication. 20 Apr. 2001 <http://www.theabc .org/history.htm>.

A Short Work from a Web Site, with a Corporate Author

General Motors. "Company Profile." General Motors. 2001. 19 Apr. 2001 <http://www.gm.com/company/corp_info/profiles/>.

A Short Work from a Web Site, with an Unknown Author

"Forgotten Inventors." Forgotten Inventors. PBS Online. 2001. 19 Apr. 2001 <http://www.pbs.org/wgbh/amex/telephone/sfeature/ index.htm>.

Article from a Database

Goldbort, Robert C. "Scientific Writing as an Art and as a Science." Journal of Environmental Health, 63.7 (2001): 22. Expanded Academic ASAP. InfoTrac. Salem State Coll. Lib., Salem, MA. 19 Apr. 2001.

Article in an Online Periodical

Ray, Tiernan. "Waiting for Wireless." SmartMoney.com 18 Apr. 2001. 19 Apr. 2001 <http://www.smartmoney.com/techmarket/index .cfm?story=20010418>.

Publication on CD-ROM

Money 99. CD-ROM. Redmond: Microsoft, 1998.

E-mail Message

Kahl, Jonathan D. "Re: Web page." E-mail to the author. 2 Oct. 2001.

MULTIMEDIA SOURCES (Print and Electronic)
Map or Chart

Wisconsin. Map. Chicago: Rand, 2000.

"Asia." Map. Maps.com. 2000. 20 Apr. 2001 <http://www.maps.com /explore/atlas/political/asia.html>.

Film or Video

Massingham, Gordon, dir., and Jane Christopher, ed. Introduction to Hazardous Chemicals. Videocassette. Edgartown, MA: Emergency Film Group, 1998.

Lawrence, Detrick, dir. Emergency Film Group: Homepage Web Video. 2001. 20 Apr. 2001 <http://www.efilmgroup.com/video1.rm>.

Radio or Television Program

"Do Americans Really Want a Tax Cut?" CNN: Crossfire. Host Robert Novak. CNN. 16 Apr. 2001.

"Energy Supplies." Host Emily Harris. All Things Considered. Natl. Public Radio. WGBH, Boston. 3 Apr. 2001. 20 Apr. 2001 <http://www.npr .org/programs/atc/>.

OTHER SOURCES
Published Interview

Gates, Bill. "The View from the Very Top." Interview. Newsweek 17 Apr. 2000: 36–39.

Personal Interview

Sariolgholam, Mahmood. Personal interview. 29 Nov. 2000.

Personal Letter

Viets, Hermann. Letter to all students, fac., and staff. University of Wisconsin, Milwaukee. 1 Sept. 1998.

Brochure or Pamphlet

Library of Congress. US Copyright Office. Copyright Registration for Online Works. Washington: GPO, 1999.

Government Document

United States. Dept. of Energy. The Energy Situation in the Next Decade. Technical Pub. 11346-53. Washington: GPO, 1998.

Report

Bertot, John Carlo, Charles R. McClure, and Douglas L. Zweizig. The 1996 National Survey of Public Libraries and the Internet: Progress and Issues: Final Report. Washington: GPO, 1996.

Lecture or Speech

McKinney, Scott. Lecture. Demarest Hall, Hobart and William Smith Colls., Geneva, NY. 3 May 2002.

Krug, Steve. "Don't Make Me Think: The Art of Designing User-Friendly Websites." The New England School of Art & Design at Suffolk U. Boston Public Lib. 3 Apr. 2001.

MLA Sample Pages

Marks 14

Author's last
name and
page num-
ber.

This report examines the nature and disposition of the 3,458 ethics

cases handled companywide by CGF's ethics officers and managers

One-inch
margins. Text
double-
spaced.

during 2002. The purpose of such reports is to provide the Ethics

and Business Conduct Committee with the information necessary for

assessing the effectiveness of the first year of CGF's Ethics Program

(Davis 142). According to Matthias Jonas, recommendations are

given for consideration "in planning for the second year of the

Ethics Program" ("Internet" 152).

The Office of Ethics and Business Conduct was created to ad-

minister the Ethics Program. The director of the Office of Ethics and

Business Conduct, along with seven ethics officers throughout CGF,

was given the responsibility for the following objectives, as de-

scribed by Rossouw:

Long quote
indented one
inch (or 10
spaces),
double-
spaced, with-
out quotation
marks.

> Communicate the values, standards, and goals of CGF's
>
> Program to employees. Provide companywide channels
>
> for employee education and guidance in resolving ethics
>
> concerns. Implement companywide programs in ethics
>
> awareness and recognition. Employee accessibility to
>
> ethics information and guidance is the immediate goal of

In-text cita-
tions give
author name
and page
number. Title
used when
multiple
works by
same author
cited.

> the Office of Business Conduct in its first year. (1543)

The purpose of the Ethics Program, according to Jonas, established

by the Committee, is to "promote ethical business conduct through

open communication and compliance with company ethics stan-

dards" ("Ethics" 89). This report examines the nature and

FIGURE D–10. MLA Sample Page

Marks 21

Heading
centered.

Works Cited

List alpha-
betized by
authors' last
names or
title and
double-
spaced.

Association for Business Communication. Aug. 1999. Assn. for Business

    Communication. 20 Apr. 2001 <http://www.theabc.org/>.

Davis, W. C., Roland Marks, and Diane Tegge. Working in the System:

    Five New Management Principles. New York: St. Martin's, 2001.

Hassab, Joseph C. Systems Management: People, Computers, Machines,

    Materials. New York: CRC, 1997.

Jonas, Matthias. "Ethics in Organizational Communication: A Review of

    the Literature." Journal of Ethics and Communication 29 (2001):

    79–99.

---. "The Internet and Ethical Communication: Toward a New

    Paradigm." Journal of Ethics and Communication 27 (1999):

Hanging-
indent style
used for
entries.

    147–77.

Library of Congress. US Copyright Office. Copyright Registration for

    Online Works. Washington: GPO, 1999.

The One-Minute Manager. CD-ROM. Boston: Bedford, 1998.

Rossouw, George J. "Business Ethics in South Africa." Journal of

    Business Ethics 16 (1997): 1539–47.

Sariolgholam, Mahmood. Personal interview. 29 Nov. 2000.

FIGURE D–11. MLA Sample List of Works Cited

Other Style Manuals

Many professional societies, publishing companies, and other organizations publish manuals that prescribe bibliographic reference formats for their publications or for publications in their fields. In addition, several general style manuals are well known and widely used.

Biology

Council of Science Editors. *Scientific Style and Format: The CBE Manual for Authors, Editors, and Publishers.* 6th ed. New York: Cambridge University Press, 1994. See also <www.council scienceeditors.org>.

Chemistry

American Chemical Society. *ACS Style Guide: A Manual for Authors and Editors.* 2nd ed. Washington, D.C.: American Chemical Society, 1998. See also <www.acs.org>.

Geology

Bates, Robert L., Rex Buchanan, and Marla Adkins-Heljeson, eds. *Geowriting: A Guide to Writing, Editing, and Printing in Earth Science.* 5th ed. Alexandria: Amer. Geological Inst., 1995. See also <www.agiweb.org>.

Government Documents

United States Government Printing Office. *Style Manual.* Washington, D.C.: U.S. Government Printing Office, 2000. See also <www.gpo.gov>.

Mathematics

American Mathematical Society. *The AMS Author Handbook: General Instructions for Preparing Manuscripts.* Rev. ed. Providence: AMS, 1996. See also <www.ams.org>.

Medicine

American Medical Association. *American Medical Association Manual of Style.* 9th ed. Baltimore: Williams, 1998. See also <www.ama-assn.org>.

Physics

American Institute of Physics, Publication Board. *Style Manual for Guidance in the Preparation of Papers.* 4th ed. New York: American Institute of Physics, 1990. See also <www.aip.org>.

D

Science and Technical Writing

National Information Standards Organization. *Scientific and Technical Reports—Elements, Organization, and Design.* Bethesda, Md.: National Information Standards Organization, 1995. See also <www.niso.org>.

Rubens, Phillip, ed. *Science and Technical Writing: A Manual of Style.* 2nd ed. New York: Routledge, 2001.

double negatives

A *double negative* is the use of an additional negative word to reinforce an expression that is already negative. In writing and speech, avoid such double negatives.

UNCLEAR	We don't have none. [This sentence literally means that we have some.]
CLEAR	We have none.

Barely, hardly, and *scarcely* cause problems because writers sometimes do not recognize that those words are already negative.

- The corporate policy ~~doesn't~~ hardly cover *s* the problem.

Not unfriendly, not without, and similar constructions are not double negatives because in such constructions two negatives are meant to suggest the gray area between negative and positive meanings. Be careful how you use such constructions; they can be confusing to the reader and should be used only if they serve a purpose.

- He is *not unfriendly.* [He is neither hostile nor friendly.]
- It is *not without* regret that I offer my resignation. [I have mixed feelings rather than only regret.]

The correlative **conjunctions** *neither* and *nor* may appear together in a clause without creating a double negative, so long as the writer does not attempt to use the word *not* in the same clause.

- It was ~~not~~ *neither*, as a matter of fact, ~~neither~~ his duty nor his desire to fire the employee.

- It was not, as a matter of fact, ~~neither~~ *either* his duty ~~nor~~ *or* his desire to fire the employee.

- She ~~did not~~ neither ~~care~~ *cared* about nor ~~notice~~ *noticed* the error.

Negative forms are full of traps that often entice writers into <u>logic errors</u>, as illustrated in the following example:

> ILLOGICAL   There is *nothing* in the book that has *not* already been published in some form, but some of it is, I believe, very little known.

In this sentence, "some of it" logically can refer only to "*nothing* in the book that has *not* already been published." The sentence can be corrected by stating the idea in more <u>positive writing</u>.

> LOGICAL   Everything in the book has been published in some form, but some of it is, I believe, very little known.

drawings

A drawing can emphasize the significant part or function of a mechanism, omit what is not significant, and focus on details or relationships that a <u>photograph</u> cannot reveal. If the actual appearance of an object is necessary, however, include a photograph.

The five types of drawings discussed in this entry are conventional line drawings, cutaway drawings, exploded-view drawings, and schematic diagrams. Each type of drawing has unique advantages, depending on your <u>purpose</u>.

A conventional drawing, like that in Figure D–12, is appropriate if

PULL
the pin: Some extinguishers require releasing a lock latch, pressing a puncture lever, or taking another first step.

FIGURE D–12. Conventional Drawing Illustrating Instructions

your <u>readers</u> need a representation of an object's general appearance or an overview of a series of steps. A cutaway drawing, like the one in Figure D–13, is used to show the internal parts of a piece of equipment and illustrate their relationship to the whole. An exploded-view drawing, like that in Figure D–14 on page 168, is used to show the proper sequence in which parts fit together or to show the details of individual parts.

D

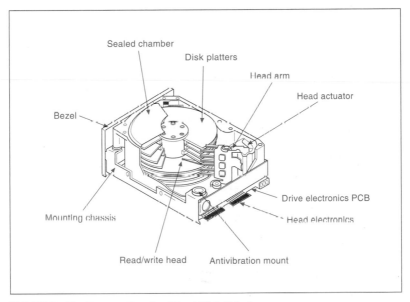

Sealed chamber
Disk platters
Head arm
Head actuator
Bezel
Drive electronics PCB
Mounting chassis
Head electronics
Read/write head   Antivibration mount

FIGURE D–13. Cutaway Drawing (Hard Disk Drive)

Schematic diagrams (or schematics) are drawings that portray the operation of electrical and mechanical systems, using lines and symbols rather than physical likeness. Schematics emphasize the relationships among parts without giving their precise proportions. The schematic diagram in Figure D–15 on page 169 depicts the process of trapping particulates from coal-fired power plants. Note that the particulate filter is depicted both symbolically and as an enlarged drawing to show relevant structural details. Schematic symbols generally represent equipment, gauges, and electrical or electronic components.

Drawings that require a high degree of accuracy and precision generally are prepared by graphics specialists. If you need only general-interest images to illustrate <u>newsletters</u> and <u>brochures</u> or to create presentation overheads, use noncopyrighted images from clip-art libraries. Such libraries contain thousands of noncopyrighted symbols,

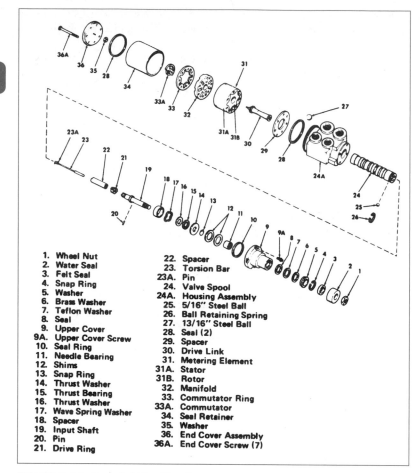

1. Wheel Nut
2. Water Seal
3. Felt Seal
4. Snap Ring
5. Washer
6. Brass Washer
7. Teflon Washer
8. Seal
9. Upper Cover
9A. Upper Cover Screw
10. Seal Ring
11. Needle Bearing
12. Shims
13. Snap Ring
14. Thrust Washer
15. Thrust Bearing
16. Thrust Washer
17. Wave Spring Washer
18. Spacer
19. Input Shaft
20. Pin
21. Drive Ring

22. Spacer
23. Torsion Bar
23A. Pin
24. Valve Spool
24A. Housing Assembly
25. 5/16" Steel Ball
26. Ball Retaining Spring
27. 13/16" Steel Ball
28. Seal (2)
29. Spacer
30. Drive Link
31. Metering Element
31A. Stator
31B. Rotor
32. Manifold
33. Commutator Ring
33A. Commutator
34. Seal Retainer
35. Washer
36. End Cover Assembly
36A. End Cover Screw (7)

FIGURE D-14. Exploded-View Drawing

shapes, and images. Figure D–16 on page 169 shows examples of clip-art images. Be careful not to use drawings from copyrighted sources without proper documentation, especially from the Web. Keep in mind that the same <u>copyright</u> laws that apply to printed material also apply to Web-based graphics. See also <u>documenting sources</u> and <u>plagiarism</u>.

Think about your need for drawings during the <u>preparation</u> and <u>research</u> stages of writing. Include them in your <u>outline</u>, indicating approximately where each should appear throughout the outline with "drawing of . . ." enclosed in brackets. See also <u>visuals</u>.

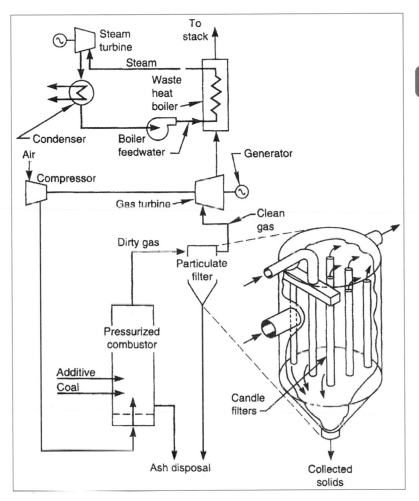

FIGURE D–15. Schematic Diagram

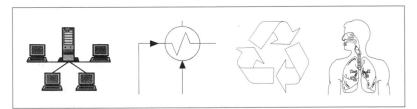

FIGURE D–16. Clip-Art Images

Writer's Checklist: Creating and Using Drawings

☑ Show equipment and other objects from the point of view of the person who will use them.

☑ When illustrating a subsystem, show its relationship to the larger system of which it is a part.

☑ Draw the different parts of an object in proportion to one another or indicate that certain parts are enlarged.

☑ When a sequence of drawings is used to illustrate a process, arrange them from left to right or from top to bottom on the page.

☑ Label parts in the drawing so that the text references to them are clear and consistent.

☑ Depending on the complexity of what is shown, label the parts themselves or use a letter/number key (see Figure D-14).

☑ For schematics, include a list of symbols in your document or <u>presentation</u> if some in your audience are not familiar with them.

due to / because of

Due to (meaning "caused by") is acceptable following a linking <u>verb</u>.

- His short temper was *due to* stress at the office.

Due to is not acceptable, however, when it is used with a nonlinking verb to replace *because of*.

- He left work ~~due to~~ illness.
  because of

E

each

When *each* is used as a subject, it takes a singular <u>verb</u> or **pronoun**.

- *Each* of the reports *is* to be submitted ten weeks after *it* is assigned.

When *each* refers to a plural subject, it takes a plural verb or pronoun. See also <u>agreement</u>.

- The reports *each have* company logos on *their* covers.

Although the phrase *each and every* is commonly used in speech in an attempt to emphasize a point, it is redundant and should be eliminated from your writing. Replace it with *each* or *every*.

- Each ~~and every~~ part was correctly labeled.

- *Every*
  ~~Each and every~~ part was correctly labeled.

economic / economical

Economic refers to the production, development, and management of material wealth. (The euro had an *economic* impact on several countries.) *Economical* simply means "not wasteful or extravagant." (Management has requested that employees be as *economical* as possible in their equipment purchases.)

editing (*see* revision *and* proofreading)

e.g. / i.e.

The abbreviation *e.g.* stands for the Latin *exempli gratia,* meaning "for example"; *i.e.* stands for the Latin *id est,* meaning "that is." Because the English expressions (*for example* and *that is*) serve a similar purpose, there is no need to use the Latin expressions or <u>abbreviations</u> except to save space in notes and <u>visuals</u>.

If you must use *i.e.* or *e.g.,* punctuate them as follows. If *i.e.* or *e.g.* connects two independent clauses, a <u>semicolon</u> should precede it and a <u>comma</u> should follow it. If *i.e.* or *e.g.* connects a <u>noun</u> and an <u>apposi-</u><u>tive</u>, a comma should precede it and follow it.

- The conference was truly international; *i.e.,* Germans, Italians, Japanese, Pakistanis, and Americans gave presentations. [independent clauses]

- The conference included speakers from five countries, *i.e.,* Germany, Italy, Japan, Pakistan, and the United States. [appositive list]

elegant variation (*see* affectation)

ellipses

An *ellipsis* is the omission of words from quoted material, and it is indicated by *ellipsis points* (or spaced periods).

ORIGINAL QUOTE	"Promotional material is sometimes charged for, particularly in high volume distribution to schools, although prices for these publications are much lower than the development costs."
WITH ELLIPSIS POINTS	"Promotional material is sometimes charged for . . . although prices for these publications are much lower than the development costs."

Notice that what remains of the <u>quotation</u> is grammatically complete.

If you are following MLA style in your writing, use <u>brackets</u> around the ellipsis points to show that the ellipsis points were not part of the original source.

- "The vast majority of the Internet's inhabitants are middle class [. . .] between the ages of eighteen and thirty-four" (5).

Do not use ellipsis points when the beginning of a quoted sentence is omitted.

- The letter states that the programmer "must create a flowchart to provide a picture of the data flow through the system."

When the ellipsis comes at the end of a sentence and the quotation continues following the omission, use four points. The first point is the period that ends the sentence, and the others indicate the omission. See also **sentence construction**.

E

- "In all departments except ours, researchers control research funds. . . . In addition, researchers control their own editorial and printing funds."

Use a full line of ellipsis points across the page to indicate the omission of one or more paragraphs from quoted material.

e-mail

DIRECTORY

E-mail (or *email*) can function in the workplace as a medium to send information, elicit discussions, collect opinions, and transmit documents and files of all types. Although e-mails may take the form of informal notes, you should generally follow the same guidelines for writing style and strategy described in the entries **correspondence** and **memos**. See also **collaborative writing** and **selecting the medium**.

Review and Confidentiality

E-mail is a quick and easy way to communicate, but avoid the temptation to dash off a first draft and send it as is. Be careful to observe the rules of netiquette in the writer's checklist that follows. As with other workplace correspondence, your message should be free of grammatical or factual errors, ambiguities, or unintended implications. It should include crucial details. Be especially careful when sending messages to superiors in your organization or to people outside the organization. Time you spend reviewing your e-mail before you send it can save a great deal of time and embarrassment sorting out misunderstandings caused by a careless message.

E

Confidentiality is another issue to keep in mind when sending e-mail. Remember, e-mail can be intercepted by someone other than the intended recipient, and e-mails are never truly deleted. Most companies back up and save all company e-mail and are legally entitled to monitor e-mail. Companies can also be compelled legally to provide e-mail in a court of law. Consider the content of all your messages in the light of these possibilities, and carefully review your text before you click "Send."

Writer's Checklist: Observing Workplace Netiquette

To maintain a high level of professionalism in workplace e-mail, observe the rules of netiquette (*Internet* + *etiquette*).

☑ Use company e-mail only for appropriate business.

- Do not send or forward jokes or humorous stories, use **biased language**, or discuss office gossip.
- Do not send *flames* (e-mails that contain abusive, obscene, or derogatory language) to attack someone.
- Do not send *spams* (mass-distributed e-mails that often promote personal projects and interests).

☑ Respond to incoming e-mail promptly. If you receive an assignment by e-mail that will take a few days or longer to complete, send a response saying so.

☑ Be scrupulous about typing e-mail addresses and otherwise ensuring that the intended recipient gets the message.

☑ Send an attachment only after verifying that your recipient wants or needs the file and that your recipient's software will accept it. Be aware that attachments can consume download time and disk space.

☑ Consider posting a large file at an Internet server and supplying the file's address so that your recipient may download the file at his or her convenience.

☑ Do not write in all-uppercase letters; such a message is difficult to read and is considered the equivalent of shouting. Likewise, do not write in all-lowercase letters; it is considered lazy and too informal for professional work.

☑ Avoid e-mail abbreviations used in personal e-mail and chat rooms (*BTW* for *by the way*, for example).

☑ Do not use emoticons (keyboard characters used to create sideways faces conveying emotions) for business and professional e-mail. For advice on providing typographic emphasis, review the next section on design considerations.

DIGITAL TIP SENDING AN E-MAIL ATTACHMENT

Sending attachments to your e-mail messages is a quick, convenient alternative to sending paper copies or disks through the regular mail. However, large files, like graphics, slow transmission speed. In addition, the recipient's software or Internet provider may not be able to accept large files. Consider using a compression software utility like WinZip, which typically reduces the file size by 80 percent or more.

A word of caution: Viruses can be embedded in e-mail attachments, so make sure you regularly update your virus-scanning software. For more on this topic, see <*www.bedfordstmartins.com/alred*> and select *Digital Tips.*

Design Considerations

Some e-mail systems allow you to use sophisticated typographical features, such as unusual fonts and bullets. Because these options increase your e-mail file size and may display unpredictably in your recipient's software, set your e-mail software to send messages in "plain text" and use alternative highlighting devices. For example, capital letters or asterisks, used sparingly, can substitute for boldface, italics, and underlines as <u>emphasis</u>:

- Dr. Wilhoit's suggestions benefit doctors AND patients.

- Although the proposal is sound in *theory,* it will never work in *practice.*

Intermittent underlining can replace solid underlining or italics when referring to published works in an e-mail message.

- My report follows the format given in the _Handbook of Technical Writing_.

Keep in mind the following additional design considerations when sending e-mail.

- Break the text into short paragraphs to avoid dense blocks of text.
- Consider providing an overview at the top in a brief paragraph for messages that run longer than a screen of text.
- Send attachments for documents containing such formatted elements as tables and bulleted lists that do not always transmit well.
- Place your response to someone else's message at the beginning (or top) of the e-mail window so that recipients can see your response immediately.

- When replying to a message, quote only relevant parts. If your system does not distinguish the quoted text, note it with a greater-than symbol (>).

- Use the "cc:" (*carbon copy*) and "bcc:" (*blind carbon copy*) address lines sparingly and thoughtfully because some recipients use their placement in the address lists to filter their e-mail.

- Provide a concise phrase in the subject line (as in memos) that describes the topic of your message; recipients use subject lines to prioritize and file their incoming messages.

Salutations, Closings, and Signature Blocks

Because e-mail can function as a letter, memo, or personal note, finding a suitable greeting and a complimentary closing can be difficult. If your employer follows a certain form, adopt that practice. Otherwise, use the following guidelines:

- When e-mail functions as a personal note to a friend, you can vary informal salutations (*Hi Mike,* or *Hello Jenny,*) and closings (*Take care,* or *Cheers,*).

- When e-mail goes outside an organization to someone with whom you have not yet corresponded, you can use the standard letter salutation (*Dear Professor Tucker:* or *Dear Docuform Customer:*) and a slightly informal closing (*Best wishes,* or *Sincerely,*).

- When e-mail functions as a memo, you may omit the salutation and closing because both your name and the name(s) of the recipient(s) appear in the "To" and "From" sections of the message. However, some e-mail users adopt a slightly more personal greeting, especially if the distribution list is relatively small or a single individual (*Project colleagues,* or *Andreas,* [recipient's first name]).

Note that in some cultures, business correspondents do not use first names as quickly as they do in American correspondence. See **international correspondence**.

Because e-mail does not provide letterhead with standard addresses and contact information, many companies and individual writers include *signature blocks* (also called *signatures*) at the bottom of their messages. Signatures, which writers can usually preprogram to appear on every e-mail they send, supply information that company letterhead usually provides as well as links to Web sites. If your organization requires a certain format, adhere to that standard. Otherwise, use the pattern shown in Figure E–1.

```
=============================
Daniel J. Vasquez, Publications Manager    ←    Name and Title
Medical Information Systems                 ←    Department or Division
TechCom Corporation                         ←    Company Name
P.O. Box 5413    Salinas CA 93962           ←    Mailing Address
Office Phone 888-229-4511 (x 341)           ←    Phone Number
General Office Fax 888-229-1132             ←    Fax Number
www.tcc.com                                 ←    Web Address (URL)
=============================
```

FIGURE E–1. E-mail Signature Block

For signature blocks, consider the following guidelines:

- Include as few lines as possible. Most netiquette guides advocate using five lines or fewer, but more are acceptable in business correspondence because it typically requires more contact information.
- Keep line length to 60 characters or fewer to avoid unpredictable line wraps.
- Test your signature block in text-only mail readers to verify your format; centered text and vertical lines may look fine in one system but chaotic in another.
- Use highlighting cues, such as <u>hyphens</u>, equal signs, and white space to separate the signature from the message.
- Avoid using quotations, aphorisms, and other messages ("May the Force be with you") in professional signatures.

DIGITAL TIP LEAVING AN AWAY-FROM-DESK MESSAGE

Many e-mail systems in the workplace allow you to create a message that is automatically sent to everyone who sends you e-mail while you are away from your desk. Your message should inform senders when you are expected back and, if necessary, whom they can contact in your absence. In the meantime, all incoming e-mail messages you receive while you are away are stored in your in box. Be sure to disable the automatic response message upon your return. For more on this topic, see <*www.bedford stmartins.com/alred*> and select *Digital Tips*.

Writer's Checklist: Managing Your E-mail

☑ Check your in box several times each day; try to clear your in box by the end of the day.

☑ Set priorities for reading e-mail by skimming sender names and subject lines.

☑ Use the search command to find topics and individual names.

☑ Review all messages on a subject before you respond, so you don't waste time responding to an issue that is no longer relevant.

☑ Learn the advanced features of your system so that you can use filters that organize messages as they arrive.

☑ Create electronic folders for e-mail, using personal names and key topics.

☑ Copy or save sent copies of important e-mails in your electronic folders.

☑ Print and file hard copies of crucial e-mail messages that are complex or that you will need for meetings.

☑ Keep an up-to-date address book.

emphasis

Emphasis is the principle of stressing the most important ideas in writing, and it can be achieved in the following ways:

- position
- sentence length
- **repetition**
- sentence type
- climactic order
- **intensifiers**
- **dashes**
- mechanical devices
- direct statements
- active **voice**

Place the idea in a particular *position;* the first and last words of a sentence, **paragraph**, or document stand out in **readers**' minds.

- Moon craters are important to understanding the earth's history because they reflect geological history.

Notice that the sentence emphasizes *moon craters* simply because the term appears at the beginning of the sentence and *geological history* because it is at the end of the sentence. See also **subordination**.

Another way to achieve emphasis is to vary *sentence length.* A very short sentence that follows a very long sentence or a series of long sentences stands out in the reader's mind, as in the short sentence ("We must cut costs") that ends the following paragraph. See sentence construction.

- We have already reviewed the problem the accounting department has experienced during the past year. We could continue to examine the causes of our problems and point an accusing finger at all the culprits beyond our control, but in the end it all leads to one simple conclusion. We must cut costs.

E

Emphasis can be achieved by the *repetition* of key words and phrases, as in the use of the word *remains* and the phrase *come and go* in the following sentence.

- Similarly, atoms come and go in a molecule, but the molecule remains; molecules come and go in a cell, but the cell remains; cells come and go in a body, but the body remains; persons come and go in an organization, but the organization remains.
  — Kenneth Boulding, *Beyond Economics*

Different emphasis can be achieved by the selection of *sentence type:* a compound sentence, a complex sentence, or a simple sentence. See sentence variety.

- The report submitted by the committee was carefully illustrated, and it covered five pages of single-spaced copy.
  [This compound sentence carries no special emphasis because it contains two coordinate independent clauses.]

- The committee's report, which was carefully illustrated, covered five pages of single-spaced copy.
  [This complex sentence emphasizes the size of the report.]

- The carefully illustrated report submitted by the committee covered five pages of single-spaced copy.
  [This simple sentence emphasizes that the report was carefully illustrated.]

Emphasis can be achieved by a *climactic order* of ideas or facts within a sentence, listing them in sequence from least to most important. See also lists.

- Over subsequent weeks the Human Resources Department worked diligently, management showed tact and patience, and the employees demonstrated remarkable support for the policy changes.

Emphasis can be achieved by the use of *intensifiers* (*most, much, very*), but this technique is so easily abused that it should be used with caution.

- The final proposal is much more persuasive than the first.

Emphasis can be achieved by setting an item apart with a *long dash* (also called an *em dash*).

- The job will be done—after we are under contract.

Emphasis also can be achieved by *mechanical devices,* such as *italics,* **bold type,** <u>underlining</u>, and CAPITAL LETTERS, but use them sparingly because overuse can create visual clutter. See also <u>capitalization</u>, <u>italics</u>, and <u>layout and design</u>.

Emphasis can be achieved by *direct statement,* such as "most important" and "foremost," as well as using someone's name in a <u>direct address</u>.

- Most important, keep in mind that everything you do affects the company's bottom line.

- John, I believe we should rethink our plans.

Finally, the *active voice* can emphasize the performer of an action by making the performer the subject of the <u>verb</u>.

- Our department designed the new system.
  [The performer, *our department,* is the subject of the verb, *designed.*]

English as a second language

Learning to write well in a second language takes a great deal of effort and practice. The most effective way to improve your command of written English is to read widely beyond the reports and professional articles your job requires, such as magazines, newspapers, novels, biographies, and any other writing that interests you. In addition, listen carefully to native speakers on television, on radio, and in person. Focus on areas of English that typically give nonnative speakers trouble, such as those discussed in this entry and referred to under "ESL Entries" on page 184. Finally, do not hesitate to consult a native speaker of English, especially for important writing tasks, such as reports, memos, and e-mails.

Count and Mass Nouns

Count nouns refer to things that can be counted (*tables, pencils, projects, reports*). *Mass nouns* (often called *non-count nouns*) identify things that cannot be counted (*electricity, water, air, loyalty, information*). This distinction can be confusing with words like *electricity* and *water*. Although we can count watt-hours of electricity and bottles of water, counting becomes inappropriate when we use the words *electricity* and *water* in a general sense, as in "Water is an essential resource." See Figure E–2 for a list of common mass nouns.

acid	education	knowledge	research
advice	electricity	loyalty	technology
air	equipment	machinery	transportation
anger	furniture	money	uranium
biology	health	news	water
clothing	honesty	oil	weather
coffee	information	precision	work

FIGURE E–2. Common Mass Nouns

The distinction between whether something can or cannot be counted determines the form of the noun to use (singular or plural), the kind of article that precedes it (*a, an, the*, or no article), and the kind of limiting adjective it requires (*fewer* or *less*, *much* or *many*, and so on). See also fewer/less.

Articles

This discussion of articles applies only to common nouns (not to proper nouns, such as the names of people) because count and mass nouns are always common nouns.

The general rule is that every count noun must be preceded by an article (*a, an*, or *the*), a demonstrative adjective (*this, that, these, those*), a possessive adjective (*my, your, her, his, its, their*), or some expression of quantity (such as *one, two, several, many, a few, a lot of, some, no*). The article, adjective, or expression of quantity appears either directly in front of the noun or in front of the whole noun phrase.

- Beth read *a* report last week. [article]
- *Those* reports Beth read were long. [demonstrative adjective]
- *Their* report was long. [possessive adjective]
- *Some* reports Beth read were long. [indefinite adjective]

E

The articles *a* and *an* are used with count nouns that refer to one item of the whole class of like items.

- Matthew has *a* pen. [Matthew could have *any* pen.]

The article *the* is used with nouns that refer to a specific item that both the reader and the writer can identify.

- Matthew has *the* pen.
  [Matthew has a *specific* pen that is known to both the reader and the writer.]

When making generalizations with count nouns, writers can either use *a* or *an* with a singular count noun or use no article with a plural count noun. Consider the following generalization using an article.

- An egg is a good source of protein. [*any egg, all eggs, eggs in general*]

However, the following generalization uses a plural count noun with no article.

- Eggs are good sources of protein. [*any egg, all eggs, eggs in general*]

When you are making a generalization with a mass noun, do not use an article in front of the mass noun.

- Sugar is bad for your teeth.

Gerunds and Infinitives

Nonnative writers of English are often puzzled by which form of a <u>verbal</u> (a <u>verb</u> used as another part of speech) to use when it functions as the direct object of a verb. No structural rule exists for distinguishing between the use of an infinitive and a gerund (a noun formed from an *-ing* verb) as the object of a verb. Any specific verb may take an infinitive as its object, others may take a gerund, and yet others take either an infinitive or a gerund. At times, even the base form of the verb is used.

- He enjoys *working*. [gerund as a complement]
- She promised *to fulfill* her part of the contract.
  [infinitive as a complement]
- The president had the manager *assign* her staff to another project.
  [basic verb form as a complement]

To make such distinctions accurately, rely on what you hear native speakers use or what you read. You might also consult a reference book for ESL students.

Adjective Clauses

Because of the variety of ways adjective clauses are constructed in different languages, they can be particularly troublesome for nonnative writers of English. The following guidelines will help you form adjective clauses correctly.

Place an adjective clause directly after the noun it modifies.

- The tall woman *who is standing across the room* is a vice president of the company ~~who is standing across the room.~~

The adjective clause *who is standing across the room* modifies *woman,* not *company,* and thus comes directly after *woman.*

Avoid using a relative pronoun with another pronoun in an adjective clause.

- The man who ~~he~~ sits at that desk is my boss.

Present Perfect Verb Tense

As a general rule, use the *present perfect __tense__* to refer to events completed in the past that have some implication for the present. When a specific time is mentioned, however, use the simple past.

PRESENT PERFECT	I *have written* the letter and I am waiting for an answer. [No specific time is mentioned, but the action, *have written,* affects the present.]
SIMPLE PAST	I *wrote* the letter yesterday. [The time when the action took place is specified, and the action, *wrote,* has no relation to the present.]

Use the present perfect tense to describe actions that occurred in the past and have some bearing on the present.

- She *has revised* that report three times. [She might revise it again.]

Use the present perfect with a *since* or *for* phrase to describe actions that began in the past and continue in the present.

- This company *has been* in business *for* seventeen years.
- This company *has been* in business *since* 1985.

Present Progressive Verb Tense

The present progressive tense is especially difficult for those whose native language does not use this tense. The *present progressive tense* is used to describe some action or condition that is ongoing (or in progress) in the present and may continue into the future.

- I *am searching* for an error in the document.
  [The search is occurring now and may continue.]

In contrast, the simple present tense more often relates to habitual actions.

- I *search* for errors in my documents.
  [I regularly search for errors, but I am not necessarily searching now.]

ESL Entries

Look under "ESL Trouble Spots" in the Topical Key for a list of entries—such as <u>idioms</u>, <u>agreement</u>, and <u>possessive case</u>—and ESL Tips boxes of particular relevance to speakers of English as a second language. See also <u>grammar</u>, <u>sentence construction</u>, and <u>usage</u>.

WEB LINK **ENGLISH AS A SECOND LANGUAGE**

For Web sites and electronic grammar exercises intended for speakers of English as a second language, see <*www.bedfordstmartins.com/alred*> and select *Links for Technical Writing* and *Exercise Central*.

English, varieties of

Written English includes two broad categories: standard and nonstandard. Standard English is used in business, industry, government, education, and all professions. It has rigorous and precise criteria for <u>capitalization</u>, <u>diction</u>, <u>punctuation</u>, <u>spelling</u>, and <u>usage</u>.

Nonstandard English does not conform to such criteria; it is often regional in origin or it reflects the special usages of a particular ethnic or social group. As a result, although it may be vigorous and colorful, the usefulness of nonstandard English as a means of communication is limited to certain contexts and to people already familiar and comfortable with it in those contexts. It rarely appears in printed material except for special effect. Nonstandard English is characterized by inexact or inconsistent punctuation, capitalization, spelling, diction, and usage choices; thus, you should avoid using it in workplace writing.

Colloquial English

Colloquial English is spoken English or writing that uses words and expressions common to casual conversation ("We need to get him up to speed"). Colloquial English is appropriate to some kinds of writing (personal letters, notes, some <u>e-mail</u>) but not to most workplace writing.

Dialectal English

Dialectal English is a social or regional variety of the language that is comprehensible to people of that social group or region but may be incomprehensible to outsiders. Dialect, which is usually nonstandard English, involves distinct <u>word choices</u>, grammatical forms, and pronunciations. For example, residents of southern Louisiana who descended from French colonists speak a dialect often referred to as Cajun.

Localisms

A *localism* is a regional wording or phrasing. For example, a large sandwich on a long split roll is variously known throughout the United States as a *hero, hoagie, grinder, poor boy, submarine,* and *torpedo.* Such words normally should be avoided in workplace writing because not all readers will be familiar with the local meanings.

Slang

Slang is an informal vocabulary composed of <u>figures of speech</u> and colorful words used in humorous or extravagant ways. There is no objective test for slang, and many standard words are given slang applications. For instance, slang may be a familiar word used in a new way ("chill" meaning relax), or it may be a new word ("wonk" meaning someone who works or studies excessively).

Most slang is short-lived and has meaning only for a narrow audience. Sometimes, however, slang becomes standard because the word fills a legitimate need. *Skyscraper* and *date* (as in "go on a date"), for example, were once considered slang expressions. Nevertheless, although slang may be valid in informal and personal writing or fiction, it generally should be avoided in workplace writing. See also <u>jargon</u> and <u>technical writing style</u>.

equal / unique / perfect

Logically, *equal* (meaning "having the same quantity or value as another"), *unique* (meaning "one of a kind"), and *perfect* (meaning "a state of highest excellence") are <u>absolute words</u> and therefore should not be compared. However, colloquial usage of *more* and *most* as <u>modifiers</u> of *equal, unique,* and *perfect* is so common that an absolute prohibition on such use is impossible.

- Our system is *more unique* (or *more perfect*) than theirs.

Some writers try to overcome the problem by using *more nearly equal* [*more nearly unique, more nearly perfect*]. When clarity and preciseness are critical, the use of comparative degrees with *equal, unique,* and *perfect* can be misleading. It is best to avoid using comparative degrees with absolute terms.

MISLEADING   Ours is a *more equal* percentage split than theirs.

PRECISE   Our percentage split is 51–49; theirs is 54–46.

-ese

The suffix *-ese* is used to designate types of <u>jargon</u> or certain languages or literary styles (computer*ese*, Chin*ese*, official*ese*, journal*ese*). Using *-ese* to coin words can become an <u>affectation</u>.

etc.

Etc. is an abbreviation for the Latin *et cetera,* meaning "and others" or "and so forth"; therefore, *etc.* should not be used with *and.*

- He purchased pencils, pads, erasers, a calculator, ~~and~~ etc.

Do not use *etc.* at the end of a list or series introduced by the phrase *such as* or *for example*—those phrases already indicate items of the same category that are not named.

- She brought ~~office supplies, such as~~ legal pads, paper clips, self-stick notes, etc., to the meeting.

- She brought office supplies, such as legal pads, paper clips, self-stick notes, ~~etc.,~~ *and* to the meeting.

In technical writing, *etc.* should be used only when there is logical progression (1, 2, 3, etc.) and when at least two items are named. It is often better to avoid *etc.* altogether because the reader may not be able to infer what other items the list might include.

- She brought legal pads, paper clips, self-stick notes, and other office supplies to the meeting.

E

ethics in writing

Ethics refers to the choices we make that affect others for good or ill. Ethical issues are inherent in writing and speaking because what we write and say can influence others. Further, how we express ideas affects both our readers' perceptions of us and our company's ethical stance. Obviously, no book can describe how to act ethically in every situation, but here are some typical ethical lapses to watch for and address during revision.*

Avoid language that attempts to evade responsibility. Some writers use the passive voice because they hope to avoid responsibility or obscure an issue.

- It has been decided. [Who has decided?]
- Several mistakes were made. [Who made them?]

Avoid deceptive language. Do not use words with more than one meaning, especially as a means to circumvent the truth. Consider the company document that stated, "A nominal charge will be assessed for using our facilities." When clients objected that the charge was actually very high, the writer pointed out that the word *nominal* means "the named amount" as well as "very small." In that situation, clients had a strong case in charging that the company was attempting to be deceptive. Various abstract words, technical and legal jargon, and euphemisms are unethical when they are used to mislead readers or to hide a serious or dangerous situation, even though technical or legal experts could interpret them as accurate. See also word choice.

Do not de-emphasize or suppress important information. Not including information that a reader would want to have, such as potential safety hazards or hidden costs for which a customer might be responsible, is unethical. Such omissions may also be illegal. (See also copyright

*Adapted from Brenda R. Sims, "Linking Ethics and Language in the Technical Communication Classroom," *Technical Communication Quarterly* 2.3 (Summer 1993): 285–99.

and plagiarism.) Use <u>layout and design</u> features like typeface size, bullets, lists, and footnotes to highlight—not hide—information that is important to readers.

Do not emphasize misleading or incorrect information. Similarly, avoid the temptation to highlight a feature or service that readers would find attractive but that is available only with some product models or at extra cost. (See also <u>logic errors</u> and <u>positive writing</u>.) Readers could justifiably object that you have given them a false impression in order to sell a product or service, especially if you also de-emphasize the extra cost or other special conditions.

In general, treat others—individuals, companies, groups—fairly and with respect. Avoid language that is biased, racist, or sexist or that perpetuates stereotypes. See also <u>biased language</u>.

In technical writing, guard against reporting false, fabricated, or plagiarized research and test results. As an author, technical reviewer, or editor, your ethical obligation is to correct or point out any misrepresentations of fact before publication, whether the publication is a technical journal article, <u>test report</u>, or product handbook. (See also <u>laboratory reports</u>.) The stakes of such ethical oversights are high because of the potential risk to the health and safety of others. Those at risk can include unwary consumers and workers injured because of faulty products or unprotected exposure to toxic materials.

Writer's Checklist: Writing Ethically

☑ *Is the document honest and truthful?* Scrutinize findings and conclusions carefully. Make sure that the data support them.

☑ *Am I acting in my employer's best interest? My client's or the public's best interest? My own best long-term interest?* Have legal counsel or someone outside your company review and comment on what you have written.

☑ *What if everybody acted or communicated in this way?* If you were the intended audience, would the message be acceptable and respectful?

☑ *Am I willing to take responsibility, publicly and privately, for what the document says?* Will you stand behind what you have written? To your employer? To your family and friends?

☑ *Does the document violate anyone's rights?* Have people from different backgrounds review your writing.

☑ *Am I ethically consistent in my writing?* Apply consistently the principles outlined here and those you have assimilated throughout your life to meet this standard.

euphemisms

A *euphemism* is an inoffensive substitute for a word or phrase that could be distasteful, offensive, or too blunt: *passed away* for *died; previously owned* or *preowned* for *used; lay off* or *downsize* for *terminate* (or *fire*). Used judiciously, a euphemism can help you avoid embarrassing or offending someone. Used carelessly, however, a euphemism can hide the facts of a situation (*incident* or *event* for *accident*) or be a form of **affectation**. Euphemisms can also confuse readers of **international correspondence**. See also **ethics in writing**.

E

everybody / everyone

Both *everybody* and *everyone* are usually considered singular and take singular **verbs** and **pronouns**.

- *Everyone* here *leaves* at 4:30 p.m.
- *Everybody* at the meeting made *his or her* proposal separately.

But the meaning can be obviously plural.

- *Everyone* thought the report should be revised, and I really couldn't blame *them*.

Although normally written as one word, *everyone* is written as two words to emphasize each individual in a group. (*Every one* of the team members contributed to this discovery.) See also **agreement**.

exclamation marks

The *exclamation mark* (!) indicates strong feeling. The most common use of an exclamation mark is after a word, phrase, clause, or sentence to indicate urgency, elation, or surprise (*Hurry! Great! Wow!*). (See also **interjections**.) In instructional writing, the exclamation mark is often used in cautions and warnings (*Danger! Stop!*).

An exclamation mark can be used after a whole sentence or an element of a sentence.

- The subject of this meeting—please note it well!—is our budget deficit.

Keep in mind that an exclamation mark cannot make an argument more convincing, lend force to a weak statement, or call attention to an intended irony.

An exclamation mark can be used after a title that is an exclamatory word, phrase, or sentence.

E

- "Our International Perspective Must Change!" is an article by Richard Moody.

When used with <u>quotation marks</u>, the exclamation mark goes outside, unless what is quoted is an exclamation.

- The manager yelled, "Get in here!" Then Ben, according to Ray, "jumped like a kangaroo"!

executive summaries

An *executive summary* consolidates the principal points of a <u>report</u> or <u>proposal</u>. Executive summaries differ from <u>abstracts</u> in that <u>readers</u> use abstracts to decide whether to read the work in full. However, an executive summary may be the only section of a longer work read by many readers, so it must accurately and concisely reflect the original document. Executive summaries tend to be about 10 percent the length of the documents they summarize and generally follow the same sequence.

Write the executive summary so that it can be read independently of the report or proposal. Do not refer by number to figures, tables, or references contained elsewhere in the document. Executive summaries may occasionally include a figure, table, or footnote if that information is integral to the summary.

The sample executive summary in Figure E–3 is from a 12-page report on the disposition of ethics cases in an aircraft corporation.

E

EXECUTIVE SUMMARY

This report examines the nature and disposition of the 3,458 ethics cases handled by the CGF Aircraft Corporation's ethics officers and managers during 2002. The purpose of this report is to provide CGF's Ethics and Business Conduct Committee with the information necessary for assessing the effectiveness of the first year of the company's Ethics Program.

Background
Effective January 1, 2002, the Ethics and Business Conduct Committee (the Committee) implemented a policy and procedures for the administration of CGF's new Ethics Program. The purpose of the Ethics Program, established by the Committee, is to "promote ethical business conduct through open communication and compliance with company ethics standards." The Office of Ethics and Business Conduct was created to administer the Ethics Program. The director of the Office of Ethics and Business Conduct, along with seven ethics officers throughout the corporation, was given the responsibility for the following objectives:

- Communicate the values and standards for CGF's Ethics Program to employees.

- Inform employees about company policies regarding ethical business conduct.

- Establish companywide channels for employees to obtain information and guidance in resolving ethics concerns.

- Implement companywide ethics-awareness and education programs.

Employee accessibility to ethics information and guidance was available through managers, ethics officers, and an ethics hotline.

Ethics Cases
Major ethics cases were defined as those situations potentially involving serious violations of company policies or illegal conduct. Examples of major ethics cases included cover-up of defective workmanship or use of defective parts in products; discrimination in hiring and promotion; involvement in monetary or other kickbacks; sexual harassment; disclosure of proprietary or company information; theft; and use of corporate Internet resources for inappropriate purposes, such as conducting personal business, gambling, or access to pornography.

1

FIGURE E–3. Executive Summary

E

Minor ethics cases were defined as including all reported concerns not classified as major ethics cases. Minor ethics cases were classified as informational queries from employees, situations involving coworkers, and situations involving management.

The effectiveness of CGF's Ethics Program during the first year of implementation is most evidenced by (1) the active participation of employees in the program and the 3,458 contacts employees made regarding ethics concerns through the various channels available to them, and (2) the action taken in the cases reported by employees, particularly the disposition of the 30 substantiated major ethics cases. Disseminating information about the disposition of ethics cases, particularly information about the severe disciplinary actions taken in major ethics violations, sends a message to employees that unethical or illegal conduct will not be tolerated.

Recommendations
Based on these conclusions, recommendations for planning the second year of the Ethics Program are (1) continuing the channels of communication now available in the Ethics Program, (2) increasing financial and technical support for the Ethics Hotline, the most highly utilized mode of contact in the ethics cases reported in 2002, (3) disseminating this report in some form to employees to ensure their awareness of CGF's commitment to uphold its Ethics Policy and Procedures, and (4) implementing some measure of recognition for ethical behavior, such as an "Ethics Employee of the Month" award to promote and reward ethical conduct.

2

FIGURE E–3. Executive Summary *(continued)*

Writer's Checklist: Writing Executive Summaries

- ☑ Write the executive summary after you have completed the original document.
- ☑ Avoid using terminology that may not be familiar to your readers.
- ☑ Spell out all uncommon symbols and **abbreviations** because executive summaries frequently are read instead of the full document.
- ☑ Make the summary concise, but do not omit transitional words and phrases (*however, moreover, therefore, for example, in summary*).
- ☑ Include only information discussed in the original document.
- ☑ Place the executive summary at the very beginning of the body of the report, as described in **formal reports**.

E

expletives

An *expletive* is a word that fills the position of another word, phrase, or clause. *It* and *there* are common expletives.

- *It* is certain that he will be promoted.

In the example, the expletive *it* occupies the position of subject in place of the real subject, *that he will be promoted.* Expletives are sometimes necessary to avoid **awkwardness,** but they are commonly overused, and most sentences can be better stated without them.

- *Many* *were*
  ~~There were many~~ orders lost because of a software error.

In addition to its usage as a grammatical term, the word *expletive* means an exclamation or oath, especially one that is obscene.

explicit / implicit

An *explicit* statement is one expressed directly, with precision and clarity.

- His directions to the Wausau facility were *explicit,* and we found it with no trouble.

An *implicit* meaning is one that is not directly expressed.

- Although the CEO did not mention the company's financial condition, the danger of overconfidence was *implicit* in her speech.

exposition

Exposition, or *expository writing,* informs readers by presenting facts and ideas in direct and concise language; it usually relies less on colorful or figurative language than does writing meant to be expressive or persuasive. Expository writing attempts to explain to readers what the subject is, how it works, and how it relates to something else. Exposition is aimed at the readers' understanding rather than at their imagination or emotions; it is a sharing of the writer's knowledge. Exposition aims to provide accurate, complete information and to analyze it for the readers.

Because it is the most effective **form of discourse** for explaining difficult subjects, exposition is widely used in **reports,** **memos,** and other types of technical writing. To write exposition, you must have a thorough knowledge of your subject. As with all writing, how much of that knowledge you pass on to your readers depends on the **readers'** needs and your **purpose.**

F

fact

Expressions containing the word *fact* ("due to the *fact* that," "except for the *fact* that," "as a matter of *fact*," or "because of the *fact* that") are often wordy substitutes for more accurate terms.

> *Because*
> • ~~Due to the fact that~~ the technical staff has a high turnover rate,
>   ^
>   our training program has suffered.

Do not use the word *fact* to refer to matters of judgment or opinion.

> *In my opinion,*
> • ~~It is a fact that~~ our research has improved because we now have a
>   ^
>   capable technical staff.

The word *fact* is, of course, valid when facts are what is meant.

> • Our tests uncovered numerous *facts* to support your conclusion.

See also <u>conciseness/wordiness</u> and <u>logic errors</u>.

fax

Fax (facsimile transmission) is used to send documents with elements, such as handwritten corrections and notes, that must be viewed as originally created. When you have to send a drawing or diagram or a document such as a contract that contains one or more signatures, a fax ensures authenticity. Keep in mind, however, that faxes almost never transmit as clearly as the original document and may be reduced in size. Therefore, use generous margins and avoid small font sizes and styles that will not transmit well.

When you fax, be aware that the document can be read by persons other than the intended recipient. If you have to fax confidential or sensitive messages, call the intended recipient first so that he or she can be waiting at the machine as you transmit your fax. Some recipients are resistant, or even hostile, to faxes they view as unnecessary or overly long. A scanned electronic document attached to an e-mail may be a better option.

feasibility reports

When organizations plan to undertake a new project—develop a new product or service, expand a customer base, purchase new equipment, or consider a move—they first try to determine the project's chances for success. A feasibility report presents evidence about the practicality of a proposed project based on specific criteria. It answers such questions as the following: Is new construction or development necessary? Is sufficient staff available? What are the costs involved? What are the legal or other special requirements? Recommendations about whether the project should be carried out are based on the findings of this analysis. Before beginning to write a feasibility report, state clearly and concisely the primary **purpose** of the study. See also **brainstorming** and **collaborative writing**.

Report Sections

Every feasibility report should contain an **introduction**, a body, a **conclusion**, and a recommendation. See also **proposals** and **formal reports** for further advice on creating these sections.

Introduction.   The introduction states the purpose of the report, describes the circumstances that led to the report, and includes any pertinent background information. It may also discuss the **scope** of the report and any procedures or methods used in the analysis of alternatives, and it notes any limitations of the study.

Body.   The body of the report presents a detailed evaluation of all the alternatives under consideration. Evaluate each alternative according to specific criteria, such as cost, availability of staff and financing, and other relevant requirements, identifying the subsections with **headings** to guide **readers**.

Conclusion.   The conclusion summarizes the evaluation of alternatives, and it usually points to one alternative as the best or most feasible.

Recommendation.   This section clearly presents the writer's opinion on which alternative best meets the criteria as summarized in the conclusion.

Typical Feasibility Report

In the typical feasibility report in Figure F–1, an engineering consulting firm conducts a feasibility study to determine how to upgrade its computer system and Internet capability.

F

Introduction
The purpose of this report is to determine which of two proposed options would best enable ACM Technology Consulting to upgrade its file servers and its Internet capacity to meet its increasing data and communication requirements.

Background. In October 2003, the Information Development and Technical Support Group at ACM put the MISSION System into operation. Since then, the volume of processing transactions has increased fivefold (from 1,000 to 5,000 updates per day). This increase has severely impaired system response time; in fact, average response time has increased from less than 10 seconds to 120 seconds. Further, our new Web-based client services system has increased exponentially the demand for processing speed and access capacity.

Scope. Two alternative solutions to provide increased processing capacity have been investigated: (1) purchase of a new ARC 98 processor to supplement the first, and (2) purchase of an HRS 60/EP with PRS enterprise software and expandable peripherals to replace the current ARC 98. The two alternatives are evaluated here according to cost and, to a lesser extent, according to expanded capacity for future operations.

Purchasing a Second ARC 98 Processor
This alternative would require additional annual maintenance costs, salary for an additional computer specialist, increased energy costs, and a one-time construction cost for necessary remodeling as well as installing Internet and other connections.

Annual maintenance costs	$35,000
Annual costs for computer specialist	75,000
Annual increased energy costs	7,500
Annual operating costs	$117,500
Construction cost (one-time)	50,000
Total first-year cost	$167,500

The costs for the installation and operation of another ARC 98 processor are expected to produce savings in system reliability and readiness.

FIGURE F–1. Feasibility Report

System Reliability.   A second ARC 98 would reduce current downtime periods from four to two per week. Downtime recovery averages 30 minutes and affects 40 users. Assuming that 50 percent of users require the system at a given time, we determined that the following reliability savings would result:

2 downtimes × 0.5 hours × 40 users × 50%
× $12.00/hour overtime × 52 weeks = $12,480 (annual savings)

Conclusion
A comparison of costs for both systems indicates that the HRS 60/EP would cost $2,200 more in first-year costs.

ARC 98 Costs
Net additional operating	$56,300
One-time (construction)	50,000
First-year total	$106,300

HRS 60/EP Costs
Net additional operating	$84,000
One-time (facility)	24,500
First-year total	$108,500

Installation of a second ARC 98 processor would permit the present information-processing systems to operate relatively smoothly and efficiently. It would not, however, provide the expanded processing capacity that the HRS 60/EP processor would for implementing new subsystems required to increase processing speed and Internet access.

Recommendation
The HRS 60/EP processor should be purchased because of the long-term savings and because its additional capacity and flexibility will allow for greater expansion in the future.

3

FIGURE F–1. Feasibility Report *(continued)*

female / male

The terms *female* and *male* are usually restricted to scientific, legal, and medical contexts (a *female* suspect, a *male* patient). Keep in mind that the terms sound cold and impersonal. *Girl, woman, lady* or *boy, man, gentleman* may be acceptable substitutes in other contexts, but they have connotations involving age, dignity, and social position. See also **biased language**.

few / a few

In certain contexts, *few* carries more negative overtones than does the phrase *a few*.

POSITIVE   There are *a few* good points in your report.

NEGATIVE   There are *few* good points in your report.

fewer / less

Fewer refers to items that can be counted (count nouns).

- *Fewer* employees took the offer than we expected.

Less refers to mass quantities or amounts (mass nouns).

- Because we had *less* rain this year, the crop yield decreased.

See also English as a second language and nouns.

figuratively / literally

These two words are often confused. *Literally* means "actually" and should not be used in place of *figuratively*, which means "metaphorically." To say that someone "literally turned green with envy" would mean that the person actually changed color.

- In the winner's circle the jockey was, *figuratively* speaking, ten feet tall.
- When he said, "Let's bury our competitors," he did not mean it *literally*.

Avoid the use of *literally* to reinforce the importance of something.

- She was ~~literally~~ the best of the applicants.

figures of speech

A *figure of speech* is an imaginative expression that often compares two things that are basically not alike but have at least one thing in common. For example, if a device is cone-shaped and has an opening at the narrow end, you might say that it looks like a volcano.

Figures of speech can clarify the unfamiliar by relating a new concept to one with which readers are familiar. In that respect, figures of speech help establish understanding between the specialist and the nonspecialist. Figures of speech can also help translate the abstract into the concrete; in the process of doing so, figures of speech also make writing more colorful and graphic. (See also **abstract/concrete words**.) A figure of speech must make sense, however, to achieve the desired effect.

F

> ILLOGICAL   Without the fuel of tax incentives, our economic engine would operate less efficiently.
> [It would not operate at all without fuel.]

Figures of speech also must be consistent to be effective.

- We must get our research program *back on track*, and we are
  engineer the effort.
  counting on you to ~~carry the ball.~~

A figure of speech should not overshadow the point the writer is trying to make. In addition, it is better to use no figure of speech at all than to use a trite one. A surprise that comes "like a bolt out of the blue" seems stale and not much of a surprise at all. See **clichés** and **trite language**.

Types of Figures of Speech

Analogies are comparisons that show the ways in which two objects or concepts are similar, often used to make one of them easier to understand. The following example explains a computer search technique by comparing it to the use of keywords in a dictionary.

- The search technique used in *indexed sequential processing* is similar to a search technique you might use to find the page on which a particular word is located in a dictionary. You might scan the keywords located at the top of each dictionary page that identify the first and last words on each page until you find the keywords that encompass the word you seek. Indexed sequential processing works the same way with computer files.

Hyperboles are gross exaggerations used to achieve an effect or **emphasis**.

- We were *dead* after working all night on the report.

Litotes are understatements, for emphasis or effect, achieved by denying the opposite of the point you are making.

- Two hundred dollars is not a small price for a book.

Metaphors are figures of speech that point out similarities between two things by treating them as though they were the same thing.

- He is the tech support department's *utility infielder.*

Metonyms are figures of speech that use one aspect of a thing to represent it, such as *the blue* for the sky and *wheels* for an automobile.

- *The hard-hat* sector of the labor force was especially hurt by the terms of the contract.

Personification is a figure of speech that attributes human characteristics to nonhuman things or abstract ideas. We might refer, for example, to the *birth* of a planet or apply emotions to machines.

- She said that she was frustrated with the *stubborn* computer system.

Similes are direct comparisons of two essentially unlike things, linking them with the word *like* or *as.*

- His feelings about his competitor are so bitter that in recent conversations with his staff he returned to the subject compulsively, *like someone scratching an itch.*

fine

When used in expressions such as "I feel *fine*" or "a *fine* day," *fine* is colloquial and, like the word *nice,* is too vague for technical writing. Use the word *fine* to mean "refined," "delicate," or "pure."

- A *fine* film of oil covered the surface of the water.
- *Fine* crystal is made in Austria.
- She made a *fine* distinction between the two possible meanings of the disputed passage.

first / firstly

First and *firstly* are both adverbs. Avoid *firstly* in favor of *first,* which sounds less stiff than *firstly.* The same is true of other ordinal **numbers**.

flammable / inflammable / nonflammable

Both *flammable* and *inflammable* mean "capable of being set on fire." Because the *in-* **prefix** usually causes the base word to take its opposite meaning (*incapable, incompetent*), use *flammable* instead of *inflammable* to avoid possible misunderstanding. (The cargo of gasoline is *flammable.*) *Nonflammable* is the opposite, meaning "not capable of being set on fire." (The asbestos suit was *nonflammable.*)

flowcharts

A *flowchart* is a diagram of a process that involves stages, shown in sequence from beginning to end. A flowchart provides an overview of a process and allows the reader to identify its essential steps quickly and easily. Flowcharts can take several forms: The steps might be represented by labeled blocks, as shown in Figure F–2; pictorial symbols, as shown in Figure F–3; or ISO (International Organization for Standardization) symbols, as shown in Figure F–4 on page 204.

Writer's Checklist: Creating Flowcharts

☑ Label each step in the process or identify each step with labeled blocks, pictorial representations, or standardized symbols.

☑ Follow the standard flow directions: left to right and top to bottom. When the flow is otherwise, indicate that with arrows.

☑ Include a key if the flowchart contains symbols your readers may not understand.

☑ Use standardized symbols for flowcharts that document computer programs and other information-processing procedures, as set forth in *Information Processing—Documentation Symbols and Conventions for Data, Program and System Flowcharts, Program Network Charts, and System Resources Charts,* ISO publication 5807-1985 (E).

For advice on integrating flowcharts into your text, see **visuals**. See also **global graphics**.

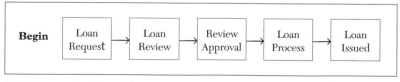

Begin → Loan Request → Loan Review → Review Approval → Loan Process → Loan Issued

FIGURE F–2. Flowchart Using Labeled Blocks

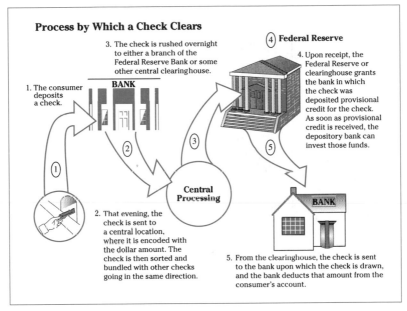

FIGURE F–3. Flowchart Using Pictorial Symbols

footnotes (*see* documenting sources)

forceful / forcible

Although *forceful* and *forcible* are both **adjectives** meaning "character-ized by or full of force," *forceful* is usually limited to persuasive ability and *forcible* to physical force.

- John made a *forceful* presentation at the committee meeting.
- The police report listed three *forcible* entries into the building.

foreign words in English

The English language has a long history of borrowing words from other languages. Most borrowings occurred so long ago that we seldom recognize the borrowed terms (also called *loan words*) as being of for-eign origin (*kindergarten* from German; *animal* from Latin; *church* from Greek).

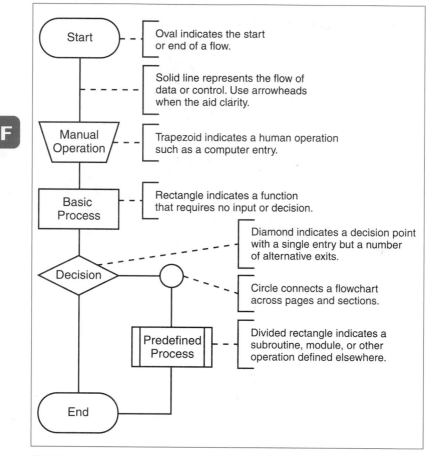

FIGURE F–4. Common ISO Flowchart Symbols (with Annotations)

Words not fully assimilated into the English language are set in italics (*sine qua non, coup de grâce, in res, in camera*). They should be underlined when italic type is not available. Words and abbreviations that have been fully assimilated need not be italicized, although they often retain their <u>diacritical marks</u> (cliché, habeas corpus, per diem, résumé, vis-à-vis). As foreign words are absorbed into English, their plural forms give way to English plurals (*agenda* becomes *agendas* and *formulae* becomes *formulas*).

Generally, foreign expressions should be used only if they serve a real need. (See also <u>e.g./i.e.</u> and <u>etc.</u>) The overuse of foreign words in an attempt to impress your reader or achieve elegance is <u>affectation</u>. Effective communication can be accomplished only if your readers understand what you write; therefore, choose foreign expressions only when they make an idea clearer.

foreword / forward

Although the pronunciation is the same, the spellings and meanings of these two words are quite different. The word *foreword* is a **noun** meaning "introductory statement at the beginning of a book or other work." (The former CEO wrote a *foreword* for the annual report.) The word *forward* is an **adjective** or **adverb** meaning "at or toward the front."

- Move the lever to the *forward* position on the panel. [adjective]

- Turn the dial until the needle begins to move *forward*. [adverb]

formal reports

DIRECTORY

Formal reports are written accounts of major projects. Most are divided into three primary parts — front matter, body, and back matter — each of which contains a number of elements.

The number and arrangement of the elements may vary, depending on the subject, the length of the **report**, and the kinds of material covered. Further, many organizations have a preferred style for formal reports and furnish guidelines for report writers to follow. If you are not required to follow a specific style, use the **format** recommended in this entry. The following list includes most of the elements a formal report might contain, in the order they typically appear. (The items shown with page numbers appear in the SAMPLE FORMAL REPORT on pages 211–27.) Often, a **cover letter** or **memo** precedes the front matter and identifies the report by title, the person or persons to whom it is being sent, and the reason for its being sent.

Front Matter
Title page (212)
Abstract (213)

Body
Back Matter

Front Matter

The *front matter* serves several functions: It gives the <u>readers</u> a general idea of the writer's <u>purpose</u>, it gives an overview of the type of information in the report, and it lists where specific information is covered in the report. Not all formal reports include every element of front matter described here. A title page and table of contents are usually mandatory, but the scope of the report and its intended audience determine whether the other elements are included.

Title Page.   Although the formats of title pages for formal reports may vary, they may include the following items:

- *The full title of the report.* Reflect the topic, scope, and purpose of the report as shown in the entry <u>titles</u>.

- *The name of the writer(s), principal investigator(s), or compiler(s).* Sometimes contributors identify themselves by their job title in the organization or by their tasks in contributing to the report (Gina Hobbs, Principal Investigator).

- *The date or dates of the report.* For one-time reports, list the date when the report is to be distributed. For periodic reports (monthly, quarterly, or yearly), list in a subtitle the period that the report covers. Elsewhere on the title page, list the date when the report is to be distributed.

- *The name of the organization for which the writer(s) works.*
- *The name of the organization to which the report is being submitted.* This information is included if the report is written for a customer or client.

The title page, although unnumbered, is considered page i. The back of the title page, which is blank and unnumbered, is considered page ii, and the abstract falls on page iii. The body of the report begins with Arabic number 1, and a new chapter or large section typically begins on a new right-hand (odd-numbered) page. Reports with printing on only one side of each sheet can be numbered consecutively regardless of where new sections begin. Center page numbers at the bottom of the page throughout the report.

Abstract.   An <u>abstract</u>, which normally follows the title page, highlights the major points of the report, enabling readers to decide whether to read the entire report.

Table of Contents.   A <u>table of contents</u> lists all the major sections or <u>headings</u> of the report in their order of appearance, along with their page numbers.

List of Figures.   When a report contains more than five figures, list them, along with their page numbers, in a separate section beginning on a new page immediately following the table of contents. Number figures consecutively with Arabic numbers. Figures include all <u>visuals</u>—<u>drawings</u>, <u>photographs</u>, <u>maps</u>, charts, and <u>graphs</u>—contained in the report.

List of Tables.   When a report contains more than five <u>tables</u>, list them, along with their titles and page numbers, in a separate section immediately following the list of figures (if there is one). Number tables consecutively with Arabic numbers.

Foreword.   A *foreword* is an optional introductory statement about a formal report or book that is written by someone other than the author(s). The foreword author is usually an authority in the field or an executive of the company. The foreword author's name and affiliation appear at the end of the foreword, along with the date it was written. The foreword generally provides background information about the publication's significance and places it in the context of other works in the field. The foreword precedes the preface when a work has both.

Preface.   The *preface,* another type of optional introductory statement, is written by the author(s) of the book or formal report. The preface may announce the work's purpose, scope, and background (including any special circumstances leading to the work). A preface may also specify the audience for a work, it may contain acknowledgments of those who helped in its preparation, and it may cite permission obtained for the use of copyrighted works. See also <u>copyright</u>.

List of Abbreviations and Symbols.   When the report uses numerous <u>abbreviations</u> and <u>symbols</u> and there is a chance that readers will not be able to interpret them, the front matter should include a list of all symbols and abbreviations and what they stand for.

Body

The *body* is the section of the report that describes in detail the methods and procedures used to generate the report, demonstrates how results were obtained, describes the results, draws conclusions, and, if appropriate, makes recommendations.

Executive Summary.   The body of the report begins with the <u>executive summary</u>, which provides a more complete overview of the report than an abstract does.

Introduction.   The <u>introduction</u> gives readers any general information, such as the report's purpose and scope, necessary to understand the detailed information in the rest of the report.

Text.   The *text* of the body presents, as appropriate, the details of how the topic was investigated, how a problem was solved, what alternatives were explored, and how the best choice among them was selected. This information is often persuasively developed by the use of illustrations and tables and supported by references to other publications.

Conclusions.   The <u>conclusions</u> section pulls together the results of the <u>research</u> and offers conclusions based on the analysis.

Recommendations.   *Recommendations,* which are sometimes combined with the conclusions, state what course of action should be taken based on the earlier arguments and results of the study. The recommendations section may state, for example, "We should pursue new markets in . . ." or "I recommend we expand our marketing efforts on the Internet."

Explanatory Notes.   Occasionally, reports contain notes that amplify terms or points for some readers. If such notes are not included as footnotes on the page where the term or point appears, they may appear in a final section, often referred to simply as "Notes."

References (or Works Cited). A list of references or works cited appears in a separate section if the report refers to material in, or quoted directly from, a published work or other research sources, including online sources. If your employer has a preferred reference style, follow it; otherwise, use the guidelines provided in the entry **documenting sources**. For a relatively short report, place the references at the end of the body of the report. For a report with a number of sections or chapters, place the reference section at the end of each major section or chapter. In either case, title the reference or works-cited section as such and begin it on a new page. If a particular reference appears in more than one section or chapter, repeat it in full in each appropriate reference section. See also **quotations** and **plagiarism**.

F

DIGITAL TIP AUTOMATING REPORT FORMATTING

Use your word-processing software to create style sheets to automate formatting for your reports. Once you've specified the format guidelines, save them as a file and use them each time you create a formal report. You can automate and make consistent elements such as fonts and font sizes for text, headings, titles, footnotes, headers, and footers; paragraphs, including indentation and margins; lists, including indentation and spacing; and columns, including the number per page. You can also create an automated table of contents for your report based on the main headings and subheadings. For more on this topic, see <*www.bedfordstmartins.com/alred*> and select *Digital Tips*. See also Digital Tip: Creating an Index on page 261.

Back Matter

The *back matter* of a formal report contains supplementary material, such as where to find additional information about the topic (bibliography), and expands on certain subjects (appendixes). Other back-matter elements clarify the terms used (glossary) and provide information on how to easily locate information in the report (index). For very large formal reports, back-matter sections may be individually numbered.

Appendixes. An **appendix** contains information that clarifies or supplements the body. It provides information that is too detailed or lengthy for the primary audience but that is relevant to secondary audiences.

Bibliography. A **bibliography** is an alphabetical list of all sources that were consulted (not just those cited) in researching the report. A bibliography is not necessary if the reference listing contains a complete list of sources.

Glossary.   A <u>glossary</u> is an alphabetical list of selected terms used in the report and their definitions.

Index.   An *index* is an alphabetical list of all the major topics and subtopics discussed in the report. It cites the page numbers where discussion of each topic can be found and allows readers to find information on topics quickly and easily. The index is always the final section of a report. See also <u>indexing</u>.

Sample Formal Report

Figures F–5 through F–13 show the typical sections of a formal report. Keep in mind that the number and arrangement of the elements vary, depending on the specific subject and requirements of an organization or client.

⚛ WEB LINK **ANNOTATED SAMPLE FORMAL REPORT**

For an annotated version of this formal report, see <*www.bedfordstmartins.com/alred*> and select *Model Documents Gallery.*

CGF Aircraft Corporation
Memo

To: Members of the Ethics and Business Conduct Committee
From: Susan Litzinger, Director of Ethics and Business Conduct   *SL*
Date: March 1, 2003
Subject: Reported Ethics Cases 2002

Enclosed is the annual Ethics and Business Conduct Report, as required by
CGF Policy CGF-EP-01, for your evaluation, covering the first year of our
Ethics Program. This report contains a review of the ethics cases handled by
CGF Ethics officers and managers during 2002.

The ethics cases reported are analyzed according to two categories:
(1) major ethics cases, or those potentially involving serious violations
of company policy or illegal conduct, and (2) minor ethics cases, or those
that do not involve serious policy violations or illegal conduct. The report
also examines the mode of contact in all of the reported cases and the
disposition of the substantiated major ethics cases.

It is my hope that this report will provide the Committee with the information needed to assess the effectiveness of the first year of CGF's Ethics Program and to plan for the coming year. Please let me know if you have any
questions about this report or if you need any further information. I may be
reached at (555) 211-2121 and by e-mail at <sl@cgf.com>.

Enc.

FIGURE F–5. Cover Letter for a Formal Report. Reprinted by permission of Susan
Litzinger, a student at Pennsylvania State University, Altoona.

F

REPORTED ETHICS CASES
Annual Report 2002

Prepared by Susan Litzinger
Director of Ethics and Business Conduct

Report Distributed March 1, 2003

Prepared for
The Ethics and Business Conduct Committee
CGF Aircraft Corporation

FIGURE F–6. Title Page for a Formal Report

F

ABSTRACT

This report examines the nature and disposition of 3,458 ethics cases handled companywide by CGF Aircraft Corporation's ethics officers and managers during 2002. The purpose of this annual report is to provide the Ethics and Business Conduct Committee with the information necessary for assessing the effectiveness of the Ethics Program's first year of operation. Records maintained by ethics officers and managers of all contacts were compiled and categorized into two main types: (1) major ethics cases, or cases involving serious violations of company policies or illegal conduct, and (2) minor ethics cases, or cases not involving serious policy violations or illegal conduct. This report provides examples of the types of cases handled in each category and analyzes the disposition of 30 substantiated major ethics cases. Recommendations for planning for the second year of the Ethics Program are (1) continuing the channels of communication now available in the Ethics Program, (2) increasing financial and technical support for the Ethics Hotline, (3) disseminating the annual ethics report in some form to employees to ensure employees' awareness of the company's commitment to uphold its Ethics Policies and Procedures, and (4) implementing some measure of recognition for ethical behavior to promote and reward ethical conduct.

iii

FIGURE F–7. Abstract for a Formal Report

TABLE OF CONTENTS

iv

FIGURE F–8. Table of Contents for a Formal Report

EXECUTIVE SUMMARY

This report examines the nature and disposition of the 3,458 ethics cases handled by the CGF Aircraft Corporation's ethics officers and managers during 2002. The purpose of this report is to provide CGF's Ethics and Business Conduct Committee with the information necessary for assessing the effectiveness of the first year of the company's Ethics Program.

Effective January 1, 2002, the Ethics and Business Conduct Committee (the Committee) implemented a policy and procedures for the administration of CGF's new Ethics Program. The purpose of the Ethics Program, established by the Committee, is to "promote ethical business conduct through open communication and compliance with company ethics standards." The Office of Ethics and Business Conduct was created to administer the Ethics Program. The director of the Office of Ethics and Business Conduct, along with seven ethics officers throughout the corporation, was given the responsibility for the following objectives:

- Communicate the values and standards for CGF's Ethics Program to employees.

- Inform employees about company policies regarding ethical business conduct.

- Establish companywide channels for employees to obtain information and guidance in resolving ethics concerns.

- Implement companywide ethics-awareness and education programs.

Employee accessibility to ethics information and guidance was available through managers, ethics officers, and an ethics hotline.

Major ethics cases were defined as those situations potentially involving serious violations of company policies or illegal conduct. Examples of major ethics cases included cover-up of defective workmanship or use of defective parts in products; discrimination in hiring and promotion; involvement in monetary or other kickbacks; sexual harassment; disclosure of proprietary or company information; theft; and use of corporate Internet resources for inappropriate purposes, such as conducting personal business, gambling, or access to pornography.

1

FIGURE F-9. Executive Summary for a Formal Report

F

Minor ethics cases were defined as including all reported concerns not classified as major ethics cases. Minor ethics cases were classified as informational queries from employees, situations involving coworkers, and situations involving management.

The effectiveness of CGF's Ethics Program during the first year of implementation is most evidenced by (1) the active participation of employees in the program and the 3,458 contacts employees made regarding ethics concerns through the various channels available to them, and (2) the action taken in the cases reported by employees, particularly the disposition of the 30 substantiated major ethics cases. Disseminating information about the disposition of ethics cases, particularly information about the severe disciplinary actions taken in major ethics violations, sends a message to employees that unethical or illegal conduct will not be tolerated.

Based on these conclusions, recommendations for planning the second year of the Ethics Program are (1) continuing the channels of communication now available in the Ethics Program, (2) increasing financial and technical support for the Ethics Hotline, the most highly utilized mode of contact in the ethics cases reported in 2002, (3) disseminating this report in some form to employees to ensure their awareness of CGF's commitment to uphold its Ethics Policy and Procedures, and (4) implementing some measure of recognition for ethical behavior, such as an "Ethics Employee of the Month" award to promote and reward ethical conduct.

2

FIGURE F–9. Executive Summary for a Formal Report *(continued)*

INTRODUCTION

This report examines the nature and disposition of the 3,458 ethics cases handled companywide by CGF's ethics officers and managers during 2002. The purpose of this report is to provide the Ethics and Business Conduct Committee with the information necessary for assessing the effectiveness of the first year of CGF's Ethics Program. Recommendations are given for the Committee's consideration in planning for the second year of the Ethics Program.

Ethics and Business Conduct Policy and Procedures

Effective January 1, 2002, the Ethics and Business Conduct Committee (the Committee) implemented Policy CGF-EP-01 and Procedure CGF-EP-02 for the administration of CGF's new Ethics Program. The purpose of the Ethics Program, established by the Committee, is to "promote ethical business conduct through open communication and compliance with company ethics standards" (CGF's "Ethical Business Conduct").

The Office of Ethics and Business Conduct was created to administer the Ethics Program. The director of the Office of Ethics and Business Conduct, along with seven ethics officers throughout CGF, was given the responsibility for the following objectives:

- Communicate the values, standards, and goals of CGF's Ethics Program to employees.

- Inform employees about company ethics policies.

- Provide companywide channels for employee education and guidance in resolving ethics concerns.

- Implement companywide programs in ethics awareness, education, and recognition.

- Ensure confidentiality in all ethics matters.

Employee accessibility to ethics information and guidance became the immediate and key goal of the Office of Ethics and Business Conduct in its first year of operation. The following channels for contact were set in motion during 2002:

3

FIGURE F-10. Introduction for a Formal Report

- Managers throughout CGF received intensive ethics training; in all ethics situations, employees were encouraged to go to their managers as the first point of contact.

- Ethics officers were available directly to employees through face-to-face or telephone contact, to managers, to callers using the ethics hotline, and by e-mail.

- The Ethics Hotline was available to all employees, 24 hours a day, 7 days a week, to anonymously report ethics concerns.

Confidentiality Issues

CGF's Ethics Policy ensures confidentiality and anonymity for employees who raise genuine ethics concerns. Procedure CGF-EP-02 guarantees appropriate discipline, up to and including dismissal, for retaliation or retribution against any employee who properly reports any genuine ethics concern.

Documentation of Ethics Cases

The following requirements were established by the director of the Office of Ethics and Business Conduct as uniform guidelines for the documentation by managers and ethics officers of all reported ethics cases:

- Name, position, and department of individual initiating contact, if available

- Date and time of contact

- Name, position, and department of contact person

- Category of ethics case

- Mode of contact

- Resolution

Managers and ethics officers entered the required information in each reported ethics case into an ACCESS database file, enabling efficient retrieval and analysis of the data.

4

FIGURE F–10. Introduction for a Formal Report *(continued)*

Major/Minor Category Definition and Examples

Major ethics cases were defined as those situations potentially involving serious violations of company policies or illegal conduct. Procedure CGF-EP-02 requires notification of the Internal Audit and the Law Departments in serious ethics cases. The staffs of the Internal Audit and the Law Departments assume primary responsibility for managing major ethics cases and for working with the employees, ethics officers, and managers involved in each case.

Examples of situations categorized as major ethics cases:

- Cover-up of defective workmanship or use of defective parts in products
- Discrimination in hiring and promotion
- Involvement in monetary or other kickbacks from customers for preferred orders
- Sexual harassment
- Disclosure of proprietary customer or company information
- Theft
- Use of corporate Internet resources for inappropriate purposes, such as conducting private business, gambling, or access to pornography

Minor ethics cases were defined as including all reported concerns not classified as major ethics cases. Minor ethics cases were classified as follows:

- Informational queries from employees
- Situations involving coworkers
- Situations involving management

5

FIGURE F-10. Introduction for a Formal Report *(continued)*

ANALYSIS OF REPORTED ETHICS CASES

Reported Ethics Cases by Major/Minor Category

CGF ethics officers and managers companywide handled a total of 3,458 ethics situations during 2002. Of these cases, only 172, or 5 percent, involved reported concerns of a serious enough nature to be classified as major ethics cases (see Figure 1). Major ethics cases were defined as those situations potentially involving serious violations of company policy or illegal conduct.

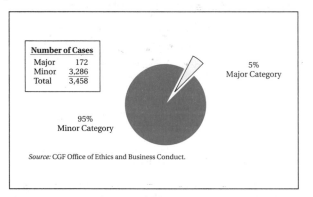

Number of Cases	
Major	172
Minor	3,286
Total	3,458

5%
Major Category

95%
Minor Category

Source: CGF Office of Ethics and Business Conduct.

Figure 1.   Reported ethics cases by major/minor category in 2002.

Major Ethics Cases

Of the 172 major ethics cases reported during 2002, 57 percent, upon investigation, were found to involve unsubstantiated concerns. Incomplete information or misinformation most frequently was discovered to be the cause of the unfounded concerns of misconduct in 98 cases. Forty-four cases, or 26 percent of the total cases reported, involved incidents partly substantiated by ethics officers as serious misconduct; however, these cases were discovered to also involve inaccurate information or unfounded issues of misconduct.

6

FIGURE F–11. Body of a Formal Report

Only 17 percent of the total number of major ethics cases, or 30 cases, were substantiated as major ethics situations involving serious ethical misconduct or illegal conduct (CGF, "2002 Ethics Hotline Results") (see Figure 2).

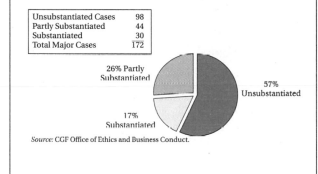

Unsubstantiated Cases	98
Partly Substantiated	44
Substantiated	30
Total Major Cases	172

26% Partly Substantiated

57% Unsubstantiated

17% Substantiated

Source: CGF Office of Ethics and Business Conduct.

Figure 2.   Major ethics cases in 2002.

Of the 30 substantiated major ethics cases, seven remain under investigation at this time, and two cases are currently in litigation. Disposition of the remainder of the 30 substantiated reported ethics cases included severe disciplinary action in five cases: the dismissal of two employees and the demotion of three employees. Seven employees were given written warnings, and nine employees received verbal warnings (see Figure 3).

7

FIGURE F–11. Body of a Formal Report *(continued)*

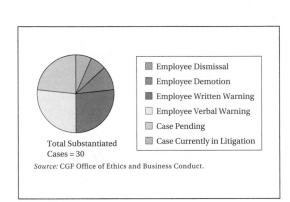

Figure 3.   Disposition of substantiated major ethics cases in 2002.

Minor Ethics Cases

Minor ethics cases included those that did not involve serious violations of company policy or illegal conduct. During 2002, ethics officers and company managers handled 3,286 such cases. Minor ethics cases were further classified as follows:

- Informational queries from employees
- Situations involving coworkers
- Situations involving management

As might be expected during the initial year of the Ethics Program implementation, the majority of contacts made by employees were informational, involving questions about the new policies and procedures. These informational contacts comprised 65 percent of all contacts of a minor nature and numbered 2,148. Employees made 989 contacts regarding ethics concerns involving coworkers and 149 contacts regarding ethics concerns involving management (see Figure 4).

8

FIGURE F–11. Body of a Formal Report *(continued)*

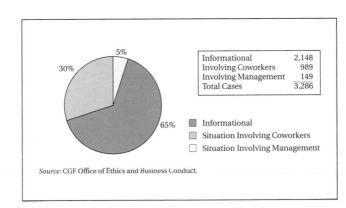

Informational	2,148
Involving Coworkers	989
Involving Management	149
Total Cases	3,286

■ Informational
▨ Situation Involving Coworkers
□ Situation Involving Management

Source: CGF Office of Ethics and Business Conduct.

F

Figure 4.   Minor ethics cases in 2002.

Mode of Contact

The effectiveness of the Ethics Program rested on the dissemination of information to employees and the provision of accessible channels through which employees could gain information, report concerns, and obtain guidance. Employees were encouraged to first go to their managers with any ethical concerns, because those managers would have the most direct knowledge of the immediate circumstances and individuals involved.

Other channels were put into operation, however, for any instance in which an employee did not feel able to go to his or her manager. The ethics officers companywide were available to employees through telephone conversations, face-to-face meetings, and e-mail contact. Ethics officers also served as contact points for managers in need of support and assistance in handling the ethics concerns reported to them by their subordinates.

The Ethics Hotline became operational in mid-January 2002 and offered employees assurance of anonymity and confidentiality. The Ethics Hotline was accessible to all employees on a 24-hour, 7-day basis. Ethics officers companywide took responsibility on a rotational basis for handling calls reported through the hotline.

9

FIGURE F–11. Body of a Formal Report *(continued)*

F

In summary, ethics information and guidance was available to all employees during 2002 through the following channels:

- Employee to manager
- Employee telephone, face-to-face, and e-mail contact with ethics officer
- Manager to ethics officer
- Employee Hotline

The mode of contact in the 3,458 reported ethics cases was as follows (see Figure 5):

- In 19 percent of the reported cases, or 657, employees went to managers with concerns.
- In 9 percent of the reported cases, or 311, employees contacted an ethics officer.
- In 5 percent of the reported cases, or 173, managers sought assistance from ethics officers.
- In 67 percent of the reported cases, or 2,317, contacts were made through the Ethics Hotline.

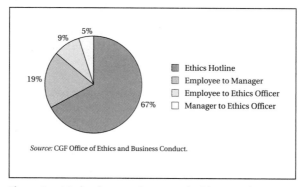

Source: CGF Office of Ethics and Business Conduct.

Figure 5.   Mode of contact in reported ethics cases in 2002.

10

FIGURE F–11. Body of a Formal Report *(continued)*

CONCLUSIONS AND RECOMMENDATIONS

The effectiveness of CGF's Ethics Program during the first year of implementation is most evidenced by (1) the active participation of employees in the program and the 3,458 contacts employees made regarding ethics concerns through the various channels available to them, and (2) the action taken in the cases reported by employees, particularly the disposition of the 30 substantiated major ethics cases.

One of the 12 steps to building a successful Ethics Program identified by Frank Navran in *Workforce* magazine is an ethics communication strategy. Navran explains that such a strategy is crucial in ensuring

> that employees have the information they need in a timely and usable fashion and that the organization is encouraging employee communication regarding the values, standards and the conduct of the organization and its members. (Navran 119)

The 3,458 contacts by employees during 2002 attest to the accessibility and effectiveness of the communication channels that exist in CGF's Ethics Program.

An equally important step in building a successful ethics program is listed by Navran as "Measurements and Rewards," which he explains as follows:

> In most organizations, employees know what's important by virtue of what the organization measures and rewards. If ethical conduct is assessed and rewarded, and if unethical conduct is identified and dissuaded, employees will believe that the organization's principals mean it when they say the values and code of ethics are important. (Navran 121)

Disseminating information about the disposition of ethics cases, particularly information about the severe disciplinary actions taken in major ethics violations, sends a message to employees that unethical or illegal conduct will not be tolerated. Making public the tough-minded actions taken in cases of ethical misconduct provides "a golden opportunity to make other employees aware that the behavior is unacceptable and why" (Ferrell and Gardiner 129).

11

FIGURE F–12. Conclusions and Recommendations for a Formal Report

With these two points in mind, I offer the following recommendations for consideration for plans for the Ethics Program's second year:

- Continuation of the channels of communication now available in the Ethics Program

- Increased financial and technical support for the Ethics Hotline, the most highly utilized mode of contact in the reported ethics cases in 2002

- Dissemination of this report in some form to employees to ensure employees' awareness of CGF's commitment to uphold its Ethics Policy and Procedures

- Implementation of some measure of recognition for ethical behavior, such as an "Ethics Employee of the Month," to promote and reward ethical conduct

To ensure that employees see the value of their continued participation in the Ethics Program, feedback is essential. The information in this annual review, in some form, should be provided to employees. Knowing that the concerns they reported were taken seriously and resulted in appropriate action by Ethics Program administrators would reinforce employee involvement in the program. While the negative consequences of ethical misconduct contained in this report send a powerful message, a means of communicating the *positive* rewards of ethical conduct at CGF should be implemented. Various options for recognition of employees exemplifying ethical conduct should be considered and approved.

Continuation of the Ethics Program's successful 2002 operations, with the implementation of the above recommendations, should ensure the continued pursuit of the Ethics Program's purpose: "to promote a positive work environment that encourages open communication regarding ethics and compliance issues and concerns."

FIGURE F–12. Conclusions and Recommendations for a Formal Report *(continued)*

WORKS CITED

CGF. "Ethics and Conduct at CGF Airlines." 5 April 2001. <www.cgfac

.com/aboutus/ethics.html>.

CGF. "2002 Ethics Hotline Investigation Results." 15 January 2003. <www

.cgfac.com/html/ethics.html>.

Ferrell, O. C., and Gareth Gardiner. In Pursuit of Ethics: Tough Choices in

the World of Work. Springfield: Smith Collins, 1991.

Kelley, Tina. "Corporate Prophets, Charting a Course to Ethical Profits."

New York Times 8 Feb. 1998: BU12.

Navran, Frank. "12 Steps to Building a Best-Practices Ethics Program."

Workforce Sept. 1997: 117–122.

13

FIGURE F–13. Works Cited for a Formal Report

format

Format refers to both the organization of information in a document and the physical arrangement of information on the page.

In one sense, format refers to the fact that some types of job-related writing, such as <u>formal reports</u> and <u>correspondence</u>, are characterized by conventions that govern the scope and placement of information. For example, in formal reports, the <u>table of contents</u> precedes the preface but follows the title page and the <u>abstract</u>. Likewise, although variations exist, parts of letters—such as inside address, salutation, and complimentary closing—typically are arranged in standard patterns.

Format also refers to the general physical appearance (as discussed in <u>layout and design</u>) of a finished document, whether printed or electronic. See also <u>e-mail</u> and <u>Web design</u>.

former / latter

Former and *latter* should be used to refer to only two items in a sentence or paragraph.

* The president and his aide emerged from the conference, the *former* looking nervous and the *latter* looking glum.

Because these terms make the reader look to previous material to identify the reference, they complicate reading and are best avoided.

forms design

Forms provide an economical and uniform way to gather and record data. Filling out a paper or online form is easier and quicker for respondents than writing a <u>memo</u>, letter, or <u>report</u>. Forms offer a standardized response that is far easier to tabulate and evaluate than longer types of writing. Figures F–14 and F–15 are two examples of forms.

Preparing a Form

An effective form makes it easy for one person to supply information and for another person to retrieve, record, and interpret that information. Ideally, a form should be self-explanatory, even to someone seeing

---

Annual Reappointment Form   July 1, 20___ to June 30, 20___

CHILDREN'S MEDICAL CENTER
**1735 Chapel Street
Toledo, Ohio 43692**

NAME:

List appointments or offices held, teaching positions, independent studies in medical or dental societies or other medical organizations, and any other professional recognitions you would like to have included in your file:

_____

_____

_____

_____

_____

_____

_____

_____

Do you wish a change in your privileges? If so, specify: _____

_____

Have there been any changes in your board specialties? Yes ☐ No ☐
     If yes: Date _____
         Specialty Board _____

Signature _____
Date _____

Return to: Chairman, Credentials & Nominating Committee

FIGURE F-14. Form (a)

it for the first time. When preparing a form, determine the kind of information you are seeking and arrange the questions in a logical order.

Choosing Online or Paper. Many organizations are moving their forms online because Web-based and other online forms both standardize respondents' interfaces and link easily to databases, thus automating the tabulation and interpretation of data. In addition, using online forms can eliminate the problems associated with distributing and collecting forms. But using online forms can also be difficult for people with lim-

F

CHILDREN'S MEDICAL CENTER

PHYSICIAN:	DEPARTMENT:
OFFICE ADDRESS:	SECTION:
OFFICE PHONE:	OHIO LICENSE NO:
STAFF STATUS:	BIRTHDATE:

1999–2000		2000–01		2001–02		2002–03	
Satisfactory Health Status		Satisfactory Health Status		Satisfactory Health Status		Satisfactory Health Status	
Satisfactory Meeting Attendance		Satisfactory Meeting Attendance		Satisfactory Meeting Attendance		Satisfactory Meeting Attendance	
Satisfactory Medical Record Completion		Satisfactory Medical Record Completion		Satisfactory Medical Record Completion		Satisfactory Medical Record Completion	
No Disciplinary Action		No Disciplinary Action		No Disciplinary Action		No Disciplinary Action	
COMMITTEE APPOINTMENTS		COMMITTEE APPOINTMENTS		COMMITTEE APPOINTMENTS		COMMITTEE APPOINTMENTS	
APPROVED	DATE	APPROVED	DATE	APPROVED	DATE	APPROVED	DATE
Chair, Credentials Committee		Chair, Credentials Committee		Chair, Credentials Committee		Chair, Credentials Committee	
Chief of Staff		Chief of Staff		Chief of Staff		Chief of Staff	
Secretary, Board of Trustees		Secretary, Board of Trustees		Secretary, Board of Trustees		Secretary, Board of Trustees	

FIGURE F–15. Form (b)

ited computer literacy or access, so consider your audience carefully before opting to collect your responses online. See also Web design.

Writing Instructions and Captions.   Place instructions at the beginning of the form and precede them with a heading designed to attract the attention of the reader, as shown in the following example.

Instructions for Completing This Form:
1. Complete the applicable blue-shaded portions on the front of pages 1, 2, and 3.
2. Itemize bills that apply to your major medical deduction on the back of page 1.
3. Mail page 1 to Securi-Med Insurance Company at the address shown above.
4. Give page 2 to your doctor.
5. Submit page 3 to a hospital, if applicable, for services provided.

F

Instructions for distributing the various copies of multiple-copy forms normally are placed at the bottom of the form and are repeated on every copy of the form.

Choosing Response Types.   Forms should ask questions in ways that are best suited to the types of data you hope to collect. In general, there are two types of questions:

- *Open-ended questions* allow respondents the freedom to choose their own words. Such questions are most appropriate if you wish to elicit answers you may not have anticipated (as in a complaint form) or if there are too many possible answers to use a multiple-choice format. However, the responses to open-ended questions are difficult to tabulate and analyze.

- *Closed-ended questions* provide a list of answers from which the respondent can select, limiting the range of possible responses. These questions are best used when you want to make sure you receive a standardized, easy-to-tabulate response. Some types of closed-ended questions are the following:
  - Multiple choice: Choose one (or sometimes more) from a preset list of options.
  - Ranked choice: Rank from best to worst, sweet to sour, etc., items in a preset list.
  - Forced choice: Choose true or false, good or bad, strong or weak, etc.

Wording Questions.   Questions are normally worded as captions. Keep captions brief and to the point; avoid unnecessary repetition by combining related information under an explanatory heading.

WORDY   What make of car do you drive? _____

What year was it manufactured? _____

What model is it? _____

What is the body style? _____

CONCISE   Vehicle Information

Make _____ Year _____

Model _____ Body Style _____

Make captions as specific as possible. If a requested date is other than the date on which the form is being filled out, the caption should read, for example, "Effective date" or "Date issued," rather than simply "Date." As in all technical writing, put yourself in your reader's place and imagine what sort of requests would be clear.

Sequence of Data. Try to arrange requests for information in an order that will be the most logical to the person filling out the form. At the top of the form, include preliminary information, such as the name of your organization, the title of your form, and any reference number. In the main portion of the form, include the entries you need to obtain the necessary data. At the end of the form, include space for a signature and a date.

The arrangement of the entries depends on several factors. First, the subject matter of the entries frequently determines the most logical order. A form requesting reimbursement for travel expenses, for instance, would logically begin with the first day of the week (or month) and end with the last day of the appropriate period. Second, if the response to one item is based on the response to another item, be sure the items appear in the correct order. Third, group requests for related information together whenever possible. Finally, entries should, in general, be arranged on the form from left to right and from top to bottom because that is the way we are accustomed to reading.

Designing a Form

When you prepare the final version of a form, pay particular attention to design details, especially to the placement of entry lines and the amount of space allowed for responses.

Entry Lines. A form can be designed so that the person filling it out provides information on a writing line, in a writing block, or in square boxes. A *writing line* is simply a rule with a caption, as shown in Figure F–16.

A *writing block* is essentially the same as a writing line, except that each entry is enclosed in a ruled block, making it impossible for the respondent to associate a caption with the wrong line, as shown in Figure F–17.

On some forms, the captions are arranged horizontally, as in Figures F–16 and F–17. On other forms, they are listed vertically, as shown in Figure F–18.

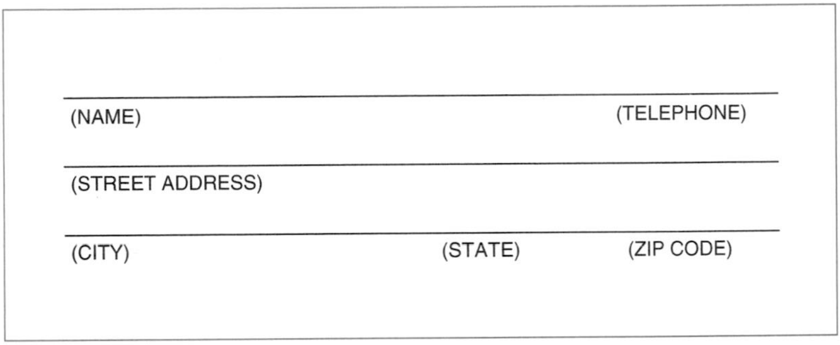

FIGURE F-16. Writing Lines for a Form

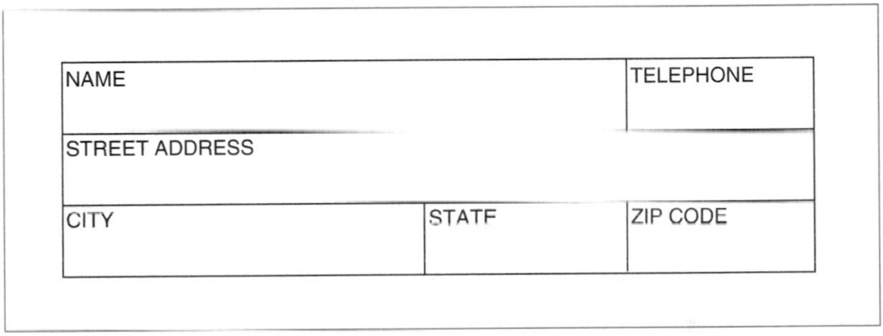

FIGURE F 17. Writing Block for a Form

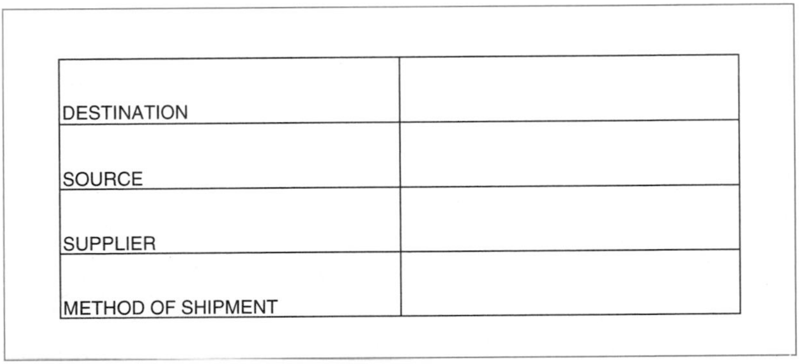

FIGURE F-18. Vertical Captions for a Form

When it is possible to anticipate all likely responses, you can make the form easy to fill out by writing the question on the form, supplying a labeled box for each possible answer, and asking the respondent to check the appropriate boxes. Such a design also makes it easy to tabulate the data. Be sure your questions are both simple and specific.

- Would you buy another MAX-PC? Yes ☐ No ☐

Spacing.   Provide enough space to enable the person filling out the form to enter the data. Insufficient writing space makes it difficult for people to respond, resulting in responses that are hard to read. Reading responses that are too tightly spaced or that snake around the side of the form can cause headaches and errors. To ensure the usability of the form, fill it out yourself before printing the final copies. See also **questionnaires**.

forms of discourse

Traditionally, the four *forms of discourse* include **exposition**, **description**, **persuasion**, and **narration**. These forms rarely exist alone in pure form; they usually appear in combination.

fortuitous / fortunate

When an event is *fortuitous,* it happens by chance or accident and without plan. Such an event may be lucky, unlucky, or neutral.

- My encounter with the general manager in Denver was entirely *fortuitous;* I had no idea he was there.

When an event is *fortunate,* it happens by good fortune or happens favorably. (Our chance meeting had a *fortunate* outcome.)

fragments (*see* sentence fragments)

functional shift

Many words shift easily from one **part of speech** to another, depending on how they are used. When they do, the process is called a *functional shift,* or a shift in function.

- It takes ten minutes to *walk* from the sales office to the accounting department. However, the long *walk* from the sales office to the accounting department reduces efficiency. [*Walk* shifts from <u>verb</u> to <u>noun</u>.]

- I talk to the Chicago office on the *phone* every day. He was concerned about the office *phone* expenses. He will *phone* the home office from London. [*Phone* shifts from <u>noun</u> to <u>adjective</u> to <u>verb</u>.]

- *After* we discuss the project, we will begin work. *After* lengthy discussions, we began work. The partners worked well together forever *after*. [*After* shifts from <u>conjunction</u> to <u>preposition</u> to <u>adverb</u>.]

<u>Jargon</u> is often the result of functional shifts. In medicine, for example, an *attending physician* is referred to as the "attending" (a shift from an adjective to a noun). Likewise, in nuclear plant construction, a *reactor containment building* is called a "containment" (a shift from an adjective to a <u>noun</u>). Do not shift the function of a word indiscriminately merely to shorten a phrase or expression. See also <u>affectation</u> and <u>conciseness/wordiness</u>.

G

garbled sentences

A *garbled sentence* is one that is so tangled with structural and grammatical problems that it cannot be repaired. Garbled sentences often result from an attempt to squeeze too many ideas into one sentence.

* My job objectives are accomplished by my having a diversified background which enables me to operate effectively and efficiently, consisting of a degree in computer science, along with twelve years of experience, including three years in Staff Engineering-Packaging sets a foundation for a strong background in areas of analyzing problems and assessing economical and reasonable solutions.

Do not try to patch such a sentence; rather, analyze the ideas it contains, list them in a logical sequence, and then construct one or more entirely new sentences.

An analysis of the preceding example yields the following five ideas:

* My job requires that I analyze problems to find economical and workable solutions.

* My diversified background helps me accomplish my job.

* I have a computer-science degree.

* I have twelve years of job experience.

* Three of these years have been in Staff Engineering-Packaging.

Using those five ideas—together with **parallel structure**, **sentence variety**, **subordination**, and **transition**—the writer might have described the job as follows:

* My job requires that I analyze problems to find economical and workable solutions. Both my training and my experience help me achieve this goal. Specifically, I have a computer-science degree and twelve years of job experience, three of which have been in the Staff Engineering-Packaging Department.

See also **clarity** and **sentence construction**.

gender

In English grammar, *gender* refers to the classification of <u>nouns</u> and <u>pronouns</u> as masculine, feminine, and neuter. The gender of most words can be identified only by the choice of the appropriate pronoun (*he, she, it*). Only these pronouns and a select few nouns reflect gender.

- *heir/heiress; duke/duchess*

Many such nouns have been replaced by single terms for both sexes. See also <u>biased language</u> and <u>he/she</u>.

Gender is important to writers because they must be sure that nouns and pronouns within a grammatical construction agree in gender. A pronoun, for example, must agree with its noun antecedent in gender. We must refer to a woman as *she* or *her*, not as *it*; to a man as *he* or *him*, not as *it*; to a barn as *it*, not as *he* or *she*. See also <u>agreement</u>.

G

ESL TIPS FOR ASSIGNING GENDER

The English language system has an almost complete lack of gender distinctions. That can be confusing for a nonnative speaker of English whose native language may be marked for gender. In the few cases in which English does make a gender distinction, there is a close connection between the assigning of gender and the sex of the subject. The few instances in which gender distinctions are made in English are summarized as follows:

Subject pronouns	he/she
Object pronouns	him/her
Possessive adjectives	his/her(s)
Some nouns	king/queen, boy/girl, bull/cow, etc.

When a noun, such as *doctor*, can refer to a person of either sex, you need to know the sex of the person to which the noun refers to determine the gender-appropriate pronoun.

- The doctor gave *her* patients lots of attention. [Doctor is female.]

- The doctor gave *his* patients lots of attention. [Doctor is male.]

When the sex of the noun antecedent is unknown, be sure to follow the guidelines for nonsexist writing in the entry <u>biased language</u>. (*Note:* Some English speakers refer to vehicles and countries as *she*, but contemporary usage tends to use *it*.)

general and specific method of development

General and specific development proceeds either from general information to specific details or from specific information to a general conclusion.

General to Specific

A general-to-specific method of development begins with a general statement and then provides facts or examples to develop and support that statement. For example, if you begin a report with the general statement "Companies that diversify are more successful than those that do not," the remainder of the report would offer examples and statistics that prove to the reader that companies that diversify are, in fact, more successful than companies that do not.

A memo or short report organized entirely in a general-to-specific sequence discusses only one point. All other information in the document supports the general statement, as in Figure G–1 from a memo about locating additional computer-chip suppliers.

Subject: Expanding Our Supplier Base for Computer Chips

General statement

On the basis of information presented at the supply meeting on April 14, we recommend that the company initiate relationships with computer-chip manufacturers. Several events make such an action necessary.

Supporting information

Our current supplier, Datacom, is experiencing growing pains and is having difficulty shipping the product on time. Specifically, we can expect a reduction of between 800 and 1,000 units per month for the remainder of this fiscal year. The number of units should stabilize at 15,000 units per month thereafter.

Domestic demand for our computers continues to grow. Demand during the current fiscal year is up 500,000 units over the last fiscal year. Our sales projections for the next five years show that demand should peak next year at about 830,000 units given the consumer demand, which will increase exponentially.

Finally, our expansion into European and Asian markets will require additional shipments of at least 750,000 units per quarter for the remainder of this fiscal year. Sales Department projections put global computer sales at double that rate, or 1,500,000 units per fiscal year, for the next five years.

FIGURE G–1. General-to-Specific Method of Development

Specific to General

The specific-to-general **method of development** begins with a specific statement and builds to a general **conclusion**. It carefully builds its case, often with examples and analogies in addition to facts or statistics, and does not actually make its point until the end. (See also **order-of-importance method of development**.) Figure G–2 is an example of the specific-to-general method of development.

G

Specific
details

> Recently, a government agency studied the use of passenger-side air bags in 4,500 accidents involving nearly 7,200 front-seat passengers of the vehicles involved. Nearly all the accidents occurred on routes that had a speed limit of at least 40 mph. Only 20 percent of the adult front seat passengers were riding in vehicles equipped with passenger-side air bags. Those riding in vehicles not equipped with passenger-side air bags were more than twice as likely to be killed as passengers riding in vehicles that were so equipped.

General
conclusion

> A conservative estimate is that 40 percent of the adult front-seat passenger-vehicle deaths could be prevented if all vehicles came equipped with passenger-side air bags. Children, however, should always ride in the backseat because other studies have indicated that a child can be killed by the deployment of an air bag. If you are an adult front-seat passenger in an accident, your chances of survival are far greater if the vehicle in which you are riding is equipped with a passenger-side air bag.

FIGURE G–2. Specific-to-General Method of Development

global communication

The prevalence of global communication technology, international trade agreements, and the emergence of Europe as a giant single market means that the ability to reach audiences from varied cultural backgrounds is essential. The audiences for such communications include clients, business partners, and colleagues.

Entries such as **meetings** and **résumés** in this book are based on U.S. cultural patterns. The treatment of such topics might be very different in other cultures where leadership styles, persuasive strategies, and even legal constraints differ. As illustrated in **international correspondence**, organizational patterns, forms of courtesy, and ideas about efficiency can vary significantly from culture to culture. What might be seen as direct and efficient in U.S. culture could be seen as blunt and even impolite in other cultures. The reasons behind these differing ways of viewing communication are complex, and those who study cultures have found various ways to measure cultural differences, such as individual versus group orientation, the importance of saving face, and conceptions of time.

Anthropologist Edward T. Hall, a pioneer in cross-cultural research, developed the concept of "context" to assess the predominant communication style of a culture.* By "context" Hall means how much or how little an individual assumes another person understands about a subject under discussion. In a very low-context communication, the participants assume they share little knowledge and must communicate in great detail. Low-context cultures tend to assume little prior knowledge on the part of those with whom they communicate; thus thorough documentation is important—written agreements (contracts) are expected, and rules are spelled out in detail.

In a high-context communication, the participants already understand the context and thus do not feel a need to exchange much background information. High-context cultures depend on shared history (or context) to relate to each other. Thus, words and written contracts are not so important while personal relationships are paramount. Of course, no culture is entirely high or low context; rather, these concepts can be helpful in understanding the complex communication style of a particular culture.†

Obviously, cultural differences and the reasons behind them are often so subtle that only someone who is very familiar with the culture can explain the effect they may have on others from that culture. For that reason, it is best to consult with someone from your intended audience's culture.‡

---

*Hall, Edward Twitchell, and Mildred Reed Hall. *Understanding Cultural Differences: Germans, French and Americans.* Yarmouth, Maine: Intercultural Press, 1990.

†Alred, Gerald J. "Teaching in Germany and the Rhetoric and Culture." *Journal of Business and Technical Communication* 11.3 (July 1997): 353–78.

‡Although many resources are available, the following books are useful starting points:

 Andrews, Deborah C. *Technical Communication in the Global Community.* Upper Saddle River, N.J.: Prentice-Hall, 1998.

 Scollon, Ron, and Suzanne Wong Scollon. *Intercultural Communication: A Discourse Approach.* Cambridge, Mass.: Basil Blackwell, 1995.

 Varner, Iris, and Linda Beamer. *Intercultural Communication in the Global Workplace.* Chicago: Irwin, 1995.

 Victor, David A. *International Business Communication.* New York: HarperCollins, 1992.

Writer's Checklist: Communicating Globally

☑ Consult with someone from your intended audience's culture. Many phrases, gestures, and visual elements are so subtle that only someone who is very familiar with the culture can explain the effect they may have on others from that culture.

☑ Acknowledge diversity within your organization. Discussing the differing cultures within your company or region will reinforce the idea that people can interpret verbal and nonverbal communications differently.

☑ Invite global and intercultural communication experts to speak to your employees. Companies in your area may have employees who could be resources for cultural discussions.

☑ Understand that the key to effective communication with global audiences is recognizing that cultural differences, despite the challenges they may present, offer growth for both you and your organization.

For more information on reaching global audiences, see **global graphics** and **presentations**.

G

WEB LINK **INTERCULTURAL RESOURCES**

Intercultural Press is a source of publications aimed at specific cultures as well as a wide variety of subjects from cross-cultural theory to international business. For this and additional links to intercultural resources, see *<www.bedfordstmartins.com/alred>* and select *Links for Technical Writing.*

---

global graphics

In the global business and technological environment, **graphs** and **visuals** require the same careful attention given to other aspects of **global communication**. The complex cultural connotations of visuals challenge writers to think beyond their own experience.

Symbols, images, and even colors are not free from cultural associations—they depend on context, and context is culturally determined. For instance, in North America, a red cross is commonly used as a symbol for first aid or hospital. In Muslim countries, however, a cross (red or otherwise) represents Christianity, whereas a crescent (usually green) signifies first aid or hospital. A manual for export to Honduras could indicate "caution" by using a picture of a person touching a finger below the eye. In France, however, that gesture would have a totally different meaning: "You can't fool me."

Careful attention to the connotations that visual elements may have for a diverse audience makes translations easier, prevents embarrassment, and earns respect for the company and its products and services. See also **presentations**.

Writer's Checklist: Communicating with Global Graphics

☑ Consult with someone from your intended audience's country who can recognize and explain the effect that visual elements will have on readers.

☑ Organize visual information for the intended audience. For example, North Americans tend to read visuals from left to right in clockwise rotation. Middle Eastern cultures read visuals from left to right in counterclockwise rotation.

☑ Be sure that the graphics you use have no religious implications.

☑ Carefully consider how you depict people in visuals. Nudity in advertising, for example, is generally acceptable in Europe but much less so in North America and Asia. In some cultures, showing even isolated bare body parts can alienate audiences.

☑ Use outlines or neutral abstractions to represent human beings. For example, use stick figures and avoid representing men and women.

☑ Examine how you display body positions in signs and visuals. Body positioning can carry unintended cultural meanings very different from your own. For example, some Middle Eastern cultures regard the display of the soles of one's shoes to be disrespectful and offensive.

☑ Try to use neutral colors in your graphics. Generally, black-and-white and gray-and-white illustrations work well. Colors can be problematic. For example, in North America, Europe, and Japan, red indicates danger. In China, however, red symbolizes joy. In Europe and North America, blue generally has a positive connotation; in Japan, blue represents villainy.

☑ Check your use of punctuation marks, which are as language specific as symbols. For example, in North America, the question mark generally represents the need for information or the help function in a computer manual or program. In many countries, that symbol has no meaning at all.

☑ Create simple visuals and use consistent labels for all visual items. In most cultures, simple shapes with fewer elements are easier to read.

☑ Explain the meaning of icons or symbols. Include a **glossary** to explain technical symbols that cannot be changed, such as company logos.

☑ Test icons and symbols in context with members of your target audience or cultural experts. See also **usability testing**.

glossaries

A *glossary* is an alphabetical list of definitions of terms used in a <u>formal report</u>, a <u>technical manual</u>, or other long document.

If you are writing a <u>report</u> that will go to readers who are not familiar with many of the terms you use, you may want to include a glossary. If you do, keep the entries concise and be sure they are written in plain language that all readers can understand.

- *Amplitude modulation*: Varying the amplitude of a carrier current with an audio-frequency signal.

G

Arrange the terms alphabetically, with each entry beginning on a new line. The definitions then follow the terms, dictionary style. In a formal report, the glossary appears after the appendix(es) and bibliography, and it begins on a new page.

Inclusion of a glossary does not relieve you of the responsibility of <u>defining terms</u> in the text that your reader may not understand when they are first mentioned.

gobbledygook

Gobbledygook is writing that suffers from an overdose of traits guaranteed to make it stuffy, pretentious, and wordy. Such traits include the overuse of big and mostly <u>abstract words</u>, <u>affectation</u> (especially long variants), <u>buzzwords</u>, inappropriate <u>jargon</u>, <u>clichés</u>, <u>euphemisms</u>, stacked <u>modifiers</u>, <u>vague words</u>, and deadwood. Gobbledygook is writing that attempts to sound official (officialese), legal (legalese), or scientific; it tries to make a "natural elevation of the geosphere's outer crust" out of a molehill. Consider the following statement from an auto repair release form.

LEGALESE   I hereby authorize the above repair work to be done along with the necessary material and hereby grant you and/or your employees permission to operate the car or truck herein described on streets, highways, or elsewhere for the purpose of testing and/or inspection. An express mechanic's lien is hereby acknowledged on above car or truck to secure the amount of repairs thereto.

DIRECT   You have my permission to do the repair work listed on this work order and to use the necessary material.

You may drive my vehicle to test its performance. I understand that you will keep my vehicle until I have paid for all repairs.

See also conciseness/wordiness and word choice.

good / well

G

Good is an adjective, and *well* is an adverb.

ADJECTIVE   Janet presented a *good* plan.

ADVERB   The plan was presented *well*.

Well also can be used as an adjective to describe someone's health.

ADJECTIVE   He is not a *well* man.

See also bad/badly.

grammar

Grammar is the systematic description of the way words work together to form a coherent language. In that sense, it is an explanation of the structure of a language. However, grammar is popularly taken to mean the set of rules that governs how a language ought to be spoken and written. In that sense, it refers to the usage conventions of a language.

Those two meanings of grammar—how the language functions and how it ought to function—are easily confused. To clarify the distinction, consider the expression *ain't*. Unless used intentionally to add colloquial flavor, *ain't* is unacceptable because its use is considered nonstandard. Yet taken strictly as a part of speech, the term functions perfectly well as a verb. Whether it appears in a declarative sentence ("I ain't going.") or an interrogative sentence ("Ain't I going?"), it conforms to the normal pattern for all verbs in the English language. Although readers may not approve of its use, they cannot argue that it is ungrammatical in such sentences.

To achieve clarity, you need to know both grammar (as a description of the way words work together) and the conventions of usage. Knowing the conventions of usage helps you select the appropriate over the inappropriate word or expression. (See also word choice.) A knowledge of grammar helps you diagnose and correct problems arising from how words and phrases function in relation to one another. For example, knowing that certain words and phrases function to modify other words and phrases gives you a basis for correcting those modifiers that are not doing their job. Understanding dangling modifiers helps you avoid or correct a construction that obscures the intended meaning. In short, an understanding of grammar and its special terminology is valuable chiefly because it enables you to recognize and correct problems so you can communicate clearly and precisely.

For a complete list of grammar entries, see the Topical Key.

> ✴ **WEB LINK GETTING HELP WITH GRAMMAR**
>
> For helpful Web sites and electronic grammar exercises, see <www .bedfordstmartins.com/alred> and select *Links for Technical Writing* and *Exercise Central*.

G

graphs

DIRECTORY

A *graph* presents numerical data in visual form and offers several advantages over presenting data within the text or in tables. Trends, movements, distributions, comparisons, and cycles are more readily apparent in graphs than they are in tables. However, although graphs often present data in a more comprehensible form than tables do, they are less precise. For that reason, graphs are often accompanied by tables that give exact data. The most common types of graphs are *line graphs, bar graphs, pie graphs,* and *picture graphs*. For additional advice, see visuals; for information about using presentation graphics, see presentations.

Line Graphs

A *line graph* shows the relationship between two variables or sets of numbers by plotting points in relation to two axes drawn at right angles. Line graphs that portray more than one set of variables are common because they allow for comparisons between two sets of statistics for the same period of time. In creating such graphs, identify each line with a label or a legend, as shown in Figure G–3. You can emphasize the difference between the two lines by shading the space between them, as also shown in Figure G–3.

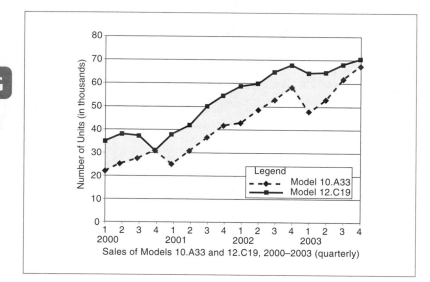

FIGURE G–3. Double-Line Graph

Writer's Checklist: Creating Line Graphs

When creating a line graph, remember that the vertical axis usually represents amounts, and the horizontal axis usually represents increments of time, as shown in Figure G-3.

☑ Indicate the zero point of the graph (the point where the two axes intersect).

☑ If the range of data shown makes it inconvenient to begin at zero, insert a break in the scale.

☑ Divide the vertical axis into equal portions, from the least amount at the bottom (or zero) to the greatest amount at the top.

☑ Divide the horizontal axis into equal units from left to right. If a label is necessary, center it directly beneath the scale.

☑ When necessary, include a key or legend that lists and explains symbols, as shown in Figure G-3.

☑ If the data comes from another source, include a source line under the graph at the lower left.

Writer's Checklist: Creating Line Graphs (continued)

☑ Place explanatory footnotes directly below the figure caption or label.

☑ Make all lettering read horizontally if possible, although the caption or label for the vertical axis is usually positioned vertically (see Figure G–3).

Be especially careful to proportion the vertical and horizontal scales so they give a precise presentation of the data that is free of visual distortion. To do otherwise is not only inaccurate but potentially unethical. (See also **ethics in writing**.) In Figure G–4, because the scale is compressed and some of the data is omitted, the graph on the left gives the appearance of a dramatic decrease; the graph on the right, which includes more data, is more accurate.

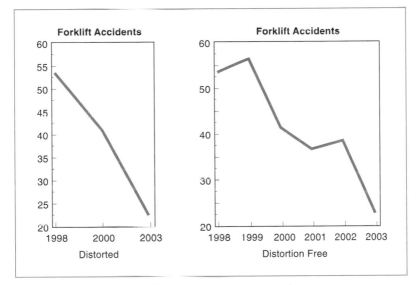

FIGURE G–4. Distorted and Distortion-Free Expressions of Data

Bar Graphs

Bar graphs consist of horizontal or vertical bars of equal width, scaled in length to represent some quantity. They are commonly used to show (1) quantities of the same item at different times, (2) quantities of different items at the same time, and (3) quantities of the different parts of an item that make up a whole — in this case, the segments of the bar graph must total 100 percent.

The horizontal graph in Figure G–5 on page 248 shows the quantities of different items for the same period of time.

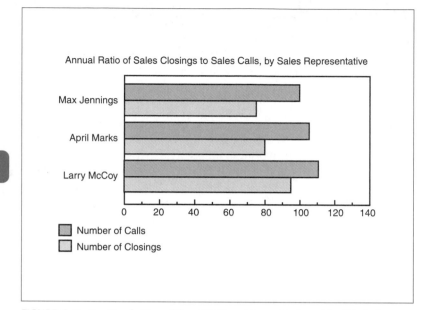

Annual Ratio of Sales Closings to Sales Calls, by Sales Representative

Max Jennings

April Marks

Larry McCoy

0   20   40   60   80   100   120   140

☐ Number of Calls
☐ Number of Closings

FIGURE G–5. Bar Graph (Quantities of Different Items During a Fixed Period)

Bar graphs can also show the different portions of an item that make up the whole, as shown in Figure G–6. Such a bar graph is divided according to the appropriate proportions of the subcomponents of the item. This type of graph can be constructed vertically or horizontally and can indicate multiple items. Where such items represent parts of a whole, as in Figure G–6, the segments in the bar graph must total 100 percent. Note that in addition to labels, each subdivision of a bar graph must be marked clearly by color, shading, or crosshatching, with a key that identifies the subdivisions represented.

Pie Graphs

A *pie graph* presents data as wedge-shaped sections of a circle. The circle equals 100 percent, or the whole, of some quantity, and the wedges represent how the whole is divided. Many times, the data shown in a bar graph could also be depicted in a pie graph. For example, Figure G–6 shows percentages of a whole in bar-graph form. In Figure G–7, the same data are converted into a pie graph, dividing "Your Municipal Tax Dollar" into wedge-shaped sections that represent percentages. Pie graphs also provide a quicker way of presenting information that can be shown in a table; in fact, a table with a more detailed breakdown of the same information often accompanies a pie graph.

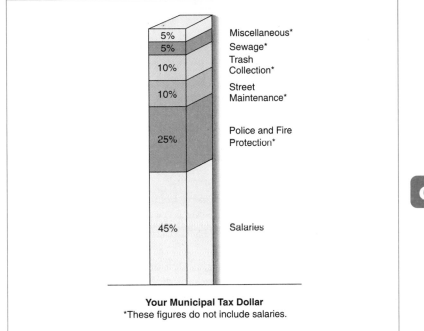

FIGURE G-6. Bar Graph (Showing the Different Parts That Make Up the Whole)

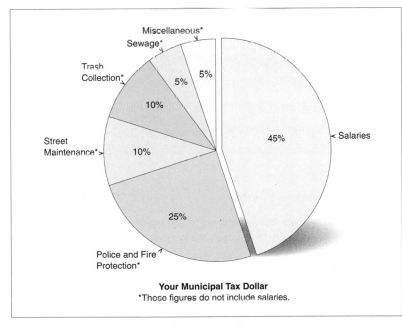

FIGURE G-7. Pie Graph (Showing Percentages of Whole)

Writer's Checklist: Creating Pie Graphs

☑ Make sure that the complete circle is equivalent to 100 percent.

☑ When possible, begin at the 12 o'clock position and sequence the wedges clockwise, from largest to smallest.

☑ Avoid presenting too many items in a pie graph; the graph can look cluttered or the slices can be too thin to be clear.

☑ Give each wedge a distinctive color, pattern, shade, or texture.

☑ Label each wedge with its percentage value and keep all callouts (labels that identify the wedges) horizontal.

☑ If you want to draw attention to a particular segment of the pie graph, detach that slice, as shown in Figure G–7.

Picture Graphs

Picture graphs are modified bar graphs that use pictorial symbols of the item portrayed. Each symbol corresponds to a specified quantity of the item, as shown in Figure G–8. Note that for precision and clarity the picture graph includes the total quantity following the symbols.

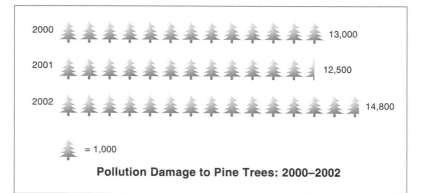

Pollution Damage to Pine Trees: 2000–2002

FIGURE G–8. Picture Graph

Writer's Checklist: Creating Picture Graphs

☑ Use picture graphs to add interest to **presentations** and documents, such as **newsletter articles**, that are aimed at wide audiences.

☑ Make sure the symbol you choose is easily recognizable. (See also **global graphics**.)

☑ Have each symbol represent a specific number of units.

☑ Because it is difficult to judge relative sizes accurately, show larger quantities by increasing the number of symbols rather than by creating a larger symbol.

H

he / she

The use of either *he* or *she* to refer to both sexes excludes half of the population. (See also **biased language**.) To avoid this problem, you could use the phrases *he or she* and *his or her*. (Whoever is appointed will find *his or her* task difficult.) However, *he or she* and *his or her* are clumsy when used repeatedly, and the *he/she* or *s/he* construction is awkward. One solution is to reword the sentence to use a plural **pronoun**; if you do, change the **noun** to which the pronoun refers to its plural form.

- The administrator cannot do his or her job until he or she understands the concept.

  Administrators ~~The administrator~~ cannot do *their jobs* ~~his or her job~~ until *they understand* ~~he or she understands~~ the concept.

In other cases, you may be able to avoid using a pronoun altogether.

- Whoever is appointed will find *the* ~~his or her~~ task difficult.

headers and footers

A header in a **report**, letter, or other document appears at the top of each page and contains identifying information; a footer appears at the bottom of each page and contains similar information. The header often contains the topic (or topic and subtopic) dealt with in that section of the document. The footer may contain the date of the document, the page number, and sometimes the document's name and section title.

Although the information included in headers and footers varies greatly from one organization to the next, the header and footer shown in Figure H–1 on page 252 are fairly typical. For headers used in letters and memos, see **correspondence**.

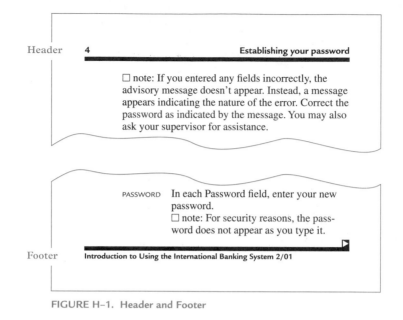

Header   4   **Establishing your password**

☐ note: If you entered any fields incorrectly, the advisory message doesn't appear. Instead, a message appears indicating the nature of the error. Correct the password as indicated by the message. You may also ask your supervisor for assistance.

PASSWORD   In each Password field, enter your new password.
☐ note: For security reasons, the password does not appear as you type it.

Footer   **Introduction to Using the International Banking System 2/01**

FIGURE H–1. Header and Footer

headings

Headings (also called *heads*) are titles or subtitles within the body of a document that help <u>readers</u> find information, divide the material into comprehensible segments, highlight the main topics, and signal topic changes. A <u>formal report</u> or <u>proposal</u> may need several levels of headings to indicate major divisions, subdivisions, and even smaller units. However, it is better to avoid using more than three levels of headings. See also <u>layout and design</u>.

Headings typically represent the major topics of a document. In a short document, you can use the major divisions of your outline as headings; in a longer document, you may need to use both major and minor divisions.

General Heading Style

There is no one correct format for headings. Often a company settles on a standard format, which everyone in the company follows. Sometimes a customer for whom a report or proposal is being prepared requires a particular format. In the absence of specific guidelines, follow the system illustrated in Figure H–2.

<table>
<tr><td>

First-
level
head

Second-
level
head

Third-
level
head

Fourth-
level
heads

</td><td>

DISTRIBUTION CENTER LOCATION REPORT

The committee initially considered 30 possible locations for the proposed new distribution center. Of these, 20 were eliminated almost immediately for one reason or another (unfavorable zoning regulations, inadequate transportation infrastructure, etc.). Of the remaining ten locations, the committee selected for intensive study the three that seemed most promising: Chicago, Minneapolis, and Salt Lake City. We have now visited these three cities and our observations and recommendations follow.

CHICAGO

Of the three cities, Chicago presently seems to the committee to offer the greatest advantages, although we wish to examine these more carefully before making a final recommendation.

Selected Location

Though not at the geographical center of the United States, Chicago is the demographic center to more than three-quarters of the U.S. population. It is within easy reach of our corporate headquarters in New York. And it is close to several of our most important suppliers of components and raw materials—those, for example, in Columbus, Detroit, and St. Louis. Several considerations were considered essential to the location, although some may not have had as great an impact on the selection. . . .

Air Transportation.   Chicago has two major airports (O'Hare and Midway) and is contemplating building a third. Both domestic and international air-cargo service are available. . . .

Sea Transportation.   Except during the winter months when the Great Lakes are frozen, Chicago is an international seaport. . . .

Rail Transportation.   Chicago is served by the following major railroads. . . .

</td></tr>
</table>

H

FIGURE H–2. Headings Used in a Document

Decimal Numbering System

The decimal numbering system uses a combination of <u>numbers</u> and decimal points to differentiate among levels of headings. The following outline shows the correspondence between different levels of headings and the decimal numbers used:

1. FIRST-LEVEL HEADING
    1.1 Second-Level Heading

1.2 Second-Level Heading
    1.2.1 Third-Level Heading
    1.2.2 Third-Level Heading
        1.2.2.1 Fourth-Level Heading
        1.2.2.2 Fourth-Level Heading
1.3 Second-Level Heading
    1.3.1 Third-Level Heading
    1.3.2 Third-Level Heading
2. FIRST-LEVEL HEADING

Although the second-, third-, and fourth-level of such headings are indented in an outline or **table of contents**, they are flush with the left margins when they function as headings in the body of a report. Every heading starts on a new line, with an extra line of space above and below the heading.

Writer's Checklist: Using Headings

☑ Use headings to signal a new topic or, if it is a lower-level heading, a new subtopic within the larger topic.

☑ Avoid too many or too few headings or levels of headings; too many clutter a document and too few fail to provide recognizable structure.

☑ Ensure that headings at the same level are of relatively equal importance and follow **parallel structure**.

☑ Subdivide sections only as needed; not every section requires lower-level headings.

☑ Subdivide higher-level headings into two or more lower-level headings whenever possible.

☑ Do not leave a heading as the final line of a page. If two lines of text cannot fit below a heading, start the section at the top of the next page.

☑ Do not allow a heading to substitute for discussion; the text should read as if the heading were not there.

hyphens

The *hyphen* (-) serves both to link and to separate words. The hyphen's most common linking function is to join **compound words** (ablebodied, self-contained, self-esteem). A hyphen is used to form compound **numbers** from twenty-one through ninety-nine and fractions when they are written out (forty-one, three-quarters).

Hyphens Used with Modifiers

Two- and three-word **modifiers** that express a single thought are hyphenated when they precede a **noun**.

- It was a *well-written* report.

However, a modifying phrase is not hyphenated when it follows the noun it modifies.

- The report was *well written.*

If each of the words can modify the noun without the aid of the other modifying word or words, do not use a hyphen.

- We purchased a *new laser* printer.

If the first word is an **adverb** ending in *-ly,* do not use a hyphen.

- We found a *newly minted* coin.

A hyphen is always used as part of a letter or number modifier.

- 9-inch, A-frame.

In a series of unit modifiers that all have the same term following the hyphen, the term following the hyphen need not be repeated throughout the series; for greater smoothness and brevity, use the term only at the end of the series.

- The third-, fourth-, and fifth-floor rooms were recently painted.

Hyphens Used with Prefixes and Suffixes

A hyphen is used with a **prefix** when the root word is a proper noun.

- pre-Columbian, anti-American, post-Newtonian

A hyphen may be used when the prefix ends and the root word begins with the same vowel.

- re-elect, re-enter, anti-inflammatory

A hyphen is used when *ex-* means "former."

- ex-president, ex-spouse

A hyphen may be used to emphasize a prefix.

- She was anti-everything.

The **suffix** *-elect* is hyphenated.

- president-elect

Hyphens and Clarity

The presence or absence of a hyphen can alter the meaning of a sentence.

AMBIGUOUS   We need a biological waste management system.

That sentence could mean one of two things.

- We need a biological-waste management system.
  [We need a system to manage "biological waste."]

- We need a biological waste-management system.
  [We need a "biological" system to manage waste.]

To avoid confusion, some words and modifiers should always be hyphenated. *Re-cover* does not mean the same thing as *recover,* for example; the same is true of *re-sent* and *resent, re-form* and *reform, re-sign* and *resign.*

Other Uses of the Hyphen

Hyphens should be used between letters showing how a word is spelled.

- In his letter, he misspelled *believed* b-e-l-i-e-v-e-d.

A hyphen can stand for *to* or *through* between letters and numbers (pp. 44-46, the Detroit-Toledo Expressway, A-L and M-Z).

Writer's Checklist: Using Hyphens to Divide Words

☑ Do not divide one-syllable words.

☑ Divide words at syllable breaks, which you can determine with a dictionary.

☑ Do not divide a word if only one letter would remain at the end of a line or if fewer than three letters would start a new line.

☑ Do not divide a word at the end of a page.

☑ If a word already has a hyphen in its spelling, divide the word at the existing hyphen.

☑ Do not use a hyphen to break a URL or an e-mail address at the end of a line because it may confuse readers who could assume that the hyphen is part of the address.

I

idioms

An *idiom* is a group of words that has a special meaning apart from its literal meaning. Someone who "runs for office" in the United States, for example, need not be an athlete. The same candidate would "stand for office" in the United Kingdom. Because such expressions are specific to a culture, nonnative speakers must memorize them. The following are typical idioms that give nonnative speakers trouble.

call off [cancel]
call on [visit a client]
cross out [draw line through]
do over [repeat task]
drop in on [visit unexpectedly]
figure out [solve a problem]
find out [discover information]
get through with [finish]
give up [quit]

hand in [submit]
hand out [distribute]
keep on [continue]
leave out [omit]
look up [seek in reference]
put off [postpone]
run into [meet by chance]
run out of [deplete supply]
watch out for [be careful]

Idioms often provide helpful shortcuts. In fact, they can make writing more natural and vigorous. Avoid them, however, if your writing is to be translated into another language or read in other English-speaking countries.

Idiom also refers to the practice of using certain **prepositions** following some **adjectives** (*similar to*), **nouns** (*need for*), and **verbs** (*approve of*). Because there is no sure system to explain such usages, the best advice is to check a dictionary. See the Web Link box below. See also **international correspondence** and **English as a second language**.

🌐 WEB LINK **PREPOSITIONAL IDIOMS**

The Arizona State University Writing Center and the Purdue University Online Writing Lab offer lists of common pairings of prepositions with nouns, verbs, and adjectives. For links to helpful Web sites and electronic grammar exercises on prepositional idioms, see <*www.bedfordstmartins.com /alred*> and select *Links for Technical Writing* and *Exercise Central*.

---

illegal / illicit

If something is *illegal*, it is prohibited by law. If something is *illicit*, it is prohibited by either law or custom. *Illicit* behavior may or may not be *illegal*, but it does violate social convention or moral codes and therefore usually has a clandestine or immoral **connotation**. (The employee's *illicit* sexual behavior caused a scandal, but the company's attorney concluded that no *illegal* acts were committed.)

illustrations (*see* visuals)

imply / infer

If you *imply* something, you hint or suggest it. (Her e-mail *implied* that the project would be delayed.) If you *infer* something, you reach a conclusion on the basis of evidence. (The manager *inferred* from the e-mail that the project would be delayed.)

in / into

In means "inside of"; *into* implies movement from the outside to the inside. (The equipment was *in* the test chamber, so she sent her lab assistant *into* the chamber to get it.)

in order to

Most often, *in order to* is a meaningless filler phrase that is dropped into a sentence without thought. See also **conciseness/wordiness**.

- ~~In order to~~ To start the engine, open the choke and throttle and then press the starter.

However, the phrase *in order to* is sometimes essential to the meaning of a sentence.

- If the vertical scale of a graph line would not normally show the zero point, use a horizontal break in the graph *in order to* include the zero point.

In order to also helps control the pace of a sentence, even when it is not essential to the meaning of the sentence.

- The committee must know the estimated costs *in order to* evaluate the feasibility of the project.

in terms of

When used to indicate a shift from one kind of language or terminology to another, the phrase *in terms of* can be useful.

- *In terms of* gross sales, the year has been relatively successful; however, *in terms of* net income, it has been discouraging.

When simply dropped into a sentence because it easily comes to mind, *in terms of* is meaningless <u>affectation</u>. See also <u>conciseness/wordiness</u>.

- She was thinking ~~in terms~~ of subcontracting much of the work.

inasmuch as / insofar as (*see* as / because / since)

increasing-order-of-importance method of development (*see* order-of-importance method of development)

indentation

Text that is indented is set in from the margin. The most common use of indentation is at the beginning of a paragraph, where the first line is usually indented five spaces, unless the full-block style is used, as shown in <u>correspondence</u>. Another use of indentation is in outlining, in which each subordinate entry is indented under its major entry.

A long quotation may be indented in a manuscript instead of being enclosed in quotation marks. The indentation varies, depending on what documentation style you are following (see <u>documenting sources</u>). If you are not following a specific style manual, you may block indent one-half inch or ten spaces from both the right and left margins for reports and other documents. See also <u>quotations</u>.

indexing

An *index* is an alphabetical list of all the major topics and sometimes subtopics in a written work. It cites the pages where each topic can be found and allows readers to find information on particular topics quickly and easily, as shown in Figure I–1. The index always comes at the very end of the work. Many Web sites also provide linked subject indexes to the content at their sites.

The key to compiling a useful index is selectivity. Instead of listing every possible reference to a topic, select references to passages where the topic is discussed fully or where a significant point is made about it. For actual index entries like those in Figure I–1, choose those words or phrases that best represent a topic. Key terms are those that a reader would most likely look for in an index. For example, the key terms in a reference to the development of legislation about environmental impact statements would probably be *legislation* and *environmental impact statement*. In selecting terms for index entries, use chapter or section titles only if they include such key terms. For index entries on tables and visuals, use the keywords in their titles. Create alphabetical Web-site indexes from links to topics in subsites throughout the larger site.

Compiling an Index

Do not attempt to compile an index until the final manuscript is completed because terminology and page numbers will not be accurate before then. The best way to manually compile a list of topics is to read through your written work from the beginning; each time a key term appears in a significant context, list the term and its page number on a 3-by-5 index card. An index entry can consist solely of a main entry and its page number.

- aquatic monitoring programs, 42

An index entry can also include a main entry, subentries, and even sub-subentries, as shown in Figure I–1. A subentry indicates pages where a specific subcategory or subdivision of the main topic can be found.

When you have completed this process for the entire work, sort the main entries alphabetically, then sort all subentries and sub-subentries alphabetically beneath their main entries. Because active Web sites add and remove content continually, their indexes must be updated regularly.

Wording Index Entries

The first word of an index entry should be the principal word because the reader will look for topics alphabetically by their main words. Se-

monitoring programs, 27–44 ————————— Main entry
  aquatic, 42
  ecological, 40 ————————————➤ Subentries
  meteorological, 37
    operational, 39
    preoperational, 37 ————————————➤ Sub-subentries
  radiological, 30
  terrestrial, 41, 43–44 ————————————➤ Subentries
  thermal, 27

FIGURE I–1. Index Entry (with Main Entry, Subentries, and Sub-subentries)

lecting the right word to list first is easier for some topics than for others. For instance, *tips on repairing electrical wire* would not be a suitable index entry because a reader looking for information on electrical wire would not look under the word *tips*. Ordinarily, an entry with two keywords, like *electrical wire*, should be indexed under each word (*electrical wire* and *wire, electrical*). An index entry should be written as a <u>noun</u> or a noun **phrase** rather than as an **adjective** alone.

- electrical wire, 20–22
    grounding, 21
    insulation, 20
    repairing, 22
    size, 21

DIGITAL TIP CREATING AN INDEX

Word-processing software can provide a quick and efficient way to create an alphabetical subject index of your document. Review the document to identify entries: the words, phrases, figure captions, or symbols that you wish to index. Then, following your software's instructions, highlight and code these entries. In Microsoft Word, for example, you can mark a keyword and the software can automatically mark all other instances of the word. Following your coding, the software sorts the entries, eliminates duplications, and arranges them alphabetically with their page numbers in a separate section at the end of the document. You can also create headings for each alphabetic grouping of the index (A, B, C, etc.).

Your draft index will still need careful review and revisions, but using the software to create the first draft can certainly save time. For more on this topic, see <*www.bedfordstmartins.com/alred*> and select *Digital Tips.*

Cross-Referencing

Cross-references in an index help readers find other related topics in the text. A reader looking up *technical writing,* for example, might find cross-references to *report* or *e-mail.* Cross-references do not include page numbers; they merely direct readers to another index entry, where they can find page numbers. There are two kinds of cross-references: *see* references and *see also* references.

See references are most commonly used with topics that can be identified by several different terms. Listing the topic page numbers by only one of the terms, the indexer then lists the other terms throughout the index as *see* references.

- economic costs. *See* benefit-cost analyses

See references also direct readers to index entries where a topic is listed as a subentry.

- L-shaped fittings. *See* elbows, L-shaped fittings

See also references indicate other entries that include additional information on a topic.

- ecological programs, 40–49 *See also* monitoring programs

Writer's Checklist: Indexing

☑ Use lowercase for the first words and all subsequent words of main entries, subentries, and sub-subentries unless they are proper nouns or would otherwise be capitalized. See **capitalization**.

☑ The cross-reference terms *see* and *see also* should appear in italics.

☑ Place each subentry in the index on a separate line, indented from its main entry. Indent sub-subentries from the preceding subentry. Indentations allow readers to scan a column quickly for pertinent subentries or sub-subentries.

☑ Separate entries from page numbers with commas.

☑ Format the index with double columns, as is done in the index to this book.

indiscreet / indiscrete

Indiscreet means "lacking in prudence or sound judgment." (His public discussion of the proposed merger was *indiscreet.*) *Indiscrete* means "not divided or divisible into parts." (The separate departments, once combined, become *indiscrete.*) See also **discreet/discrete**.

individual

Individual is most appropriate when used as an <u>adjective</u> to distinguish a single person from a group. (The *individual* employee's obligations are detailed in the policy manual.) Using *individual* as a <u>noun</u> is an <u>affectation</u>. Use *people* or another appropriate term. See also <u>persons/people</u>.

members of
- Several ~~individuals on~~ the committee did not vote.
  ^

ingenious / ingenuous

Ingenious means "marked by cleverness and originality." (Seon Ju's *ingenious* solution led to her promotion.) *Ingenuous* means "straightforward" or "characterized by innocence and simplicity." (The *ingenuous* co-op students bring fresh perspectives to the company.)

inquiries and responses

An inquiry letter or <u>e-mail</u> may be as simple as a request for a free brochure or as complex as asking a consultant to define the requirements for establishing a usability testing lab. See also <u>correspondence</u>.

There are two broad categories of inquiries. One kind provides a benefit to the <u>reader</u> (for instance, you may ask for information about a product that a company has recently advertised); the other kind primarily benefits the writer (you may request from a public utility information on an energy-related project you are developing). The second kind of letter requires the use of <u>persuasion</u> and special consideration of your reader's needs.

Writing Inquiries

Your purpose in writing an inquiry will probably be to obtain answers to specific questions, as shown in Figure I–2 on page 265. You will be more likely to receive a prompt, helpful reply if you follow these guidelines:

- Keep your questions specific and clear but concise.
- Phrase your questions so that the reader will immediately know the type of information you are seeking, why you need it, and how you will use it.

- If possible, present your questions in a numbered list to make it easy for your reader to respond to them.
- Keep the number of questions to a minimum.
- Offer some inducement for the reader to respond, such as promising to share the results of what you are doing.
- Promise to keep responses confidential, if appropriate.

In the closing, thank the reader for taking the time to respond. In addition, make it convenient for the recipient to respond by providing contact information, such as a phone number or an e-mail address, as shown in Figure I–2.

Responding to Inquiries

When you receive an inquiry, determine whether you have both the information and authority to respond. If you are the right person in your organization to respond, answer as promptly as you can, and be sure to answer every question asked. How long and how detailed your response should be depends on the nature of the question and the information provided by the writer.

If you receive an inquiry that you feel you cannot answer, find out who can and forward it to that person. Notify the letter writer that you have forwarded the inquiry, as shown in Figure I–3 on page 266.

When an inquiry is forwarded, the person who replies should state in the first paragraph of the response why someone else is answering the original inquiry, as shown in Figure I–4 on page 267.

inside / inside of

In the phrase *inside of,* the word *of* is redundant and should be omitted.

- The switch is just inside ~~of~~ the door.

Using *inside of* to mean "in less time than" is colloquial and should be avoided in writing.

- They were finished ~~inside of~~ an hour.
  in less than

insoluble / insolvable

The words *insoluble* and *insolvable* are sometimes used interchangeably to mean "incapable of being solved." *Insoluble* also means "incapable of being dissolved."

University of Dayton
P.O. Box 113
Dayton, OH 45409
March 11, 2003

Jane E. Metcalf
Engineering Services
Miami Valley Power Company
P.O. Box 1444
Miamitown, OH 45733

Dear Ms. Metcalf:

Could you please send me some information on heating systems for a computerized, energy-efficient house that a team of engineering students at the University of Dayton is designing?

The house, which contains 2,000 square feet of living space (17,600 cubic feet), meets all the requirements stipulated in your brochure "Insulating for Efficiency." We need the following information:

1. The proper-size heat pump to use in this climate for such a home.
2. The wattage of the supplemental electrical heating units that would be required for this climate.
3. The estimated power consumption and current rates of those units for one year.

We will be happy to send you a copy of our preliminary design report. If you have questions or suggestions, please contact me at kjp@fly.ud.edu or call 513-229-4598.

Thank you for your help.

Sincerely,

Kathryn J. Parsons

Kathryn J. Parsons

FIGURE I–2. Inquiry

- Until yesterday, the production problem seemed *insolvable*.
- Rubber gloves are *insoluble* in most household solvents.

instructions

Instructions that are clear and easy to follow prevent miscommunication and help <u>readers</u> complete tasks effectively and safely. To write ef-

---

Subject: Report Received
   Date: Mon, 19 March 2003 11:42:25 -0500 (EDT)
   From: Jane E. Metcalf <metcalf@mvpc.org>
     To: Kathryn J. Parsons <kjp@fly.ud.edu>

Dear Kathryn Parsons:

Thank you for inquiring about the heating system we recommend for use in homes designed according to the specifications outlined in our brochure "Insulating for Efficiency."

Because I cannot answer your specific questions, I have
forwarded your inquiry to Michael Wang, Engineering Assistant in our Development Group. He should be able to answer
the questions you have raised. You should be hearing from
him shortly.

Best wishes,

Jane E. Metcalf

===================================
Jane E. Metcalf, Director of Public Information
Miami Valley Power Company
P.O. Box 1444 ~ Miamitown, OH 45733
Office 513-264-4800 ~ Fax 513-264-4889
Web ~ http://www.enersaving.com
===================================

FIGURE I-3. **Response to an Inquiry (Indicating That It Has Been Forwarded)**

fective instructions, you must thoroughly understand the process, system, or device you are describing. Often you must observe someone as
he or she completes the task and perform the steps yourself before you
begin to write. See also **process explanation**.

Writing Instructions

Keep in mind your readers' level of knowledge. If you know that all
your readers have good backgrounds in the topic, use fairly specialized
terms. If that is not the case, use plain language or include a **glossary**
for specialized terms that you cannot avoid.

Clear and easy-to-follow instructions are written as commands in
the imperative **mood**, active **voice**, and (whenever possible) simple
present **tense**.

MIAMI VALLEY POWER COMPANY
P.O. BOX 1444
MIAMITOWN, OH 45733
(513) 264-4800

March 24, 2003

Ms. Kathryn J. Parsons
University of Dayton
P.O. Box 113
Dayton, OH 45409

Dear Ms. Parsons:

Jane Metcalf forwarded to me your inquiry of March 11 about the house
that your engineering team is designing. I can estimate the insulation requirements of a typical home of 17,600 cubic feet as follows:

1. For such a home, we would generally recommend a heat pump capable
   of delivering 40,000 BTUs. Our model AL-42 (17 kilowatts) meets that
   requirement.
2. With the AL-42's efficiency, you don't need supplemental heating units.
3. Depending on usage, the AL-42 unit averages between 1,000 and 1,500
   kilowatt-hours from December through March. To determine the current
   rate for such usage, check with Dayton Power and Light Company.

I can give you an answer that would apply specifically to your house only
with information about its particular design (such as number of stories, windows, and entrances). If you send me more details, I will be happy to provide more precise figures. Your project sounds interesting.

Sincerely,

Michael Wang

Michael Wang
Engineering Assistant
mwang@mvpc.org

http://www.enersaving.com

FIGURE I–4. Response to an Inquiry

Raise the .
- ~~The~~ access lid ~~will be raised by the operator.~~
       ^   ^

Although <u>conciseness</u> is important in instructions, <u>clarity</u> is essential. You can make sentences shorter by leaving out some <u>articles</u> (*a, an, the*), some <u>pronouns</u> (*you, this, these*), and some <u>verbs</u>, but such sentences may result in <u>telegraphic style</u> and be harder to understand. For example, the first version of the following instruction for placing a document in a scanner tray is confusing.

CONFUSING   Place document in tray with printed side facing opposite.

CLEAR   Place the document in the document tray with the printed side facing away from you.

One good way to make instructions easy to follow is to divide them into short, simple steps in their proper sequence. Steps can be organized with words (*first, next, finally*) that indicate time or sequence.

- *First,* determine the problem the customer is having with the computer. *Next,* observe the system in operation. *At that time,* question the operator until you are sure that the problem has been explained completely. *Then* analyze the problem and make any necessary adjustments.

Or you can use numbers, as in the following:

- 1. Connect each black cable wire to a brass terminal.
  2. Attach one 4-inch green jumper wire to the back.
  3. Connect the jumper wire to the bare cable wire.

Consider using the numbered- or bulleted-list feature of your word-processing software to create sequenced steps.

Plan ahead for your reader. If the instructions in step 2 will affect a process in step 9, say so in step 2. Sometimes your instructions have to make clear that two operations must be performed simultaneously. Either state that fact in an <u>introduction</u> to the specific instructions or include both operations in one step.

CONFUSING   1. Hold down the CONTROL key.
             2. Press the RETURN key before releasing the CONTROL key.

CLEAR   1. While holding down the CONTROL key, press the RETURN key.

If your instructions involve a great many steps, break them into stages, each with a separate heading so that each stage begins again with step 1.

Using <u>headings</u> as dividers is especially important if your reader is likely to be performing the operation as he or she reads the instructions.

Using Illustrations

Illustrations or <u>visuals</u> can simplify even complex instructions by reducing the number of words needed to explain a process or procedure. <u>Drawings</u> and <u>photographs</u> enable your reader to identify parts and the relationships between parts more easily than long explanations.

The instructions in Figure I–5 on page 270 guide the reader through the steps of streaking a saucer-sized disk of material (called *agar*) used to grow bacteria colonies. The purpose is to thin out the original specimen (the *inoculum*) so that the bacteria will grow in small, isolated colonies.

Warning Readers

Alert your readers to any potentially hazardous materials (or actions) before they reach the step for which the material is needed. Caution readers handling hazardous materials about any requirements for special clothing, tools, equipment, and other safety measures. Highlight warnings, cautions, and precautions to make them stand out visually from the surrounding text. Present warning notices in a box, in all uppercase letters, in large and distinctive fonts, or in color. Experiment with font style, size, and color to determine which devices are most effective. See also <u>layout and design</u>.

Figure I–6 on page 271 shows a warning from an instruction manual for the use of a gas grill. Notice that a drawing supports the text of the warning.

Testing Instructions

Finally, to test the accuracy and clarity of your instructions, ask someone who is not familiar with the task to follow your directions. A first-time user can spot missing steps or point out passages that should be worded more clearly. As you observe your tester, note any steps that seem especially confusing and revise accordingly. See also <u>usability testing</u>.

Writer's Checklist: Writing Instructions

- ☑ Use the imperative mood and the active voice.
- ☑ Use short sentences and simple present tense as much as possible.
- ☑ Avoid technical terminology and <u>jargon</u> that your readers might not know, including undefined <u>abbreviations</u>.
- ☑ Do not use elegant variation (two different words for the same thing). See also <u>affectation</u>.
- ☑ Eliminate any <u>ambiguity</u>.

Writer's Checklist: Writing Instructions *(continued)*

☑ Use effective visuals, and place them properly.

☑ Include appropriate warnings and cautions.

☑ Verify that measurements, distances, times, and relationships are precise and accurate.

☑ Test your instructions by having someone else follow them while you observe.

STREAKING AN AGAR PLATE

Distribute the inoculum over the surface of the agar in the following manner:

1. Beginning at one edge of the saucer, thin the inoculum by streaking back and forth over the same area several times, sweeping across the agar surface until approximately one-quarter of the surface has been covered. *Sterilize the loop in an open flame.*

2. Streak at right angles to the originally inoculated area, carrying the inoculum out from the streaked areas onto the sterile surface with only the first stroke of the wire. Cover half of the remaining sterile agar surface. *Sterilize the loop.*

3. Repeat as described in Step 2, covering the remaining sterile agar surface.

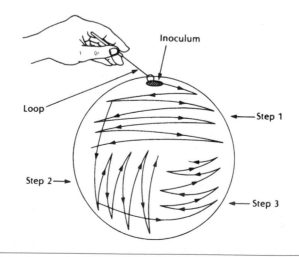

FIGURE I–5. Illustrated Instructions

WARNING

- It is the responsibility of assembler/owner to assemble, install, and maintain gas grill.
- Use the grill outdoors only.
- Do not let children operate or play near your grill.
- Keep the grill area clear and free from materials that burn, gasoline, bottled gas in any form, and other flammable vapors and liquids.
- Do not block holes in bottom and back of grill.
- Visually check burner flames on a regular basis.
- Use the grill in a well-ventilated space. Never use in an enclosed space, carport, garage, porch, patio, or building made of combustible construction, or under overhead construction.
- Do not install your grill in or on recreational vehicles and/or boats.
- Keep grills a distance of 36" or 3 ft. (approximately 1 m) from buildings to ensure there is no fire or melting of mate rials on the building.

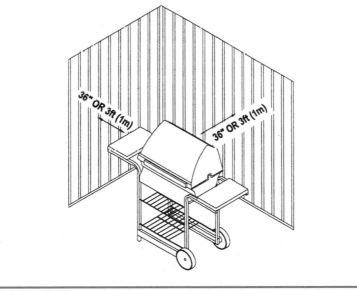

FIGURE I–6. Warning in a Set of Instructions   *Source:* "Char-Broil Use and Care Manual," Courtesy of W. C. Bradley Company, 1997, Columbus, GA.

insure / ensure / assure

Insure, ensure, and *assure* all mean "make secure or certain." *Assure* refers to people, and it alone has the connotation of setting a person's mind at rest. (I *assure* you that the equipment will be available.) *Ensure* and *insure* mean "make secure from harm." Only *insure* is widely used in the sense of guaranteeing the value of life or property.

- We need all the data to *ensure* the success of the project.
- We should *insure* the contents of the building.

intensifiers

Intensifiers are <u>adverbs</u> that emphasize degree, such as *very, quite, rather, such,* and *too.* Although they serve a legitimate and necessary function, they are often overused. Too many intensifiers weaken your writing. Eliminate intensifiers that do not make a definite contribution or replace them with specific details.

- The team was ~~quite~~ happy to ~~receive the very good news~~ *learn* that it had been awarded a ~~rather substantial monetary~~ *$5,000* prize for its design.

Some words (such as *perfect, impossible,* and *final*) do not logically permit intensification because, by definition, they do not allow degrees of comparison. Although <u>usage</u> often ignores that logical restriction, to ignore it is, strictly speaking, to defy the basic meanings of such words. See also <u>absolute words</u>, <u>conciseness/wordiness</u>, and <u>equal/unique/perfect</u>.

interface

An *interface* is a surface that provides a common boundary between two bodies or areas. The bodies or areas may be physical (the interface of a piston and a cylinder or a person and a computer screen) or conceptual (the interface of mathematics and economics). Do not use *interface* as a substitute for the verb *cooperate, interact,* or even *work.* See also <u>affectation</u> and <u>buzzwords</u>.

interjections

An *interjection* is a word or phrase standing alone or inserted into a sentence to exclaim or command attention. Grammatically, it has no connection to the sentence. An interjection can be strong (*Hey! Ouch! Wow!*) or mild (*oh, well, indeed*). A strong interjection is followed by an <u>exclamation mark</u>.

- *Wow!* Profits more than doubled last quarter!

A weak interjection is followed by a comma.

- *Well,* we need to rethink the proposal.

An interjection inserted into a sentence usually requires a comma before it and after it.

- We must, *indeed,* rethink the proposal.

Because they get their main expressive force from sound, interjections are more common in speech than in writing. Use them sparingly in technical writing.

I

international correspondence

Business <u>correspondence</u> varies among cultures. Organizational patterns, persuasive strategies, forms of courtesy, and ideas about efficiency vary from country to country. For example, in the United States, direct, concise correspondence may demonstrate courtesy by not wasting another person's time; in other cultures (in countries such as Spain and India), such directness and brevity may suggest that the writer dislikes the <u>reader</u> so much that he or she wishes to make the communication as short as possible. Further, where an American writer might consider one brief letter sufficient to communicate a request, a writer in another culture may expect an exchange of three or four longer letters to pave the way for action.

Cultural Differences in Correspondence

When you read correspondence from people in other cultures or countries, be alert to differences in such features as customary expressions, openings, and closings. Japanese business writers, for example, have traditionally used indirect openings that reflect on the season, compliment the reader's success, and offer hopes for the reader's continued prosperity. Consider as well deeper issues, such as how writers from other cultures express bad news. Traditionally, Japanese writers often express negative messages indirectly to avoid embarrassing the recipient. (See also **refusal letters**.) Such differences are often based on cultural perceptions of time, face-saving, and expectations. For more information and resources for cross-cultural study, see **global communication** and **global graphics**.

✺ WEB LINK **GOOGLE'S INTERNATIONAL DIRECTORY**

Google's International Business and Trade Directory provides an excellent starting point for searching the Web for information related to customs, communication, and international standards. See <*www.bedfordstmartins .com/alred*> and select *Links for Technical Writing.*

Cross-Cultural Examples

Figures I–7 and I–8 are two versions of a letter written by an American businessman to a Japanese businessman. The first letter in Figure I–7 does not include enough politeness strategies important to Japanese culture in the opening and closing, including the salutation (*Dear Ichiro*). The letter contains an **affectation** (*limited capacity*) and is filled with slang (*heads up, powers that be*), **idioms** (*fruits of your labor, burning the midnight oil*), **jargon** (*temp*), and inappropriately informal language (*stuff, folks*).

Compare that letter to the one in Figure I–8, which is written in language that is literal and specific. The letter begins with concern about the recipient's family and prosperity because that opening honors Japanese traditional patterns in business correspondence. The letter is free of slang, idioms, and jargon. The sentences are shorter, bulleted **lists** are used to break up the paragraphs, **contractions** are not used, and months are spelled out.

When you are writing for international readers, rethink the ingrained habits that define how you express yourself, learn as much as you can about the cultural expectations of others, and focus on politeness strategies that demonstrate your respect for readers. Doing so will help you achieve clarity and mutual understanding with international readers.

Sun West Corporation, Inc.

2565 North Armadillo
Tucson, AZ 85719
Phone: (602) 555-6677
Fax: (602) 555-6678   sunwest@aol.com

February 27, 2003

Ichiro Katsumi, Investment Director
Toshiba Investment Company
1-29-10 Ichiban-cho   *Abrupt opening: add personal greeting*
Tokyo 105, Japan

Informal

Dear Ichiro: *Avoid contractions*

Slang

I've just received a heads up that you'll be coming to visit us in Tucson next
month. That's great, we've been looking forward to seeing you for some time now,
especially since we heard you're interested in investing in our company because
as you know, cash flow is very important to any company, especially a small one   *Weak
like ours.   strategy*

Slang

I've been asked by the powers that be here to confirm your flight reservations. A
Jargon — temp took the original information, but because of his limited capacity, I need to *Affectation*
confirm it again. You'll be coming in on 3/20/03 on Delta, flight 435 at 2:00 p.m.
U.S. date / And we'll send someone to pick you up at the airport. I'm sure you'll be tired,
format   you'll probably have some computer equipment with you and lots of luggage, so
be sure to tell the skycap to help you and we'll reimburse you for that and any-
Jargon — thing else you spend money on. When you get off the plane, just go to the baggage claim, then get your stuff, then go outside to the limo area and our driver
will be there with a sign with your name on it.   *Informal
slang*

Now, to the important stuff. In all honesty, we are very excited that you are coming to invest in our company. I think this will provide us with a much-needed in-
fusion of funds with which to not only stabilize but spur growth of our little company. Our products are unique and we could never expand without your help   *Informal*
Contraction — since Tucson's a growing place with lots of folks moving here. And, of course,
you'd end up being the recipients of the fruits of your labor, too. So if all works
out well, we should realize immense profits in two years or so. And despite what   *Idioms*
other people around the world say about Americans, we really are hard workers,
especially my boss who heads up the company — nose to the grindstone every
day and burning the midnight oil every night!

FIGURE I–7. Inappropriate International Correspondence (Marked for Revision)

Ichiro Katsumi   2   February 27, 2003

Informal — (Anyway,) I've enclosed a guidebook and map of Tucson and material on our company. If you see anything you'd like to do in town, let me know, and if you have any questions about the company before we see you, just drop us a quick e-mail or fax (I don't think (snail mail) will get to us in time).

Jargon

Have a safe trip, —— *Abrupt: add more goodwill*

Ty Smith

Ty Smith

———————— *Position title needed*

FIGURE I–7. Inappropriate International Correspondence (Marked for Revision) *(continued)*

Sun West Corporation, Inc.

2565 North Armadillo
Tucson, AZ 85719
Phone: (602) 555-6677
Fax: (602) 555-6678   sunwest@aol.com

February 27, 2003

Ichiro Katsumi
Investment Director
Toshiba Investment Company
1-29-10 Ichiban-cho
Tokyo 105, Japan

Dear Mr. Katsumi:

I hope that you and your family are well and prospering in the new year. We at Sun Corporation are very pleased that you will be coming to visit us in Tucson this month. It will be a pleasure to meet you, and we are very gratified and honored that you are interested in investing in our company.

So that we can ensure that your stay will be pleasurable, we have taken care of all of your travel arrangements. You will

- Leave Narita-New Tokyo International Airport on Delta Airlines flight #75 at 5:00 p.m. on March 20, 2003
- Arrive at Los Angeles International Airport at 10:50 a.m. local time and depart for Tucson on Delta flight #186 at 12.05 p.m.
- Arrive at Tucson International Airport at 1:30 p.m. local time on March 20
- Depart Tucson International Airport on Delta flight #123 at 6:45 a.m. on March 27
- Arrive in Salt Lake City, Utah, at 10:40 a.m. and depart at 11:15 a.m. on Delta flight #34 and arrive in Portland, Oregon, at 12:10 p.m. local time
- Depart Portland, Oregon, on Delta flight #254 at 1:05 p.m. and arrive in Tokyo at 3:05 p.m. local time on March 28

If this information is not accurate or if you need additional information about your travel plans or information on Sun West Corporation, please call, fax, or e-mail me directly. That way, we will receive your message in time to make the appropriate changes or additions.

FIGURE I–8. Appropriate International Correspondence

Mr. Ichiro Katsumi   2   February 27, 2003

After you arrive in Tucson, a chauffeur from Skyline Limousines will be waiting for you at Gate 12. He or she will be carrying a card with your name, will help you collect your luggage from the baggage claim area, and will then drive you to the Loews Ventana Canyon Resort. This resort is one of the most prestigious in Tucson, with spectacular desert views, high-quality amenities, and one of the best golf courses in the city. The next day, the chauffeur will be back at the Ventana at 9:00 a.m. to drive you to Sun West Corporation.

We at Sun West Corporation are very excited to meet you and introduce you to all the members of our hard-working and growing company family. After you meet everyone, you will enjoy a catered breakfast in our conference room. At that time, you will receive a schedule of events planned for the remainder of your trip. Events include presentations from the president of the company and from departmental directors on

- The history of Sun West Corporation
- The uniqueness of our products and current success in the marketplace
- Demographic information and benefits of being located in Tucson
- The potential for considerable profits for both our companies with your company's investment

We encourage you to read through the enclosed guidebook and map of Tucson. In addition to events planned at Sun West Corporation, you will find many natural wonders and historical sites to see in Tucson and in Arizona in general. If you see any particular event or place that you would like to visit, please let us know. We will be happy to show you our city and all it has to offer.

Again, we are very honored that you will be visiting us, and we look forward to a successful business relationship between our two companies.

Sincerely,

Ty Smith

Ty Smith
Vice President

Enclosures (2)

FIGURE I–8. Appropriate International Correspondence *(continued)*

Writer's Checklist: Writing International Correspondence

If at all possible, consult with someone from your intended audience's culture, and keep in mind the following. See also **English as a second language**.

- ☑ Avoid humor, irony, and sarcasm; they are easily misunderstood outside their cultural context.

- ☑ Avoid idioms ("it's a slam dunk," "give a heads up"), unusual **figures of speech**, and allusions to events or attitudes particular to American life.

- ☑ Consider whether jargon or technical terminology can be found in abbreviated English-language dictionaries.

- ☑ Write clear and complete sentences: Unusual word order or rambling sentences will frustrate and confuse readers. See **clarity**, **sentence construction**, **sentence fragments**, and **garbled sentences**.

- ☑ Avoid using an overly simplified style that will potentially offend the reader.

- ☑ Avoid using first names and **direct address** too quickly.

- ☑ Write out the date and name of the month to make the entire date immediately clear (January 11, 2003, *not* 1/11/03). See also **numbers**.

- ☑ Specify time zones or refer to international standards, such as Greenwich Mean Time (GMT) or Universal Time Coordinated (UTC).

- ☑ Where possible, use international measurement standards, such as the metric system (18°C, 14 cm, 45 kg, and so on).

- ☑ Proofread your correspondence. See **proofreading**.

Internet

The Internet is composed of public and private computer networks that allow people to communicate globally, find and share information, and offer commercial services. Internet services permit users to communicate with others; exchange documents, data, and software; and connect to computers in different locations. These resources include **e-mail**, discussion groups, chat environments, the World Wide Web, file transfer protocol (FTP), and Telnet. Guidelines for locating and evaluating Internet sources can be found in **Internet research**. See also **Web design** and **writing for the Web**.

WEB LINK ONLINE REFERENCES FOR INTERNET
TERMINOLOGY

The Library of Congress offers a Web glossary that contains links to other glossaries and terminology resources. See <*www.bedfordstmartins.com/alred*> and select *Links for Technical Writing.*

Internet research

DIRECTORY

The <u>Internet</u> provides access to a staggering amount of information, including access to many public and university library catalogs and databases. You can also conduct primary <u>research</u> by participating in discussion groups and newsgroups and by using e-mail to request information from specific audiences.

WEB LINK RESEARCH RESOURCES

For research-related links and resources, see <*www.bedfordstmartins.com/ alred*> and select *Links for Technical Writing, Diana Hacker's Research and Documentation Online,* and *Research Room.*

Using Search Engines and Directories

Search engines and directories enable you to find what you need at your library and on the Internet. Search engines—like Google or AltaVista—are machine-generated databases created by "spiders" or "bots" that travel the Web looking for new or updated pages; search engines typically have databases of billions of pages. Directories—like Yahoo! or Lycos—are human generated, so they index fewer pages but offer directory trees to help organize their content, as shown in Figure I–9. In ad-

Web Directory

Arts Movies, Music, Television, ...	**Home** Consumers, Homeowners, Family, ...	**Regional** Asia, Europe, North America, ...
Business Industries, Finance, Jobs, ...	**Kids and Teens** Computers, Entertainment, School, ...	**Science** Biology, Psychology, Physics, ...
Computers Internet, Hardware, Software, ...	**News** Media, Newspapers, Current Events, ...	**Shopping** Autos, Clothing, Gifts, ...
Games Board, Roleplaying, Video, ...	**Recreation** Food, Outdoors, Travel, ...	**Society** Issues, People, Religion, ...
Health Alternative, Fitness, Medicine, ...	**Reference** Education, Libraries, Maps, ...	**Sports** Basketball, Football, Soccer, ...

World
Deutsch, Español, Français, Italiano, Japanese, Korean, Nederlands, Polska, Svenska ...

FIGURE I-9. Google's Main Subject Directory

dition to the subject directories offered by many search engines, the following directories will help you to conduct selective, scholarly research on the Web:

> Argus Clearinghouse <www.clearinghouse.net>
> Infomine <http://infomine.ucr.edu>
> World Wide Web Virtual Library <www.vlib.org>

Increasingly, search sites are using a hybrid approach. For example, some search engines, such as Northern Light <www.northern light.com>, search not only the Web but also their own database of articles — content that is edited and compiled by staff librarians and not available elsewhere on the Web. Others combine search engines with directories; for example, Google offers both a standard search engine and directory as well as a special contributor-generated directory referred to as an "Open Directory" <http://dmoz.org>.

As comprehensive as search engines and directories may seem, keep in mind that none is complete or objective. Most search sites are incomplete, carrying only a preselected range of content. Many, for example, do not index Adobe Portable Document Format (PDF) files or Usenet Newsgroups and many cannot index databases and other non-HTML-based content. Further, search sites' ranking of the sites they believe will be relevant to you is based on a number of different strategies. Some sites base relevance on how high on the given page your search term appears, on the number of appearances of your term, or on the number of other sites that link to the page. Almost all major search sites now sell high rankings to the highest bidders, so your results may not highlight the pages most relevant to your search.

Your best strategy is to research how your favorite search engines work; nearly all provide detailed methodologies on their help pages. Many search engines also give you more control through the option of conducting an advanced search, which provides you with a number of ways to obtain more selective results. Figure I–10 shows an advanced search conducted on Google for technical writing programs — a search limited to results in English and to sites within the ".edu" domain. This particular advanced search resulted in ten selective hits. Although search engines vary in what and how they search, you can use some basic strategies (see Writer's Checklist: Searching the Web below).

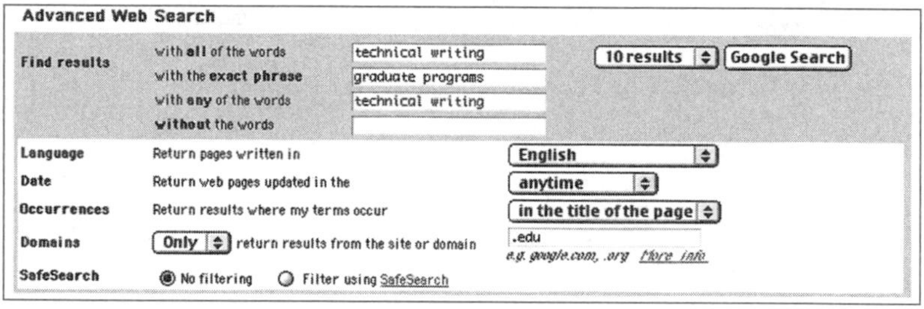

FIGURE I–10. Advanced Google Search

Writer's Checklist: Searching the Web

☑　Enter words and phrases that are as specific to your topic as possible. For example, if you are looking for information about *nuclear power* and enter only the term *nuclear,* the search will also yield listings for *nuclear family, nuclear medicine,* and *nuclear winter.*

☑　Use Boolean operators (AND, OR, NOT) to narrow your search. For example, if you're searching for information on breast cancer and are finding references to nothing but prostate cancer, try "breast AND cancer NOT prostate."

☑　Consider conducting an advanced search (see Figure I-10).

☑　Check any search tips available at the engine you use. For example, some engines allow you to narrow your search by combining phrases with double quotation marks: "usability testing" will only return pages that have the full compound phrase.

☑　Use field specifiers on search engines that support them:

SITE: *www.eserver.org* limits the results to pages on the *eserver* Web site.

LINK: *www.eserver.org* will return only sites that link to *eserver.org.*

URL: *eserver* will return only sites that contain the word *eserver* in their Web addresses.

Writer's Checklist: Searching the Web (continued)

☑   Use a variety of search engines.

☑   If you are interested in obtaining as many hits as possible, consider using a metasearch engine.

The following search engines are used widely on the Web:

AltaVista <www.altavista.com>
Excite <www.excite.com>
Google <www.google.com>
Hotbot <www.hotbot.lycos.com>
Lycos <www.lycos.com>
Northern Light <www.northernlight.com>
WebCrawler <www.webcrawler.com>
Yahoo! <www.yahoo.com>

The Writer's Checklist: Searching the Web suggests using a metasearch engine that searches the Web using multiple search engines at the same time. For example, a metasearch could allow you to search AltaVista, Google, and public and university libraries at the same time. Keep in mind, however, that metasearches result in numerous hits, so be prepared to refine and narrow your results. Dogpile <www.dogpile.com> and Meta crawler <www.metacrawler.com> are two useful metasearch engines.

> **WEB LINK SEARCH-ENGINE RESOURCE**
>
> Search Engine Watch is a site that provides up-to-date information about search engines — classifying, evaluating, and summarizing the current features of each. For this and additional research-related links and resources, see <*www.bedfordstmartins.com/alred*>.

Locating Business and Government Sites

The Web includes numerous sites devoted to specific subject areas. Following are some suggested resources for researching a technical topic.

TECHNICAL RESOURCES

IEEE Spectrum Online <www.spectrum.ieee.org>

National Science Foundation (NSF) <www.nsf.gov>

WWW Virtual Library, engineering <www.eevl.ac.uk/wwwvl.html>

SCIENCE LIBRARIES ON THE WEB

<directory.google.com/Top/Reference/Libraries/Subject_Specific/Science>

GOVERNMENT RESOURCES

Federal Government Agencies Directory <www.lib.lsu.edu/gov /fedgov.html>

FedStats <www.fedstats.gov>

Evaluating Internet Sources

Evaluate the usefulness and reliability of information on the Internet by the same standards that you use to evaluate information from other sources. For Internet sources, be especially concerned about the validity of the information provided.

The easiest way to ensure that information is valid is to obtain it from a reputable source. For example, American businesses rely on and widely use the compilations of data from the Bureau of Labor Statistics, the Securities and Exchange Commission, and the Bureau of the Census. Likewise, the Internet versions of established, reputable journals in medicine, management, engineering, computer software, and the like merit the same level of trust as the printed versions. However, as you move away from established, reputable sites, exercise more caution. Be especially wary of unmoderated discussion groups on Usenet and other public Web forums. Use the following cues for determining the sponsor of an Internet site:

.aero	aerospace industry	.mil	U.S. military
.biz	business	.museum	museum
.com	company or individual	.name	individual
.coop	business cooperative	.net	network provider
.edu	college or university	.org	nonprofit organization
.gov	federal government	.pro	professional, such as a
.int	international organization		doctor or lawyer

In addition, keep in mind the following four criteria when evaluating Internet sources: authority, accuracy, bias, and currency.*

Authority.   Because anyone can publish on the Web, it is sometimes difficult to determine authorship of a document, and frequently a person's qualifications for speaking on a topic are absent or questionable. If you do not recognize the author as well known and respected in the field, here are some possible ways you can determine authority:

- The author's document was listed or linked from a reliable source or document.

*From Ryan, Leigh. *The Bedford Guide for Writing Tutors.* 3rd ed. Boston: Bedford/St. Martin's, 2002.

- The document gives substantive biographical information about the author so you can evaluate his or her credentials, or you can get this information by linking to another document.
- The author is referenced or mentioned positively by another author or organization whose authority you trust.

If the publisher or sponsor is an organization, you may generally assume that the document meets the standards and aims of the group. Consider also

- the suitability of the organization to address this topic
- whether this organization or agency is recognized and respected in the field
- the relationship of the author to the publisher or sponsor   does the document tell you something about the author's expertise or qualifications?

Accuracy.   Criteria for evaluating accuracy might include the following:

- Other sources that the document relies on are linked or are included in a bibliography.
- Background information can be verified.
- Methodology is appropriate for the topic.
- With a research project, data that was gathered includes explanations of research methods and interpretations.
- The graphs and visuals are free of distortion.
- The site is modified or updated regularly.

Bias.   To determine bias, consider how the context reveals the author's knowledge of the subject and his or her stance on the topic. Check the site for the following:

- The site identifies in some form the audience it targets.
- The site was developed by a recognized academic institution; government agency; or national, international, or commercial organization with an established reputation in the subject area.
- The author shows knowledge of theories, techniques, or schools of thought usually related to the topic.
- The author shows knowledge of related sources and attributes them properly.
- The author discusses the value and limitations of the approach, if it is new.

- The author acknowledges that the subject itself or his or her treatment of it is controversial, if you know that to be the case.

Currency. If currency is important, consider whether the document has a publication or "last updated" date or includes date of copyright, gives dates showing when information was gathered, or gives information about new material when appropriate.

See also Writer's Checklist: Evaluating Library Resources. For information about citing Internet sources, see **documenting sources**.

WEB LINK EVALUATING ONLINE SOURCES

Purdue University, UC Berkeley, and Widener University offer resources for evaluating content on the Web. For these and additional research-related links and resources, see *<www.bedfordstmartins.com/alred>*.

interviewing for information

Interviewing others is often a crucial part of your **research**. The process of interviewing includes determining the proper person to interview, preparing for the interview, conducting the interview, and expanding your notes immediately after the interview.

Determining Who to Interview

Many times, your subject or **purpose** logically points to the proper person to interview for information. If you were writing a proposal to a government agency, you might want to interview someone who has experience both in writing government proposals and with the agency involved. The following sources can help you determine the appropriate person to interview: (1) workplace colleagues or faculty in appropriate academic departments, (2) information from **Internet research**, (3) local chapters of professional societies, and (4) yellow or business pages of the local telephone directory.

Preparing for the Interview

Before the interview, learn as much as possible about the person you are going to interview and the organization for which he or she works. When you contact the prospective interviewee, explain who you are, why you would like to interview him or her, the subject and purpose of the interview, and how much time it will take, and let your interviewee know that you will allow him or her to review your draft.

After you have made the appointment, prepare a list of questions to ask your interviewee. Avoid vague, general questions. A question such as "What do you think of the Internet?" is too general to elicit useful information. It is more helpful to ask specific but open-ended questions; for example, "Some local professionals in your field are making extensive use of the Internet for helping clients. How do you use the Internet? How has it helped your organization?" Such questions prompt interviewees to provide specific information. The information gained from an interview is also useful for preparing a **questionnaire**.

Writer's Checklist: Conducting an Interview

☑ Arrive promptly for the interview and set the interviewee at ease.

☑ Be pleasant but purposeful. Do not be timid about asking leading questions on the subject. Use your prepared list of questions as your guide: Begin with the least complex aspects of the topic, then move to the more complex aspects.

☑ Don't get sidetracked. If the interviewee strays too far from the subject, ask a specific question to direct the conversation back on track.

☑ Avoid being rigid; if a prepared question is no longer suitable, move to the next question.

☑ Some answers prompt additional questions; ask them as they arise.

☑ Let your interviewee do most of the talking. Remember that the interviewee is the expert. See also **listening**.

☑ Take only memory-jogging notes that will help you recall the conversation later. Concentrate on key facts and figures.

☑ Use a tape recorder if both you and your interviewee are comfortable with it.

☑ As the interview is reaching a close, take a few minutes to skim your notes. If time allows, ask the interviewee to clarify anything that is ambiguous.

☑ After thanking the interviewee, ask permission to telephone to clarify a point or two as you complete your interview notes.

☑ Immediately after leaving the interview, expand your memory-jogging notes to help you mentally review the interview. Do not postpone this step. See also **note-taking**.

☑ A day or two following the interview, send the interviewee a note of thanks in a brief letter or e-mail.

interviewing for a job

A job interview may last thirty minutes, an hour, several hours, or more. Sometimes, an initial interview is followed by a series of job interviews that can last a half or full day. Often, just one person or a few people conduct a job interview, but, at times, a group of four or more might do so. Job interviews can take place in person, by phone, or by teleconference. Because it is impossible to know exactly what to expect, it is important that you be well prepared. See also **application letters**, **job searches**, and **résumés**.

Before the Interview

The interview is not a one-way communication. It presents you with an opportunity to ask questions of your potential employer. In preparation, learn everything you can about the company before the interview. Use the following questions as a guide.

- What kind of organization is it (e.g., nonprofit, government)?
- How diversified is the organization?
- Is it locally owned?
- Does it provide a product or service? If so, what kind?
- How large is the business? How large are its assets?
- Is the owner self-employed? Is the company a subsidiary of a larger operation? Is it expanding?
- How long has it been in business?
- Where will you fit in?

You can obtain information from current employees, the Internet, company publications, and the business section of back issues of local newspapers (available in the library or online). You may be able to learn the company's size, sales volume, product line, credit rating, branch locations, subsidiary companies, new products and services, building programs, and other such information from its annual reports; publications such as *Moody's Industrials, Dun and Bradstreet, Standard and Poor's,* and *Thomas' Register;* and other business reference sources a librarian might suggest. Ask your interviewer about what you cannot find through your own research. Now is your chance to demonstrate your interest and make certain that you are considering a healthy and growing company.

Try to anticipate the questions your interviewer might ask, and prepare your answers in advance. Be sure you understand a question before answering it, and avoid responding too quickly with a canned answer—be prepared to answer in a natural and relaxed manner. Interviewers typically ask the following questions:

- What are your short-term and long-term occupational goals?
- Where do you see yourself five years from now?
- What are your major strengths and weaknesses?
- Do you work better with others or alone?
- How do you spend your free time?
- What are your personal goals?
- Describe an accomplishment you are particularly proud of.
- Why are you leaving your current job?
- Why do you want to work for this company?
- Why should I hire you?
- What salary and benefits do you expect?

Many employers use behavioral interviews. Rather than traditional, straightforward questions, the behavioral interview focuses on asking the candidate to provide examples or respond to hypothetical situations. Interviewers who use behavioral-based questions are looking for specific examples from your experience. Prepare for the behavioral interview by recollecting challenging situations or problems that were successfully resolved. Examples of behavior-based questions include the following:

- Tell me about a time when you experienced conflict on a team.
- If I were your boss and you disagreed with a decision I made, what would you do?
- How have you used your leadership skills to bring about change?
- Tell me about a time when you failed.

Arrive for your interview on time, or even ten or fifteen minutes early—you may be asked to fill out an application or other paperwork before you meet your interviewer. Always bring extra copies of your résumé and samples of your work (if applicable). If you are asked to complete an application form, read it carefully before you write and proofread it when you are finished. The form provides a written record for company files and indicates to the company how well you follow directions and complete a task.

During the Interview

The interview actually begins before you are seated: What you wear and how you act make a first impression. In general, dress simply and conservatively and avoid extremes in fragrance and cosmetics. Be well-groomed.

Behavior.   First, thank the interviewer for his or her time, express your pleasure at meeting him or her, and remain standing until you are offered a seat. Then sit up straight (good posture suggests self-assurance), look directly at the interviewer, and try to appear relaxed and confident. During the interview, you may find yourself feeling a little nervous. Use that nervous energy to your advantage by channeling it into the alertness that you will need to listen and respond effectively. Do not attempt to take extensive notes during the interview, although it is acceptable to jot down a few facts and figures. See also listening.

Responses.   When you answer questions, do not ramble or stray from the subject. Say only what you must to answer each question properly and then stop, but avoid giving just yes or no answers — they usually don't allow the interviewer to learn enough about you. Some interviewers allow a silence to fall just to see how you will react. The burden of conducting the interview is the interviewer's, not yours — and he or she may interpret your rush to fill a void in the conversation as a sign of insecurity. If such a silence makes you uncomfortable, be ready to ask an intelligent question about the company.

If the interviewer overlooks important points, bring them up. However, let the interviewer mention salary first, if possible. Doing so yourself may indicate that you are more interested in the money than the work. However, make sure that you are aware of prevailing salaries and benefits in your field, and review the entry salary negotiations.

Interviewers look for a degree of self-confidence and an applicant's understanding of the field, as well as genuine interest in the field, the company, and the job. Ask questions to communicate your interest in the job and company. Interviewers respond favorably to applicants who can communicate and present themselves well.

Conclusion.   At the conclusion of the interview, thank the interviewer for his or her time. Indicate that you are interested in the job (if true), and try to get an idea of the company's hiring timeline. Reaffirm friendly contact with a firm handshake.

After the Interview

After you leave the interview, jot down the pertinent information you obtained, as it may be helpful in comparing job offers. A day or two following a job interview, send the interviewer a note of thanks in a brief letter or e-mail. Such notes often include the following:

- Your thanks for the interview and to individuals or groups that gave you special help or attention during the interview
- The name of the specific job for which you interviewed

- Your impression that the job is attractive
- Your confidence that you can perform the job well
- An offer to provide further information or answer further questions

Figure I–11 shows a typical follow-up letter.

Dear Mr. Vallone:

Thank you for the informative and pleasant interview we had last Wednesday. Please extend my thanks to Mr. Wilson of the Media Group as well.

I came away from our meeting most favorably impressed with Calcutex Industries. I find the position of junior ACR designer to be an attractive one and feel confident that my qualifications would enable me to perform the duties to everyone's advantage.

If I can answer any further questions, please let me know.

Sincerely yours,

FIGURE I–11. Follow-up Letter

If you are offered a job you want, accept the offer verbally and write a brief letter of acceptance as soon as possible — certainly within a week — or if you do not want the job, you will need to write a refusal letter, as described in <u>acceptance/refusal letters</u>.

introductions

DIRECTORY

Introductions are used for short and routine types of <u>correspondence</u>, such as letters and <u>e-mail</u>, and large writing projects, such as <u>formal reports</u> and major <u>proposals</u>. See also <u>conclusions</u>.

Routine Openings

Not every document needs a fully developed introduction or opening. When your <u>readers</u> are already familiar with your subject, or if what you are writing is short, a brief or routine opening, as shown in the following examples, is adequate.

CORRESPONDENCE
Dear Mr. Ignatowski:
You will be happy to know that we have corrected the error in your bank balance. The new balance shows . . .

PROGRESS REPORT LETTER
Dear Dr. Chang:
To date, 18 of the 20 specimens you submitted for analysis have been examined. Our preliminary analysis indicates . . .

LONGER PROGRESS REPORT
PROGRESS REPORT ON REWIRING THE SPORTS ARENA
The rewiring program at the Sports Arena is proceeding ahead of schedule. Although the costs of certain equipment are higher than our original bid, we expect to complete the project without exceeding our budget because the speedy completion will save labor costs.
Work Completed
As of August 15, we have . . .

E-MAIL
Jane, as I promised in my earlier e-mail, I've attached the personnel budget estimates for fiscal year 2003.

Opening Strategies

Opening strategies are aimed at focusing the readers' attention and motivating them to read the entire document.

Objective.   In reporting on a project, you might open with a statement of the project's objective to give the readers a basis for judging the results.

- The primary goal of this project was to develop new techniques to solve the problem of waste disposal. Our first step was to investigate . . .

Problem Statement.   One way to give readers the perspective of your report is to present a brief account of the problem that led to the study or project being reported.

- Several weeks ago a manager noticed a recurring problem in the software developed by Datacom Systems. Specifically, error messages repeatedly appeared when, in fact, no specific trouble. . . . After an extensive investigation, we found that Datacom Systems . . .

Of course, for **proposals** or **formal reports,** problem statements may be more elaborate and a part of the full-scale introduction discussed later in this entry.

Scope.   You may want to present the scope of your document in your opening. By providing the parameters of your material, the limitations of the subject, or the amount of detail to be presented, you enable your readers to determine whether they want to or need to read your document.

- This pamphlet provides a review of the requirements for obtaining an FAA pilot's license. It is not intended as a textbook to prepare you for the examination itself; rather, it outlines the steps you need to take and the costs involved.

Background.   The background or history of a subject may be interesting and lend perspective and insight to a subject. Consider the following example from a newsletter describing the process of oil drilling:

- From the bamboo poles the Chinese used when the pyramids were young to today's giant rigs drilling in a hundred feet of water, there has been considerable progress in the search for oil. But whether in ancient China or a modern city, under water or on a mountaintop, the object of drilling has always been the same — to manufacture a hole in the ground, inch by inch.

Summary.   You can provide a summary opening by describing in abbreviated form the results, conclusions, or recommendations of your article or report. Be concise: Do not begin a summary by writing "This report summarizes . . .".

CHANGE   This report summarizes the advantages offered by the photon as a means of examining the structural features of the atom.

TO   As a means of examining the structure of the atom, the photon offers several advantages.

Interesting Detail.   Often an interesting detail will gain the readers' attention and arouse their curiosity. Readers of an annual report for a manufacturer of telescopes and scientific instruments, for example, may be persuaded to invest if they believe that the company is developing innovative, cutting-edge products.

- The rings of Saturn have puzzled astronomers ever since they were discovered by Galileo in 1610 using the first telescope. Recently, even more rings have been discovered. . . .
  Our company's Scientific Instrument Division designs and manufactures research-quality, computer-controlled telescopes that promise to solve the puzzles of Saturn's rings by enabling scientists to use multicolor differential photometry to determine the rings' origins and compositions.

Definition.   Although a definition can be useful as an opening, do not define something with which the reader is familiar or provide a definition that is obviously a contrived opening (such as "Webster defines *technology* as . . ."). A definition should be used as an opening only if it offers insight into what follows.

- *Risk* is often a loosely defined term. For the purposes of this report, risk refers to a qualitative combination of the probability of an event and the severity of the consequences of that event.

Anecdote.   An anecdote can be used to attract and build interest in a subject that may otherwise be mundane; however, this strategy is best suited to longer documents and presentations.

- In his poem "The Calf Path," Sam Walter Foss tells of a wandering, wobbly calf trying to find its way home at night through the lonesome woods. It made a crooked path, which was taken up the next day by a lone dog. Then "a bellwether sheep pursued the trail over vale and steep, drawing behind him the flock, too, as all good bellwethers do." At last the path became a country road; then a lane that bent and turned and turned again. The lane became a village street, and at last the main street of a flourishing city. The poet ends by saying, "A hundred thousand men were led by a calf, three centuries dead."
  Many companies today follow a "calf path" because they react to events rather than planning . . .

Quotation.   Occasionally, you can use a quotation to stimulate interest in your subject. To be effective, however, the quotation must be pertinent—not some loosely related remark selected from a book of quotations.

- According to Deborah Andrews, "technical communicators in the twenty-first century must reach audiences and collaborate across borders of culture, language, and technology." One way of accomplishing that goal is to make sure our training includes cross-cultural experiences that provide . . .

Forecast.   Sometimes you can use a forecast of a new development or trend to arouse the reader's interest.

- In the not-too-distant future, we may be able to use a hand-held medical diagnostic device similar to those in science fiction to assess the complete physical condition of accident victims. This project and others are now being developed at The Seldi Group, Inc.

Persuasive Hook.   While all opening strategies contain persuasive elements, the hook is the most overtly persuasive. A brochure touting the newest innovation in tax-preparation software might address readers as follows:

- Welcome to the newest way to do your taxes! TaxPro EZ ends the headache of last-minute tax preparation with its unique Web-Link feature.

Full-Scale Introductions

The purpose of a full-scale introduction is to give readers enough general information about the subject to enable them to understand the details in the body of the document. An introduction should accomplish the following:

- *State the subject.* Provide background information, such as definition, history, or theory to provide context for your readers.
- *State the purpose.* Make your readers aware of why the document exists and whether the material provides a new perspective or clarifies an existing perspective.
- *State the scope.* Tell readers the amount of detail you plan to cover.
- *Preview the development of the subject.* Especially in a longer document, outline how you plan to develop the subject. Providing such information allows readers to anticipate how the subject will be presented and helps them evaluate your conclusions or recommendations.

Consider writing the introduction last. Many writers find that it is only when they have drafted the body of the document that they have a full enough perspective on the subject to introduce it adequately.

Technical Manuals and Specifications

You may find that you need to write one kind of introduction for re-ports, academic papers, or trade journal articles and a different kind for technical manuals or specifications. When writing an introduction for a technical manual or set of specifications, identify the topic and its primary purpose or function in the first sentence or two. Be specific, but do not go into elaborate detail. Your introduction sets the stage for the entire document, and it should provide readers with a broad frame of reference and an understanding of the overall topic. Then the reader is ready for technical details in the body of the document.

How technical your introduction should be depends on your read-ers: What are their technical backgrounds? What kind of information are they seeking in the manual or specification? The topic should be in-troduced with a specific audience in mind—a computer user, for ex-ample, has different interests in an application program and a different technical vocabulary than a programmer. Whether you need to provide explanations or definitions of terminology will depend on your in-tended audience. The following example is written for readers who un-derstand such terms as "constructor" and "software modules."

- The System Constructor is a program that can be used to create operating systems for a specific range of microcomputer systems. The constructor selects requested operating software modules from an existing file of software modules and combines those modules with a previously compiled application program to create a functional operating system designed for a specific hardware configuration. It selects the requested software modules, estab-lishes the necessary linkage between the modules, and generates the control tables for the system according to parameters specified at run time.

You may encounter a dilemma that is common in technical writ-ing: Although you cannot explain topic A until you have explained topic B, you cannot explain topic B before explaining topic A. The solution is to explain both topics in broad, general terms in the introduction. Then, when you need to write a detailed explanation of topic A, you will be able to do so because your reader will know just enough about both topics to be able to understand your detailed explanation.

- The NEAT/3 programming language, which treats all peripheral units as file storage units, allows your program to perform data input or output operations depending on the specific unit. Periph-eral units from which your program can only input data are re-ferred to as *source units;* those to which your program can only out-put data are referred to as *destination units.*

investigative reports

Investigative **reports** may be written for a variety of reasons—most often in response to a request for information. You might be asked, for instance, to research the Web sites of competing companies in your industry or to conduct an opinion survey among your customers. An investigative report gives a precise analysis of a topic and offers conclusions and recommendations.

Open the report with a statement of its primary and (if any) secondary **purposes**, then define the **scope** of your investigation. If the report is on a survey of opinions, indicate the number of people surveyed, income categories, occupations, and other identifying information. (See also **questionnaires**.) Include any information that is pertinent in defining the extent of the investigation. Then report your findings and, if necessary, discuss their significance. End the report with your conclusions and any recommendations. An example of an investigative report is shown in Figure I–12 on page 298.

irregardless / regardless (see *regardless*)

italics

Italics is a style of type used to denote emphasis and to distinguish foreign expressions, book titles, and certain other elements. Italic type is signaled by underlining in manuscripts submitted for publication or where italic font is not available (see also **e-mail**). *This sentence is printed in italics.* You may need to italicize words that require special emphasis in a sentence. (Contrary to projections, sales have *not* improved.) Do not overuse italics for emphasis, however. (This will hurt *you* more than *me.*)

Foreign Words and Phrases

Foreign words and phrases that have not been assimilated into the English language are italicized (*sine qua non, coup de grâce, in res, in camera*). Foreign words that have been fully assimilated into the language need not be italicized, although they often retain their **diacritical marks**. A word may be considered assimilated if it appears in most standard dictionaries and is familiar to most readers (cliché, etiquette, vis-à-vis, de facto, siesta). See also **foreign words in English**.

Memo

To: Noreen Rinaldo, Training Manager
From: Charles Lapinski, Senior Instructor   *CL*
Date: February 14, 2003
Subject: Adler's Basic English Program

As requested, I have investigated Adler Medical Instruments' (AMI's) Basic English Program to determine whether we might adopt a similar program.

The purpose of AMI's program is to teach medical technologists outside the United States who do not speak or read English to understand procedures written in a special 800-word vocabulary called *Basic English*. This program eliminates the need for AMI to translate its documentation into a number of different languages. The Basic English Program does not attempt to teach the medical technologists to be fluent in English but, rather, to recognize the 800 basic words that appear in Adler's documentation.

Course Requirements
The course does not train technologists. Students must know, in their own language, what a word like *hemostat* means; the course simply teaches them the English term for it. As prerequisites, students must have basic knowledge of their specialty, must be able to identify a part in an illustrated parts book, must have used AMI products for at least one year, and must be able to read and write in their own language.

Students are given an instruction manual, an illustrated book of equipment with parts and their English names, and pocket references containing the 800 words of the Basic English vocabulary plus the English names of parts. Students can write the corresponding word in their language beside the English word and then use the pocket reference as a bilingual dictionary. The course consists of 30 two-hour lessons, each lesson introducing approximately 27 words. No effort is made to teach pronunciation; the course teaches only recognition of the 800 words, which include 450 nouns, 70 verbs, 180 adjectives and adverbs, and 100 articles, prepositions, conjunctions, and pronouns.

Course Outcomes
The 800-word vocabulary enables the writers of documentation to provide medical technologists with any information that might be required because the subject areas are strictly limited to usage, troubleshooting, safety, and operation of AMI medical equipment. All nonessential words (such as *apple, father, mountain,* and so on) have been eliminated, as have most synonyms (for example, *under* appears, but *beneath* does not).

Conclusions
I see two possible ways in which we could use some or all of the elements of AMI's Program: (1) in the preparation of our student manuals or (2) as AMI uses the program.

I think it would be unnecessary to use the Basic English methods in the preparation of manuals for *all* of our students. Most of our students are English speakers to whom an unrestricted vocabulary presents no problem.

As for our initiating a program similar to AMI's, we could create our own version of the Basic English vocabulary and write our instructional materials in it. Because our product lines are much broader than AMI's, however, we would need to create illustrated parts books for each of the different product lines.

FIGURE I–12. Investigative Report

Titles

Italicize the <u>titles</u> of separately published documents, such as books, periodicals, newspapers, pamphlets, brochures, legal cases, movies, and television programs.

- *Turning Workplace Conflicts into Collaboration* by Joyce Richards was reviewed in the *New York Times*.

Abbreviations of such titles are italicized if their spelled-out forms would be italicized.

- The *NYT* is one of the nation's oldest newspapers.

Italicize the titles of compact discs, videotapes, plays, long poems, paintings, sculptures, and musical works.

CD-ROM	*Computer Security Tutorial on CD-ROM*
PLAY	Arthur Miller's *Death of a Salesman*
LONG POEM	T. S. Eliot's *The Wasteland*
MUSICAL WORK	Gershwin's *Porgy and Bess*

Use <u>quotation marks</u> for parts of publications, such as book chapters and articles in periodicals.

- "Clarity and Conciseness: The Writer's Tightrope" was an article in *Intercom*.

Titles of reports, short poems, musical works, and songs are also enclosed in quotation marks.

REPORT	"Analysis of Alternative Technology Options in the Commercial and Automotive Sectors"
SHORT POEM	Yusef Komunyakaa's "Elegy for Thelonious"
SONG	Bob Dylan's "Like a Rolling Stone"

Exceptions are titles of holy books and legislative documents, which are not italicized or placed in quotation marks (Old Testament and Magna Carta).

Proper Names

The names of ships, trains, and aircraft (but not the companies or governments that own them) are italicized (Dutch Submarine *Walrus (2)*; U.S. Space Shuttle *Endeavour*). Craft that are known by model or serial designations are exceptions and are not italicized (DC-7; Boeing 747).

Words, Letters, and Figures

Words, letters, and figures discussed as such are italicized.

- The word *inflammable* is often misinterpreted.
- I need a new keyboard because the *S* and *6* keys on my old one do not function.

Subheads

Subheads in a report are sometimes italicized.

- *Training Writers.* We are certainly leading the way in developing first-line managers who not only are professionally competent but . . .

See also <u>headings</u> and <u>layout and design</u>.

its / it's

Its is the possessive case form of *it*; *it's* is a contraction of *it is*. (*It's* important that the sales department meet *its* quota.) Although nouns normally form the possessive by the addition of an apostrophe and an *s*, the contraction of *it is* (*it's*) already uses that device. Therefore, the possessive form of the pronoun *it* is formed by adding only the *s*.

J

jargon

Jargon is a highly specialized slang that is unique to an occupational or professional group. Jargon is at first understood only by insiders; later, it may become known more widely. For example, computer programmers coined the term *debugging* to describe the discovery and correction of errors in software. If all your readers are members of a particular occupational group, jargon may provide an efficient means of communicating. However, if you have any doubt that your entire reading audience is part of such a group, avoid using jargon. See also <u>affectation</u> and <u>gobbledygook</u>.

job descriptions

Most large companies and many small ones use formal *job descriptions* to specify the duties of and requirements for many of the jobs in the firm. Job descriptions fulfill several important functions: They provide information on which equitable salary scales can be based; they help management determine whether all functions within a company are adequately supported; and they let both prospective and current employees know exactly what is expected of them. Together, all the job descriptions in a firm present a picture of the organization's structure.

Sometimes middle managers are given the task of writing job descriptions for their employees. In many organizations, though, employees are required to draft their own job descriptions, which supervisors then check and approve.

Although job-description formats vary from organization to organization, they commonly contain the following sections:

- The *accountability* section identifies, by title only, the person to whom the employee reports.
- The *scope of responsibilities* section provides an overview of the primary and secondary functions of the job and states, if applicable, who reports to the employee.

- The *specific duties* section gives a detailed account of the specific duties of the job as concisely as possible.
- The *personal requirements* section lists the required or preferred education, training, experience, and licensing for the job.

Writer's Checklist: Writing Job Descriptions

☑ Before attempting to write your job description, make a list of all the different tasks you do in a week or a month. Otherwise, you will almost certainly leave out some of your duties.

☑ Focus on content. Remember that you are describing your job, not yourself.

☑ List your duties in decreasing order of importance. Knowing how your various duties rank in importance makes it easier to set valid job qualifications.

☑ Begin each statement of a duty with a <u>verb</u> and be specific. Write "Greet and assist customers" rather than "Take care of customers."

☑ Review existing job descriptions that are considered well written.

The job description shown in Figure J–1 is typical. It never mentions the person holding the job described; it focuses, instead, on the job and the qualifications required to fill the position.

job search

Whether you are trying to land your first job or you want to change careers entirely, begin by assessing your skills, interests, and abilities, perhaps through <u>brainstorming</u>.* Next, ask yourself what are your career goals and values. For instance, do you prefer working independently or collaboratively? Do you enjoy public settings? meeting people? How important are career stability and location? Finally, ask yourself what you would most like to be doing in the immediate future, in two years, and in five years.

Once you have reflected and brainstormed about the job that is right for you, a number of sources can help you locate the job you want. Of course, you should not rely on any one of these sources exclusively:

*A good source for stimulating your thinking is the most recent edition of *What Color Is Your Parachute? A Practical Manual for Job-Hunters & Career-Changers* by Richard Nelson Bolles, published by Ten Speed Press.

Manager, Technical Publications
Dakota Electrical Corporation

Accountability

Reports directly to the Vice President, Customer Service.

Scope of Responsibilities

The Manager of Technical Publications plans, coordinates, and supervises the design and development of technical publications and documentation required in the support of the sale, installation, and maintenance of Dakota products. The manager is responsible for the administration and morale of the staff. The supervisor for instruction manuals and the supervisor for parts manuals report directly to the manager.

Specific Duties

Directs an organization presently composed of 20 people (including two supervisors), over 75 percent of whom are writing professionals and graphic artists.

Screens, selects, and hires qualified applicants for the department.

Prepares a formal program designed to orient writing trainees to the production of reproducible copy and graphic arts.

Evaluates the performance of departmental members and determines salary adjustments for all personnel in the department.

Plans documentation to support new and existing products.

Determines the need for subcontracted publications and acts as a purchasing agent when needed.

Offers editorial advice to supervisors.

Develops and manages an annual budget for the Technical Publications Department.

Cooperates with the Engineering, Parts, and Service Departments to provide the necessary repair and spare parts manuals upon the introduction of new equipment.

Serves as a liaison between technical specialists, the publications staff, and field engineers.

Recommends new and appropriate uses for the department within the company.

Keeps up with new technologies in printing, typesetting, art, and graphics and uses them to the advantage of Dakota Electrical Corporation where applicable.

Requirements

B.A. in technical communication desired.

Minimum of three years' professional writing experience and a general knowledge of graphics, production, and Web design.

Minimum of two years' management experience with a knowledge of the general principles of management.

Strong interpersonal skills.

FIGURE J–1. Job Description

- Networking
- Campus career services
- Internet resources
- Advertisements
- Trade and professional journal listings
- Private (or temporary) employment agencies
- Letters of inquiry

Keep a file during your job search of dated job ads, copies of <u>application letters</u> and <u>résumés,</u> notes requesting interviews, and the names of important contacts. This collection can serve as a future resource and reminder. See also <u>interviewing for a job</u> and <u>salary negotiations.</u>

Networking

Networking involves communicating with people who might provide useful advice or might be able to connect you with potential jobs in your interest areas. They may include people already working in your chosen field, contacts in professional organizations, professors, family members, or friends. Use your contacts to develop a network of even more contacts. Keep in mind that of all open positions, an estimated 80 percent are filled through networking.*

Campus Career Services

A visit to a college career-development center is another good way to begin your job search. Government, business, and industry recruiters often visit campus career offices to interview prospective employees; recruiters also keep career counselors aware of their companies' current employment needs and submit <u>job descriptions.</u> Not only can career counselors help you select a career, they can also put you in touch with the best, most current resources—identifying where to begin your search and saving you time. Career development centers often hold workshops on résumé preparation and offer other job-finding resources on their Web sites.

Internet Resources

Using the Web can enhance your job search in a number of ways. First, you can consult sites that give advice to college graduates about careers, job seeking, and résumé preparation. Second, you can learn about businesses and organizations that may hire employees in your

*From *JobStar Central,* an online job-search guide hosted by the *Wall Street Journal* at <http://jobstar.org>.

area by visiting their Web sites. Such sites often list job openings and provide instructions for applicants and offer other information, such as employee benefits. Third, you can learn about jobs in your field and post your résumé for prospective employers at employment databases, such as Monster.com or America's Job Bank. (See the Web Link box "Finding a Job," page 306.) Fourth, you can post your résumé at your personal Web site. Although posting your résumé at an employment database will undoubtedly attract more potential employers, including your résumé at your own site has benefits. For example, you might provide a link to your site in e-mail correspondence or provide your Web site's URL in an inquiry letter to a prospective employer.

Employment specialists suggest that you spend time on the Web in the evening or early morning so that you can focus on in-person contacts during working hours. See also Internet research.

Advertisements

Many employers advertise in the classified sections of newspapers and on their own Web sites. For the widest selection of help-wanted listings, look in the Sunday editions or the help-wanted Web pages of local and big-city newspapers. An item-by-item check is necessary because many times a position can be listed under various classifications. A clinical medical technologist seeking a job, for example, might find the specialty listed under "Medical Technologist" or "Clinical Laboratory Technologist." Depending on a hospital's or a pathologist's needs, the listing could be even more specific, such as "Blood Bank Technologist" or "Hematology Technologist." As you read the ads, take notes on salary ranges, job locations, job duties and responsibilities, and even the terminology used in the ads to describe the work. A knowledge of keywords and expressions that are generally used to describe a particular type of work can be helpful when you prepare your résumé and letters of application.

Trade and Professional Journal Listings

In many industries, associations publish periodicals of interest to people working in the industry. Such periodicals (print and online) often contain job listings. To learn about the trade or professional associations for your occupation, consult resources on the Web, such as Google's Directory of Professional Organizations <http://directory .google.com/Top/Society/Organizations/Professional> or online resources offered by your library or campus career office. You may also consult the following references at a library: *Encyclopedia of Associations, Encyclopedia of Business Information Sources,* and *National Directory of Employment Services.* See also library research.

Private (or Temporary) Employment Agencies

Private employment agencies are profit-making organizations that are in business to help people find jobs—for a fee. Reputable agencies provide you with job leads, help you organize your job search, and supply information on companies doing the hiring. A staffing agency, or temporary placement agency, could match you with an appropriate temporary or permanent job in your field. Temporary work for an organization for which you might want to work permanently is an excellent way to build your network while continuing your job search.

Choose an employment or temporary placement agency carefully. Some are well established and reputable; others are not. Check with your local Better Business Bureau <www.bbb.org> and your college career office before you sign an agreement with a private employment agency. Further, be sure you understand who is paying the agency's fee. Often the employer pays the agency's fee; however, if you have to pay, make sure you know exactly how much. As with any written agreement, read the fine print carefully.

Letters of Inquiry

If you would like to work for a particular firm, write and ask whether it has any openings for people with your qualifications. Normally, you can send the letter to the department head, the director of human resources, or both; for a small firm, however, write to the head of the firm. For more information and examples of inquiry letters, see **application letters**.

Other Sources

Local, state, and federal government agencies offer many employment services. Local government agencies are listed in telephone and Web directories under the name of your city, county, or state.

WEB LINK FINDING A JOB

For job-hunting sites, tips, sample documents, and more, see <*www.bedfordstmartins.com/alred*> and select *Links for Technical Writing* and *Finding an Internship or Job*.

---

journal articles (*see* **trade journal articles**)

judicial / judicious

Judicial is a term that pertains only to law. (The *judicial* branch is one of the three branches of the U.S. government.) *Judicious* refers to careful or wise judgment. (The reorganization was accomplished smoothly by *judicious* management.)

J

K

kind of / sort of

In writing, *kind of* and *sort of* should be used only to refer to a class or type of things.

- They used a special *kind of* metal in the process.

Do not use *kind of* or *sort of* to mean "rather," "somewhat," or "somehow." That usage can lead to vagueness; it is better to be specific.

VAGUE   It was *kind of* a bad year for the company.

SPECIFIC   The company's profits fell 10 percent last year.

know-how

An informal term for "special competence or knowledge," *know-how* should be avoided in formal writing <u>style</u>.

- He has a reputation for technical ~~know-how.~~
  skill.

L

laboratory reports

A laboratory report should begin by stating the reason that a laboratory investigation was conducted; it should also list the equipment and procedures used during the test, the problems encountered, the conclusions reached, and the recommendations made. Test reports are similar but briefer and less formal.

A laboratory report often emphasizes the equipment and procedures used in the investigation because those two factors can be critical in determining the accuracy of the data. Present the results of the laboratory investigation clearly and precisely. If your report requires graphs or tables, integrate them into your report as described in the entry visuals. The example in Figure L–1 on page 310 shows a typical laboratory report.

lay / lie

Lay is a transitive verb—a verb that requires a direct object to complete its meaning—that means "place" or "put."

- We will *lay* the foundation of the building one section at a time.

The past tense form of *lay* is *laid*.

- We *laid* the first section of the foundation on the 27th of June.

The perfect tense form of *lay* is also *laid*.

- Since June, we *have laid* all but two sections of the foundation.

Lay is frequently confused with *lie*, which is an intransitive verb—a verb that does not require an object to complete its meaning—that means "recline" or "remain."

- Injured employees should *lie* down and remain still until the doctor arrives.

The past tense form of *lie* is *lay*. This form causes the confusion between *lie* and *lay*.

- The injured employee *lay* still for approximately five minutes.

The perfect tense form of *lie* is *lain*.

- The injured employee *had lain* still for approximately five minutes before the doctor arrived.

PCB Exposure from Oil Combustion

Wayne County Firefighters Association

Submitted to:
Mr. Philip Landowe
President, Wayne County Firefighters Association
Wandell, IN 45602

Submitted by:
Analytical Laboratories, Incorporated
1220 Pfeiffer Parkway
Indianapolis, IN 46223
February 28, 2003

INTRODUCTION

Waste oil used to train firefighters was suspected by the Wayne County Firefighters Association of containing polychlorinated biphenyls (PCBs). According to information provided by Mr. Philip Landowe, President of the Association, it has been standard practice in training firefighters to burn 20–100 gallons of oil in a diked area of approximately 25–50 m³. Firefighters would then extinguish the fire at close range. Exposure would last several minutes, and the exercise would be repeated two or three times each day for one week.

Oil samples were collected from three holding tanks near the training area in Englewood Park on November 11, 2002. To determine potential firefighter exposure to PCBs, bulk oil analyses were conducted on each of the samples. In addition, the oil was heated and burned to determine the degree to which PCB is volatized from the oil, thus increasing the potential for firefighter exposure via inhalation.

TESTING PROCEDURES

Bulk oil samples were diluted with hexane, put through a cleanup step, and analyzed in electron-capture gas chromatography. The oil from the underground tank that contained PCBs was then exposed to temperatures of 1008°C without ignition and 2008°C with ignition. Air was passed over the enclosed sample during heating, and volatized PCB was trapped in an absorbing medium. The absorbing medium was then extracted and analyzed for PCB released from the sample.

RESULTS

Bulk oil analyses are presented in Table 1. Only the sample from the underground tank contained detectable amounts of PCB. Aroclor 1260, containing 60 percent chlorine, was found to be present in this sample at 18 mg. Concentrations of 50 mg PCB in oil are considered hazardous. Stringent storage and disposal techniques are required for oil with PCB concentrations at these levels.

TABLE 1. Bulk Oil Analyses

Source	Sample #	PCB Content (mg/g)
Underground tank (11' deep)	6062	18*
Circle tank (3' deep)	6063a	<1
	6063b	<1
Square pool (3' deep)	6064a	<1
	6064b	<1

*Aroclor 1260 is the PCB type. This sample was taken for volatilization study.

FIGURE L–1. Laboratory Report

DISCUSSION AND CONCLUSIONS

At a concentration of 18 mg/g, 100 gallons of oil would contain approximately 5.5 g of PCB. Of the 5.5 g of PCB, about 0.3 g would be released to the atmosphere under the worst conditions.

The American Conference of Governmental Industrial Hygienists has established a threshold limit value (TLV)* of 0.5 mg/m^3 air for a PCB containing 54 percent Cl as a time-weighted average over an 8-hour work shift and has stipulated that exposure over a 15-minute period should not exceed 1 mg/m^3. The 0.3 g of released PCB would have to be diluted to 600 m^3 air to result in a concentration of 0.5 mg/m^3 or less. Because the combustion of oil lasted several minutes, a dilution to more than 600 m^3 is likely; thus, exposure would be less than 0.5 mg/m^3.

In summary, because exposure to this oil was limited and because PCB concentrations in the oil were low, it is unlikely that exposure from inhalation would be sufficient to cause adverse health effects. However, we cannot rule out the possibility that excessive exposure may have occurred under certain circumstances, based on factors such as excessive skin contact and the possibility that oil with a higher-level PCB concentration could have been used earlier. The practice of using this oil should be terminated.

*The safe average concentration that most individuals can be exposed to in an 8-hour day.

L

FIGURE L-1. Laboratory Report *(continued)*

layout and design

The layout and design of a document can make even the most complex information look accessible and give <u>readers</u> a favorable impression of the writer and the organization. To accomplish those goals, a design should help readers find information easily; offer a simple and uncluttered presentation; and highlight structure, hierarchy, and order. The design should also reinforce an organization's image. For example, if clients are paying a high price for consulting services, they may expect a sophisticated, polished design; if employees inside an organization expect management to be frugal, they may accept—even expect—economical and standard company design.

Effective design is based on visual simplicity and harmony and can be achieved through the selection of fonts, the choice of devices to highlight information, and the arrangement of text and visual components on a page.

Typography

Typography refers to the style and arrangement of type on a printed page. A complete set of all the letters, numbers, and symbols available in one typeface (or style) is called a *font*. The letters in a typeface have a number of distinctive characteristics, as shown in Figure L–2.

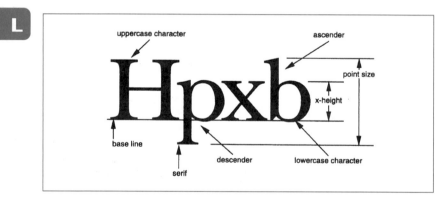

FIGURE L–2. Primary Components of Letter Characters

Typeface and Type Size.   For most on-the-job writing, select a typeface primarily for its legibility. Avoid typefaces that may distract readers. Instead, choose popular typefaces with which readers are familiar, such as Times Roman, Garamond, or Gill Sans. Do not use more than two typefaces in the text of a document. For certain documents, such as **brochures** and **newsletters,** you may wish to create a contrast between headlines and text. To do so, use typefaces that are distinctively different. Always experiment before making final decisions.

One way typefaces are characterized is by the presence or absence of *serifs*. Serif typefaces have projections, as shown in Figure L–2; sans serif styles do not. (*Sans* is French for "without.") The text of this book is set in Plantin, a serif typeface. Although sans serif type has a modern look, serif type is easier to read, especially in the smaller sizes. Sans serif, however, works well for headings (like the ones in this book). The simpler design of sans serif letters also makes them ideal for use on Web sites and other documents read on-screen.

A font size that is too small will cause eyestrain and make the text look crammed and intimidating. Type that is too large uses more space than necessary and makes reading difficult and inefficient. Figure L–3 illustrates various type sizes.

---

6 pt. This size might be used for dating a source.

8 pt. This size might be used for footnotes.

10 pt. This size might be used for figure captions.

12 pt. This size might be used for main text.

14 pt. This size might be used for headings.

FIGURE L–3. Type Sizes (6- to 14-Point)

Ideal font sizes for the main text of paper documents range from 8 to 13 points; 11- and 12-point type are the most common sizes. Your readers and the distance from which a document will be read should help determine type size. For example, instructions that will rest on a table at which the reader stands require a larger typeface than a document that will be read up close. For presentations and Web documents, preview your document to see the effectiveness of your choice of point sizes and typefaces. See writing for the Web.

Left- or Full-Justified Margins. Left-justified margins are generally easier to read than full-justified margins that can produce irregular spaces between words and unwanted white space in blocks of text. However, because left-justified (ragged-right) margins look informal, full-justified text is more appropriate for publications aimed at a broad readership that expects a more formal, polished appearance. Further, full justification is often useful with multiple-column formats because the spaces between the columns (called *alleys*) need the definition that full justification provides.

Highlighting Devices

When thoughtfully used, highlighting devices—typography; headings and captions; headers and footers; rules, icons, and color—give a document visual logic and organization. For example, rules and boxes can set off steps and illustrations from surrounding explanations. Consistency and moderation are important: Use the same technique to highlight a particular feature throughout your document and be careful not to overuse any single device.

Typographical Devices.   One method of achieving <u>emphasis</u> through typography is to use capital letters. HOWEVER, LONG STRETCHES OF ALL UPPERCASE LETTERS ARE DIFFICULT TO READ. (See also <u>e-mail</u>.) Use all uppercase letters only in short spans, such as in headings. Likewise, use <u>italics</u> sparingly because *continuous italic type reduces legibility and thus slows readers.* Of course, italics are useful if your aim is to slow readers, as in cautions and warnings. **Boldface,** used in moderation, may be the best cuing device because it is visually different yet retains the customary shapes of letters and numbers.

Headings and Captions.   <u>Headings</u> (or heads) reveal the organization of a document and help readers decide which sections they need to read. Headings appear in many typeface variations (boldface being the most common) and often use sans serif typefaces. Never leave a heading as the final line on a page—the heading is disconnected from its text and thus ineffective. Instead, move the heading to the start of the next page.

Captions are titles that highlight or describe illustrations or blocks of text. Captions often appear below or above figures and <u>tables</u> and in the left or right margins next to blocks of text.

Headers and Footers.   A header appears at the top of each page and contains identifying information; a footer appears at the bottom of each page and contains similar information. Document pages may have headers or footers or both that include the topic or subtopic of a section, identifying numbers, the date the document was written, page numbers, and the document name. Keep your <u>**headers and footers**</u> concise—too much information in them can create visual clutter.

Rules, Icons, and Color.   *Rules* are vertical or horizontal lines used to divide one area of the page from another or to create boxes. They highlight elements and make information more accessible.

An *icon* is a pictorial representation of an idea. Commonly used icons include the small envelopes used in <u>Web design</u> to symbolize e-mail links and national flags to symbolize different language versions of a document. To be effective, icons must be simple and easily recognized or defined.

Color and screening (shaded areas on a page) can distinguish one part of a document from another or unify a series of documents. They can set off sections within a document, highlight examples, or emphasize warnings. In tables, screening can highlight column titles or sets of data to which you want to draw the reader's attention.

Page Design

Page design is the process of combining the various design elements on a page to make a coherent whole. The flexibility of your design is based on the capabilities of your word-processing software, how the document will be reproduced, and the budget.

Thumbnail Sketches.   Before you spend time positioning actual text and visuals on a page, you may want to create a *thumbnail sketch,* in which blocks indicate the placement of elements. You can go further by roughly assembling all the thumbnail pages, showing size, shape, form, and general style of a large document. Such a mock-up, called a *dummy,* allows you to see how a finished document will look.

Columns.   As you design pages, consider how columns may improve the readability of your document. A single-column format works well with larger typefaces, double spacing, and left-justified margins. For smaller typefaces and single-spaced lines, the two-column structure keeps text columns narrow enough so readers need not scan back and forth across the width of the entire page for every line.

A word on a line by itself at the end of a column is called a *widow.* A single word carried over to the top of a column or page is called an *orphan.* Avoid both.

White Space.   White space visually frames information and breaks it into manageable chunks. For example, white space between paragraphs helps readers see the information in each paragraph as a unit. Use extra white space between sections as a visual cue to signal that one section is ending and another is beginning.

Lists.   Lists are an effective way to highlight words, phrases, and short sentences. Lists are particularly useful for certain types of information, such as steps in sequence, materials or parts needed, concluding points, and recommendations.

Illustrations.   Readers notice illustrations before they notice text, and they notice larger illustrations before they notice smaller ones.

Thus, the size of an illustration suggests its relative importance. For newsletter articles and publications aimed at wide audiences, consider especially the proportion of the illustration to the text. Magazine designers often use the three-fifths rule: Page layout is more dramatic and appealing when the major element (<u>photograph</u>, <u>drawing</u>, or other <u>visual</u>) occupies three-fifths rather than half the available space. The same principle can be used to enhance the visual appeal of a <u>report</u>.

Remember that clarity and usefulness take precedence over aesthetics in many business and technical documents. Illustrations can be gathered in one place (for example, at the end of a report), but placing them in the text closer to their accompanying explanations makes them more effective. Using illustrations in the text also provides visual relief. For advice on the placement of illustrations, see the Writer's Checklist: Using and Integrating Visuals (page 568).

DIGITAL TIP LAYING OUT A PAGE

Word-processing programs offer a variety of features you can use to improve the layout and other elements affecting the appearance of text on your page. For more on these features, see *<www.bedfordstmartins.com /alred>* and select *Digital Tips*. See also Digital Tip: Creating an Index on page 261.

---

lend / loan

Both *lend* and *loan* can be used as <u>verbs</u>, but *lend* is more common. (You can *lend* [or *loan*] them the money if you wish.) Unlike *lend, loan* can be a <u>noun</u>. (We made arrangements at the bank for a *loan.*)

---

letters (*see* **correspondence** and **e-mail**)

---

liable / libel

Liable means "legally subject to" or "responsible for."

- Employers are held *liable* for their employees' actions.

Libel refers to "anything circulated in writing or pictures that injures someone's good reputation." When someone's reputation is injured in speech, the term is *slander.*

- When an editorial charged our board of directors with bribing a representative, the board sued the newspaper for *libel.*
- If the mayor supports the bribery charge in her speech tonight, we will accuse her of *slander.*

In workplace writing, *liable* should retain its legal meaning. Where a condition of probability is intended, use *likely.*

- Rita is ~~liable~~ *likely* to be promoted.

---

library research

L

This entry, along with the <u>Internet research</u> entry, is intended to provide you with a starting point for conducting research; the Web Link boxes and references throughout provide resources for more detailed information. See also <u>research</u> for other methods of gathering information, <u>note-taking</u> for advice about recording notes during research, and <u>documenting sources</u> for guidance on citing research material.

The first step for conducting library research is to develop a search strategy appropriate to the information needed for your topic. You may want to begin by meeting with a research librarian. Research librarians are information specialists who can help you quickly find the best print or online resources for your topic—a brief conversation can focus your research and save you time. In addition, use your library's homepage for access to its catalog, databases of articles, subject directories to the Web, and more. A sample of a library's homepage is shown in Figure L–4.

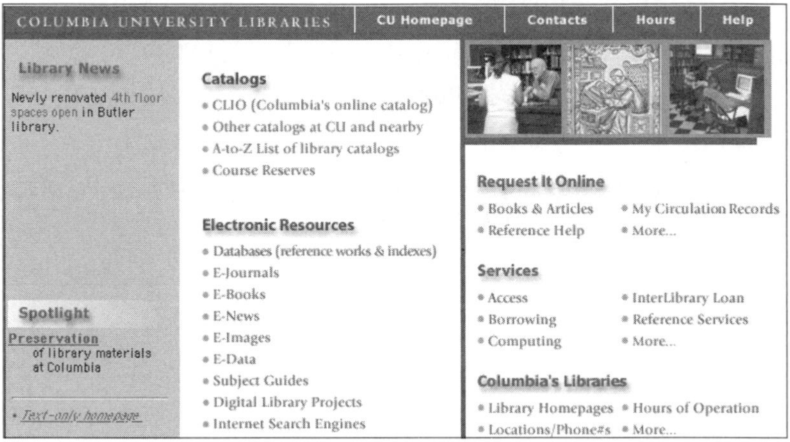

FIGURE L–4. Library Homepage

L

Your search strategy depends on the kind of information you are seeking. For example, if you need the latest data offered by government research, check the Web. Likewise, if you need a current article on a topic, you might search an online database—such as InfoTrac—subscribed to by your library. For an overview of a subject, you might turn to an encyclopedia; for historical background, your best resources are books, journals, and primary documents.

> WEB LINK USING COMMON LIBRARY TERMS
>
> For glossaries of common research-related library and Internet terms, see <www.bedfordstmartins.com/alred> and select *Diana Hacker's Research and Documentation Online* and *Research Room*.

Online Catalogs (Locating Books)

An online catalog—accessed through a library terminal and through the Internet—allows you to search a library's holdings, indicates an item's location and availability, and may allow you to arrange an interlibrary loan.

You can search a library's online catalog by author, title, or subject. The most typical way of searching the catalog is to search by subject. Most catalogs allow you to search by subject in two ways: by keyword or by subject. If your search turns up too many results, you can usually narrow your search by using the "limit search" or "advanced search" option offered by many catalogs. An example of an advanced search by keyword is shown in Figure L–5.

1. Select database(s) to search:

```
Metro Boston Library Network
BPL Research Lib, 1974-Present
Minuteman
CLAMS
Fenway Libraries Online
```

2. Enter a word or phrase to search:
(Searching by "Keyword" permits the use of operators and qualifiers *.)*

```
Business AND environmental protection
```

Pressing <Enter> without selecting a search type searches by the **highlighted** *search button.*

3. Search records by:

[Author] [Title] [Subject] [Keyword]

Press Button to Submit Search

4. Limit the search (optional):

By language: By material type: By date:

[English   ◆]   [All Materials   ◆]   [All Dates   ◆]

FIGURE L–5. Advanced Search of a Catalog

🌐 **WEB LINK ACCESS TO LIBRARY CATALOGS**

You can gain access via the Internet to public, government, and college and university online libraries. In some cases, you may need to have either a public or an academic library card to use online databases and other library resources.

The LibDex Library Index is a worldwide directory of library homepages and Web-based Online Public Access Catalogs (OPACs). The Library of Congress offers access to its online catalog holdings and special collections. See also the Google Directory of Libraries. For these and other research-related links and resources, see <*www.bedfordstmartins.com/alred*>.

Databases and Indexes (Locating Articles)

Most college, university, and public libraries subscribe to online data-bases—such as collections of online articles—many of which are avail-able through a library's Web site.

- *InfoTrac:* a collection of databases of articles, with many available in full text; can also provide specialized databases in business, health, and other fields
- *ProQuest:* a database of articles, with many available in full text; can also provide specialized databases for nursing, biology, and psychology
- *EBSCOhost:* a database of articles, with many available in full text; can also provide specialized databases in a range of subjects
- *FirstSearch:* a collection of specialized databases such as WorldCat (library collections) and ArticleFirst (articles, some in full text)
- *Lexis/Nexis Universe:* a collection of databases containing news, business, legal, and congressional information, with most available in full text

These databases, sometimes called *periodical indexes*, are excellent re-sources for articles published within the last ten to twenty years. Some include descriptive abstracts and full texts of articles. To find older arti-cles, you may need to consult a print index, such as the *Readers' Guide to Periodical Literature* and the *New York Times Index*, or a reference li-brarian. (For more information on print indexes, see Reference Works below.)

To locate articles in a database, conduct a keyword search, as shown in Figure L–6. If your search turns up too many results, narrow your search by connecting two search terms with *AND*—"technical writing AND employment"—or use other options offered by the data-base, such as a limited, modified, or advanced search.

Reference Works

In addition to articles, books, and Web sources (see Internet research), you may want to consult reference works such as encyclopedias, dic-tionaries, and atlases for a brief overview of your subject. *Bibliographies,* which are lists of works written about a topic, can direct you to more specialized sources. Ask your reference librarian to recommend refer-ence works and bibliographies that are most relevant to your topic.

Encyclopedias.   Encyclopedias are comprehensive, multivolume collections of articles arranged alphabetically. Some, such as the *En-carta Encyclopedia* <www.encarta@msn.com>, cover a wide range of

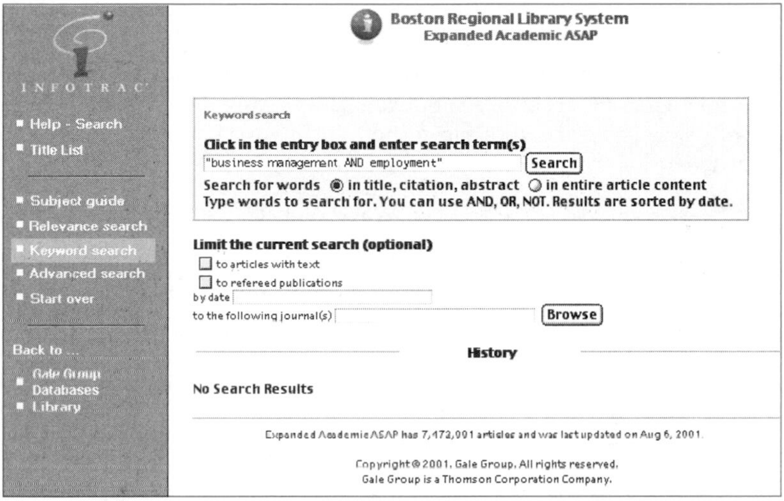

FIGURE L–6. InfoTrac Search Page

subjects, while others, such as *The Encyclopedia of Careers and Vocational Guidance*, 11th ed., edited by William Hopke (Chicago: Ferguson, 1999), and *McGraw Hill Encyclopedia of Science and Technology*, 8th ed., edited by Sybil P. Parker (New York: McGraw-Hill, 1997), focus on specific areas.

Dictionaries.   General and specialized dictionaries are available in print, on CD-ROM, and on the Web. General dictionaries can be compact or comprehensive, unabridged publications. Specialized dictionaries define terms used in a particular field, such as engineering, computers, architecture, or consumer affairs, and offer detailed definitions of field-specific terms, usually written in straightforward language.

Handbooks and Manuals.   Handbooks and manuals are typically one-volume compilations of frequently used information in a particular field. They offer brief definitions of terms or concepts, standards for presenting information, procedures for documenting sources, and visuals such as graphs and tables. For examples of handbooks and manuals for citing sources, see documenting sources.

Bibliographies.   Bibliographies list books, periodicals, and other research materials published in areas such as engineering, medicine, the humanities, and the social sciences. One example is *The St. Martin's Bibliography of Business and Technical Communication* by Gerald J. Alred (New York: St. Martin's, 1997).

General Guides.   The annotated *Guide to Reference Books,* 11th ed., by Robert Balay (Chicago: American Library Association, 1996), can help you locate reference books, indexes, and other research materials. The following are specialized indexes. Check your library's homepage or with your reference librarian to find out if your library subscribes to a particular index and whether it is available online.

Applied Science and Technology Index, 1958–. Alphabetical subject listing of articles from about 300 periodicals; issued monthly.

Engineering Index, 1934–. Alphabetical subject listing containing brief abstracts; issued monthly.

Government Reports Announcements and Index, 1965–. Semimonthly index of reports, arranged by subject, author, and report number.

Index to the Times (London), 1790–. Monthly.

Monthly Catalog of U.S. Government Publications, 1895–. Unclassified publications of all federal agencies, listed by subject, author, and report number; issued monthly.

New York Times Index, 1851–. Alphabetical list of subjects covered in *New York Times* articles; issued bimonthly.

Readers' Guide to Periodical Literature, 1900–. Monthly index of about 200 general U.S. periodicals, arranged alphabetically by subject.

Safety Sciences Abstracts Journal, 1974–. Listing of literature on industrial and occupational safety, transportation, and environmental and medical safety; issued quarterly.

Atlases and Statistical Sources.   *Atlases* provide representations of the physical and political boundaries of countries, climate, population, or natural resources. *Statistical sources* are collections of numerical data. They are the best source for such information as the U.S. gross domestic product, the consumer price index, or the demographic breakdown of the general population.

ATLAS

Microsoft® Encarta® World Atlas 2001. CD-ROM for Windows®. Microsoft Corporation, 2001.

STATISTICAL SOURCES

American Statistics Index. Washington: Congressional Information Service, 1978–. Monthly, quarterly, and annual supplements.

U.S. Bureau of the Census. *Statistical Abstract of the United States.* Washington: Government Printing Office, 1879–. Annual. <www.census.gov>

Writer's Checklist: Evaluating Library Resources

FOR A BOOK

☑ Is the text recent enough and relevant to your topic? Is it readily available?

☑ Who is the author? Does the preface or introduction indicate the author's purpose?

☑ Does the table of contents relate to your topic? Does the index contain terms related to your topic? Does the text contain a bibliography, reference list, and footnotes?

☑ Are the chapters useful? (Skim through one that seems related to your topic — notice especially the introduction, headings, and closing.)

☑ Does the author present information in an unbiased way? Are the language, tone, and style inviting?

FOR AN ARTICLE

☑ Is the article recent enough and relevant to your topic? Is it readily available?

☑ Is the publisher of the magazine or other periodical well known? Who is the publication's main audience? (The mainstream public? A small group of professionals?) Does the publication target or show bias toward a particular audience?

☑ What is the article's purpose? (For a journal article, read the abstract; for a newspaper article, read the headline and the lead sentences.)

☑ Does the article contain informative diagrams or other visuals that indicate its scope?

L

-like (see suffixes)

like / as

To avoid confusion between *like* and *as*, remember that *like* is a **preposition** and *as* (or *as if*) is a **conjunction**. Use *like* with a **noun** or **pronoun** that is not followed by a **verb**.

• The new supervisor behaves *like* a novice.

Use *as* before **clauses**, which contain verbs.

• She responded *as* we expected she would.

- It seemed *as if* the presentation would never end.

Like may be used in elliptical constructions that omit the verb. See also **figures of speech**.

- He adapted to the new system *like* a duck ~~adapts~~ to water.

listening

Effective listening is essential for communication. It enables the listener to understand the directions of an instructor, the message in a speaker's **presentation**, the goals of a manager, and the needs and wants of customers. Above all, it lays the foundation for cooperation. Productive communication occurs when both the speaker and the listener focus clearly on the content of the message and attempt to eliminate as much interference as possible.

Fallacies About Listening

Most people assume that because they can hear they know how to listen. In fact, *hearing* is passive, whereas *listening* is active. Hearing voices in a crowd or a ringing telephone requires no analysis and no active involvement. We hear such sounds without choosing to listen to them — we have no choice but to hear them. Listening, however, requires taking action, interpreting the message, and assessing its worth.

Many people believe that words have absolute meanings; however, words can have multiple meanings that are determined by the context in which they are used. Differences in meaning may be the result of differences in the speaker's and the listener's occupation, education, culture, sex, race, or other factors. The use of **idioms** can result in misunderstanding. See also **biased language, English as a second language, jargon,** and **international correspondence**.

Active Listening

To listen actively, you should (1) make a conscious decision to listen, (2) define your purpose for listening, (3) take specific actions to listen more efficiently, and (4) adapt to the situation.

Step 1: Make a Conscious Decision.   The first step to active listening is simply making up your mind to do so. Active listening requires a conscious effort, something that does not come naturally. To listen actively, "seek first to understand and *then* to be understood."

Step 2: Define Your Purpose. Knowing why you are listening can go a long way toward managing the most common listening problems: drifting attention, formulating your response while the speaker is still talking, and interrupting the speaker. To help you define your purpose for listening, ask yourself these questions:

- What kind of information do I hope to get from this exchange and how will I use it?
- What kind of message do I want to send while I am listening? (Do I want to portray understanding, determination, flexibility, competence, or patience?)
- What might interfere with listening during the interaction—boredom, daydreaming, anger, impatience? How can I keep these factors from placing a barrier between the speaker and me?

Step 3: Take Specific Actions. Becoming an active listener requires a willingness to become a responder rather than a reactor. A *responder* is a listener who slows down the communication to be certain that he or she is accurately receiving the message sent by the speaker. A *reactor* simply says the first thing that comes to mind, without checking to make sure that he or she accurately understands the message. Take the following actions to help you become a responder and not a reactor.

- Make a conscious effort to be impartial when evaluating a message. For example, do not dismiss a message because you dislike the speaker or are distracted by the speaker's appearance or mannerisms.
- Slow down the communication by asking for more information or by paraphrasing the message received before you offer your thoughts. Paraphrasing lets the speaker know you are listening, gives the speaker an opportunity to clear up any misunderstanding, and keeps you focused. See also **paraphrasing**.
- Listen with empathy by putting yourself in the speaker's position. When people feel they are being listened to empathetically, they tend to respond with appreciation and cooperation, thereby improving the communication.
- To help you stay focused on what the speaker is saying, take notes while you are listening. **Note-taking** not only communicates your attentiveness to the speaker, it also reinforces the message and helps you remember it.

Step 4: Adapt to the Situation. The requirements of active listening differ from one situation to another. For example, when you are listening to a lecture, you may be listening only for specific information.

However, if you are on a team project that depends on everyone's contribution, you need to listen at the highest level so you can gather information as well as pick up on nuances the other speakers may be communicating. See also **collaborative writing**.

lists

Lists can save readers time by allowing them to see at a glance specific items, questions, or directions. Lists also help readers by breaking up complex statements and by allowing key ideas to stand out, as in Figure L–7. Notice in Figure L–7 that all the list items in the example have **parallel structure**. In addition, all the items are balanced; that is, all points are relatively equal in importance and are of the same general length. To ensure that your readers understand how a list fits with the surrounding sentences, provide adequate **transition** before and after any list.

Before we agree to hold the technical staff meeting at the Brent Hotel, we should make sure the hotel facilities provide the following:

- Service center with phones, faxes, Internet hookup, PCs, and copying services for the conference committee
- Ground-floor exhibit area large enough for thirty 8-by-15-foot booths
- Eight meeting rooms to accommodate 25 people each
- Internet hookups, projection screens for presentations, and overhead projectors in each room
- Ballroom and dining facilities for 250 people

To confirm that the Brent Hotel is our best choice, we should tour the facilities during our stay in Kansas City.

FIGURE L–7. Bulleted List

In an attempt to avoid writing paragraphs, some writers tend to overuse lists. Memos or reports that consist almost entirely of lists can be difficult to understand because the reader is forced to connect the separate items mentally.

Writer's Checklist: Using Lists

- ☑ List only comparable items, using parallel structure throughout.
- ☑ Use only words, phrases, or short sentences of the same general length.
- ☑ Provide context by introducing each list, typically with a complete sentence followed with a **colon**.
- ☑ Ensure **coherence** by providing adequate transition before and after a list.
- ☑ Use bullets, as opposed to numbers, when rank or sequence is not important.
- ☑ Do not include too many items, as in entire pages of lists.

literature reviews

A *literature review* is a summary **report** on the relevant literature (printed and electronic) available on a particular subject over a specified period of time. For example, a literature review might describe significant material published in the past five years on a technique for improving emergency medical diagnostic procedures, or it might describe all reports written in the past ten years on efforts to improve safety procedures at a particular company. A literature review tells readers what has been published on a particular subject and gives them an idea of what material they should read in full.

Some **trade journal articles** or theses begin with a brief literature review to bring the reader up-to-date on current research in the field. The writer then uses the review as background for his or her own discussion of the subject. Figure L–8 on page 328 shows a two-paragraph literature review that serves as an **introduction** for a medical article on facial clefts in newborns. A literature review also could be a whole document in itself. Fully developed literature reviews are good starting points for detailed **research**.

To prepare a literature review, you must begin by researching published material on your topic. Because your reader may begin research on the basis of your literature review, be especially careful to accurately cite all bibliographic information. As you review each source, note the scope of the book or article and judge its value to the reader. Save all printouts of computer-assisted searches — you may want to incorporate the sources in a bibliography. See also **documenting sources**, **Internet research**, and **library research**.

Assessment of attractiveness of newborn infants with unrepaired facial clefts by different groups of individuals was studied to investigate the validity of a surgeons' rule of thumb system for this based upon cleft severity.

Recent studies have attempted to construct objective rating scales to quantify the severity of facial disfigurement in children and young people (Eliason et al., 1991; Poole et al., 1991; Roberts-Harry et al., 1992; Tobiasen and Hiebert, 1993). The medical community increasingly recognizes the need for a holistic approach to congenital disfigurement, including such aspects as parents' reactions, feelings, and attitudes (Bull and Rumsey, 1988). Results of these studies suggest that there may be a mistaken assumption of consensus between plastic surgeons and people unfamiliar with facial clefts in how each group respectively rates attractiveness of facial appearance. The studies all refer, however, to repaired facial clefts and other anomalies; none has assessed unrepaired facial clefts in neonates. Tobiasen (1991) has concluded that it is premature to suggest that surgeons rate impairment differently from others. Surgeons commonly use a rule of thumb system to classify severity on the basis of dichotomous categorizations (i.e., bilateral or unilateral clefting along the axis of symmetry; complete or incomplete extension to the nasal alar, and with or without palate involvement). The proposed system used by plastic surgeons and derived from Freedlander et al. (1990) is shown in Table 1.

Source: Slade, Pauline. "Relationships Between Cleft Severity and Attractiveness of Newborns with Unrepaired Clefts." *Cleft Palate-Craniofacial Journal,* vol. 32, no. 4 (July 1995): 318–322.

FIGURE L–8. Literature Review (in Introductory Paragraphs)

Begin a literature review by defining the area to be covered and the types of works to be reviewed. For example, a literature review may be limited to articles and reports and not include any books. You can arrange your discussion chronologically, beginning with a description of the earliest relevant literature and progressing to the most recent (or vice versa). You can also subdivide the topic, discussing works in various subcategories of the topic.

Related to literature reviews are annotated <u>bibliographies,</u> which also give readers information about published material. Rather than discussing the literature in paragraphs, however, an annotated bibliography lists each bibliographic item and then briefly describes it. The description (or *annotation*) may include the purpose of the article or book, its scope, the main topics covered, its historical importance, and anything else the writer thinks the reader should know.

logic errors

In persuasive writing, logic is essential to convincing readers that your conclusions are valid. (See also **persuasion**.) Errors in logic can undermine the point you are trying to communicate and your credibility.

Lack of Reason

When a statement is contrary to the reader's common sense, that statement is not reasonable. If, for example, you stated "Los Angeles is a small town," your reader might immediately question your logic. If, however, you stated "Although Los Angeles is a large sprawling city, it is composed of many areas that are much like small towns," your reader could probably accept the statement as reasonable.

Sweeping Generalizations

Sweeping generalizations are statements that are either too broad or all-inclusive to be supportable; they generally enlarge an observation about a small group to refer to an entire population. A flat statement—such as "All engineers are poor writers"—ignores any possibility that some engineers may be good writers. Using such generalizations weakens your credibility.

Non Sequiturs

A *non sequitur* is a statement that does not logically follow a previous statement.

- I arrived at work early today, so the weather is calm.

Common sense tells us that arriving early for work is unrelated to the weather; thus, the second part of the sentence does not follow logically from the first. In your own writing, be careful that all points stand logically connected.

False Analogies

A *false analogy* (also called *post hoc, ergo propter hoc*) refers to the logical fallacy that because one event followed another event, the first somehow caused the second.

- I didn't bring my umbrella today. No wonder it is now raining.

In on-the-job writing, such an error in reasoning can happen when the writer hastily concludes that two events are related without examining the logical connection between them.

Biased or Suppressed Evidence

A conclusion reached as a result of self-serving data, questionable sources, suppressed evidence, or incomplete facts is both illogical and unethical. Suppose you are preparing a report on the acceptance of a new policy among employees. If you distribute <u>questionnaires</u> only to those who think the policy is effective, the resulting evidence will be biased. If you purposely ignore employees who do not believe the policy is effective, you will be suppressing evidence. Intentionally ignoring relevant data that might not support your position not only produces inaccurate results, but, perhaps more importantly, is unethical. See also <u>ethics in writing</u>.

Fact Versus Opinion

Distinguish between fact and opinion. Facts include verifiable data or statements, whereas opinions are personal conclusions that may or may not be based on facts. For example, it is verifiable that distilled water boils at 100 degrees Centigrade; that it tastes better than tap water is an opinion. Distinguish the facts from your opinions in your writing so that your readers can draw their own conclusions.

Loaded Arguments

When you include an opinion in a statement and then reach conclusions that are based on that statement, you are loading the argument. Consider the following opening for a memo.

- I have several suggestions to improve the poorly written policy manual. First, we should change . . .

Unless everyone agrees that the manual is poorly written, readers may reject a writer's entire message because they disagree with this loaded premise. Do not load arguments in your writing; conclusions reached

---

WEB LINK UNDERSTANDING AN ARGUMENT

Dr. Frank Edler of Metropolitan Community College in Omaha, Nebraska, offers a tutorial in recognizing the logical components of an argument and thinking critically. See <*www.bedfordstmartins.com/alred*> and select *Links for Technical Writing*.

with loaded statements are weak and can produce negative reactions in readers who detect the loading.

long variants (*see* affectation)

loose / lose

Loose is an <u>adjective</u> meaning "not fastened" or "unrestrained." (He found a *loose* wire.) *Lose* is a <u>verb</u> meaning "be deprived of" or "fail to win." (I hope we do not *lose* the Emerson account.)

lowercase and uppercase letters

Lowercase letters are small letters, as distinguished from capital letters (known as uppercase letters). The terms were coined in the early history of printing, when printers kept the small letters in a case, or tray, below the case where they kept the capital letters.

Avoid using all capital letters (all caps) to emphasize important information, especially in <u>e-mail</u>. A passage set in all caps is hard to read because the shapes of capital letters are not as distinctive as the shapes of lowercase letters or a combination of uppercase and lowercase letters. See also <u>capitalization</u> and <u>layout and design</u>.

L

M

malapropisms

A *malapropism* is a word that sounds similar to the one intended but is ludicrously wrong in the context.

> **INCORRECT**   Our employees are no longer *sedimentary* now that we have a fitness center.

> **CORRECT**   Our employees are no longer *sedentary* now that we have a fitness center.

Intentional malapropisms are sometimes used in humorous writing; unintentional malapropisms can embarrass a writer. See also <u>figures of speech</u>.

male (see *female / male*)

manuals

Manuals—whether printed or electronic—help customers and technical specialists use and maintain products. The manuals are usually written by professional technical writers, although in smaller companies, engineers and technicians may write them.

Types of Manuals

Following are descriptions of typical technical manuals.

User Manuals.   User manuals are aimed at skilled or unskilled users of equipment and provide instructions for the setup, operation, and maintenance of a product. User manuals also typically include safety precautions and troubleshooting charts and guides. See also <u>instructions</u>.

Tutorials.   Tutorials are self-study guides for users of a product or system. Either packaged with user manuals or provided electronically, tutorials walk novice users through the operation of a product or system.

Training Manuals.   Training manuals are used to train individuals in some procedure or skill, such as operating equipment, flying an airplane, or processing an insurance claim. In many technical and vocational fields, training manuals are the primary teaching devices. Training manuals are often accompanied by videos, CD-ROMs, or other audiovisual material.

Operators' Manuals.   Written for skilled operators of construction, manufacturing, computer, or military equipment, operators' manuals contain essential instructions and safety warnings. They are often published in a convenient format that allows operators to use them at a work site.

Service Manuals.   Service manuals help trained technicians repair equipment or systems, usually at the customer's location. Such manuals often contain troubleshooting guides for locating technical problems.

Special-Purpose Manuals.   A number of manual types, including programmer reference manuals, overhaul manuals, handling and setup manuals, and safety manuals, have very specialized and limited uses.

Designing and Writing Effective Manuals

Identify and Write for Your Audience.   Will you be writing for novice users, intermediate users, or experts? Or, will you be instructing a combination of users with different levels of technical knowledge and experience? Identify your audience before you begin writing. Depending on your audience, you will make the following decisions:

- Which details to include (fewer for experts, more for novices)
- What level of technical vocabulary to use (necessary technical terminology for experts, plain language for novices)
- Whether to include a summary list of steps (experts will probably prefer using this summary list and not the entire manual; once novices and intermediate users gain more skill, they will also prefer referring just to this summary list)

Design your manual so that <u>readers</u> can use the equipment, software, or machinery while they are also reading your instructions.

- Readers should be able to easily find particular sections and instructions. Use generous amounts of white space between sections and actions, and use headings, subheadings, and words that readers will find familiar.
- Write with precision and accuracy so that readers can perform the procedures easily.
- Use clearly drawn and labeled illustrations, in addition to written text, to show readers exactly what equipment, online screens, or other items in front of them should look like.

Provide an Overview. An overview at the beginning of a manual should explain the overall purpose of the procedure, how the procedure can be useful to the reader, and any cautions or warnings the reader should know about before starting (see **instructions**). If readers know the purpose of the procedure, along with its specific workplace applications, they will be more likely to learn, recall the material that follows, and pay close attention to that material.

Create Major Sections. Divide any procedure into separate goals, create major sections to cover those goals, and state them in the section headings. If your manual has chapters, you can divide each chapter into specific subsections. In a manual about designing a Web page, an introductory chapter might include these subsections: (1) obtaining Web space, (2) viewing the new space, (3) using the Pico Editor to create text.

Indicate the Goals of Actions. Within each section, explain *why* readers must follow each step or each related set of steps. The conventional way to indicate a goal is to use either the infinitive form of a **verb** ("*To turn* on the machine") or the gerund form ("*Turning* on the machine"). Whichever verb form you choose, use it consistently throughout your manual, each time you want to indicate a goal. In the following example, the manual writer decided to indicate the goal of a set of actions (Copying Files) with a boldfaced, upper- and lowercase, flush-left heading using the gerund form of the verb.

Copying Files
Action 1   At the Filer, type F (for filecopy)
Action 2   Type the file's old volume and filename in the "VOLUME:FILENAME" format, then hit <ENTER>.

Use the Imperative Verb Form for Actions. The conventional way to indicate an action is by using the imperative form of verbs (for example, "*Type* your name"; "*Press* the Control Key"). Use the imperative form consistently, each time you want to indicate an action. In the example about copying files, the manual writer designated actions by placing subheads (Action 1, Action 2, and so on) in the left margin and imperative verbs ("type" and "hit") in the instructions to the right of the subheads.

Use Simple and Direct Verbs. Simple and direct verbs are most meaningful, especially for novice readers. Avoid jargon and terms known only to intermediate readers or experts, unless you know that they constitute all of the reading audience.

POOR VERB CHOICE	BETTER VERB CHOICE
Attempt	Try
Depress	Press
Discontinue	Stop
Display	Show
Employ	Use
Enumerate	Count
Execute	Do
Observe	Watch
Segregate	Divide

Indicate the Response of Actions. When appropriate, indicate the expected response of an action to reassure readers that they are performing the procedure correctly (for example, "A blinking light will appear" or "You will see a red triangle"). Use the form you choose consistently throughout a set of instructions to designate a response. In the following example, you can see how the response is indented farther to the right than the action, begins with a subhead ("Response"), and is indicated by a full sentence.

Action 3   Type the file's old volume and filename in the "VOLUME:FILENAME" format, then press <ENTER>.

Response The system asks for the old volume and filename of the file you want to copy: **Filecopy what file?**

Writer's Checklist: Preparing Manuals

☑ Determine the best medium for your manual: Web, online document, CD-ROM, spiral binding, loose-leaf binding, and so on. See also **layout and design**.

☑ Pay attention to **organization** and **outlining** because of the complexity of the products and systems for which manuals are written.

☑ Use a consistent format for each part of a manual (headings, subheadings, goals, actions, responses, warnings, cautions, and tips).

Writer's Checklist: Preparing Manuals (continued)

☑ Use **visuals**, such as schematics and exploded-view **drawings**, **flow-charts**, **photographs**, and **tables**, placing them where readers would benefit most from seeing them.

☑ Include **indexes** to help readers find information and use warning statements and standard symbols for potential dangers in **instructions**.

☑ Have manuals reviewed by your peers as well as by technical and legal experts to ensure that the manuals are helpful and accurate.

☑ Conduct **usability testing** on manual drafts so that you can detect errors and other problems readers are likely to have.

☑ Carefully review the entries **process explanation**, **revision**, **proofreading**, and **technical writing style**.

maps

Maps are often used to show specific geographic features (roads, mountains, rivers, and the like). They can also illustrate geographic distributions of population, climate patterns, corporate branch offices, and so forth. Note that the map in Figure M–1 contains a figure number and title, scale of distance, key (or legend), compass, and distinctive highlighting for emphasis.

Writer's Checklist: Creating and Using Maps

☑ Follow the general guidelines discussed in **visuals** for placement of maps.

☑ Label each map clearly, and assign each map a figure number if it is one of a number of illustrations.

☑ Clearly identify all boundaries in the map. Eliminate unnecessary boundaries.

☑ Eliminate unnecessary information that may clutter a map. For example, if the purpose of the map is to show population centers, do not include mountain elevations, rivers, or other physical features.

☑ Include a scale of miles/kilometers or feet/meters to give your readers an indication of the map's proportions.

☑ Indicate which direction is north.

☑ Emphasize key features by using color, shading, dots, cross-hatching, or other appropriate symbols.

☑ Include a key, or legend, that explains what the different colors, shadings, or symbols represent.

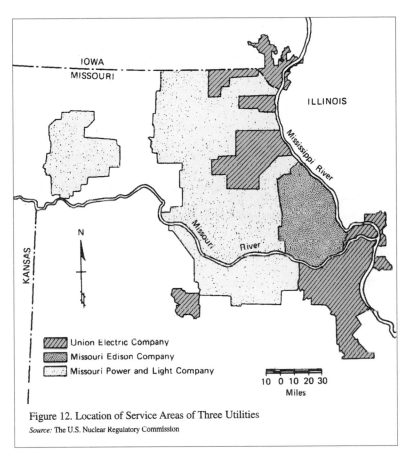

Figure 12. Location of Service Areas of Three Utilities

Source: The U.S. Nuclear Regulatory Commission

FIGURE M–1. Map

M

☀ WEB LINK **MAPS AND MAPPING INFORMATION**

The University of Iowa Libraries offer a useful site with links to online maps and other types of mapping and cartographic resources. See <*www.bedford stmartins.com/alred*> and select *Links for Technical Writing.*

---

mathematical equations

You can accurately prepare material with mathematical equations and make it easy to read by following consistent standards throughout a document. Unless you need to follow a specific style manual or specifications, the following guidelines should serve you well.

Set short and simple equations, such as $x(y) = y^2 + 3y + 2$, as part of the running text rather than displaying them on separate lines, as long as an equation does not appear at the beginning of a sentence. If a document contains multiple equations, identify them with numbers, as the following example shows:

$$x(y) = y^2 + 3y + 2 \tag{1}$$

Number displayed equations consecutively throughout the work. Place the equation number, in parentheses, at the right margin of the same line as the equation (or of the first line if the equation runs longer than one line). Leave at least four spaces between the equation and the equation number. Refer to displayed equations by number, for example, as "Equation 1" or "Eq. 1."

Positioning Displayed Equations

Equations that are set off from the text need to be surrounded by space. Triple-space between displayed equations and normal text. Double-space between one equation and another and between the lines of multiline equations. Count space above the equation from the uppermost character in the equation; count space below from the lowermost character.

Type displayed equations either at the left margin or indented five spaces from the left margin, depending on their length. When a series of short equations is displayed in sequence, align them on the equal signs.

$$p(x,y) = \sin(x + y) \tag{2}$$
$$p(x,y) = \sin x \cos y + \cos x \sin y \tag{3}$$
$$p(x_0, y_0) = \sin x_0 \cos y_0 + \cos x_0 \sin y_0 \tag{4}$$
$$q(x,y) = \cos(x + y) \tag{5}$$
$$= \cos x \cos y - \sin x \sin y$$
$$q(x_0, y_0) = \cos x_0 \cos y_0 - \sin x_0 \sin y_0 \tag{6}$$

Break an equation that requires two lines at the equal sign, carrying the equal sign over to the second portion of the equation.

$$_0\!\int^1 (f_n - \tfrac{u}{r}f_n)^2 \, r \, dr + 2n \, _0\!\int^1 f_n f_n dr = \tag{7}$$
$$_0\!\int^1 (f_n - \tfrac{u}{r}f_n)^2 \, r \, dr + nf_n^2(1)$$

If you cannot break an equation at the equal sign, break it at a plus or minus sign that is not in **parentheses** or **brackets**. Bring the plus or minus sign to the next line of the equation, which should be positioned to end near the right margin of the equation.

$$\emptyset(x, y, z) = (x^2 + y^2 + z^2)^{1/2} \, (x - y + z) \, (x + y - z)^2 \tag{8}$$
$$- [f(x, y, z) - 3x^2]$$

The next best place to break an equation is between parentheses or brackets that indicate multiplication of two major elements.

For equations that require more than two lines, start the first line at the left margin, end the last line at the right margin (or four spaces to the left of the equation's number), and center intermediate lines between the margins. Whenever possible, break equations at operational signs, parentheses, or brackets.

Omit punctuation after displayed equations, even when they end a sentence and even when a key list defining terms follows (for example, P = pressure, psf; V = volume, cu ft; T = temperature, °C). Punctuation may be used before an equation, however, depending on the grammatical construction.

The term $(n)_r$ may be written in a more familiar way by using the following algebraic device:

$$(n)_r = \frac{(n)(n-1)(n-2) \ldots (n-r+1)(n-r)(n-r-1) \ldots 3 \cdot 2 \cdot 1}{(n-r)(n-r-1) \ldots 3 \cdot 2 \cdot 1}$$
$$= \frac{n!}{(n-r)!} \tag{9}$$

maybe / may be

M

Maybe (one word) is an **adverb** meaning "perhaps." (*Maybe* the legal staff can resolve this issue.) *May be* (two words) is a **verb** phrase. (It *may be* necessary to hire a specialist.)

media / medium

Media is the plural of *medium* and should always be used with a plural verb.

- Many communication *media* are available today.

- The Internet is a multifaceted *medium*.

meetings

Meetings allow people to share information and collaborate to produce better results than exchanges of **memos** or **e-mails** or other means. (See also **selecting the medium**.) Like an oral **presentation**, a successful meeting requires planning and preparation.

Planning a Meeting

For a meeting to be successful, determine the focus of the meeting, decide who should attend, and choose the best time and place to hold the meeting. Prepare an agenda for the meeting and determine who should take the minutes.

Determine the Purpose of the Meeting.   The first step in planning a meeting is to focus on the desired outcome. Ask yourself the following question to help you determine the purpose of the meeting: What should participants *know, believe, do,* or *be able to do* as a result of attending the meeting?

Once you have focused your desired outcome, use the information to write a *purpose statement* for the meeting that answers the questions *what* and *why*.

- The purpose of this meeting is to gather ideas from the sales force [what] to create a successful sales campaign for our new scanner [why].

Decide Who Should Attend.   There is really no sense in having a meeting if not enough of the key people are present. If a meeting must be held without some key participants, e-mail those people prior to the meeting for their contributions. Of course, the meeting minutes should be distributed to everyone, including significant nonattendees.

Choose the Meeting Time.   The time of day and the length of the meeting can influence the outcome of the meeting. Consider the following when you are planning a meeting:

- People need Monday morning to focus on the week's work after the weekend.
- People need Friday afternoon to complete tasks that must be finished before the weekend.
- Long meetings should include adequate breaks to allow participants to check their messages, make phone calls, and refresh themselves.
- A meeting held during the last 15 minutes of the day will be quick, but few people will remember what happened.

Choose the Meeting Location.   Having a meeting on your own premises can give you an advantage: You feel more comfortable, which, along with your guests' newness to their surroundings, may help you get an edge. Holding the meeting on others' premises, however, can signal cooperation. For balance, especially when people are meeting for the first time or are discussing sensitive issues, having a meeting at a neutral site may be the best solution. No one gains an advantage in off-site meetings, and attendees often feel freer to participate.

Establish the Agenda. A tool for focusing the group, the *agenda* is an outline of what the meeting will address. Never begin a meeting without an agenda, even if it is only a handwritten list of topics. Ideally, the agenda should be distributed to attendees a day or two before the meeting. For a longer meeting in which participants are required to make a presentation, try to distribute the agenda a week or more in advance.

The agenda should list the attendees, the meeting time and place, and the topics you plan to discuss. If the meeting includes presentations, list the time allotted for each speaker. Finally, indicate an approximate length for the meeting so that participants can plan the rest of their day. Figure M-2 shows a typical agenda.

Sales Meeting Agenda

Purpose: To get input for a sales campaign for the new software
Date: May 12, 2003
Place: Conference Room E
Time: 9:30 a.m.–11:00 a.m.
Attendees: New Products Advertising Manager, Equipment Sales Reps, Customer-Service Staff, and Service Managers

Topic	Presenter	Time
The Software	Bob Arbuckle	Presentation, 9:30–9:45
The Campaign	Maria Lopez	Presentation, 9:45–10:00
The Sales Strategy	Mary Winifred	Presentation, 10:00–10:15
Discussion	Led by Dave Crimes	Discussion, 10:15–11:00

FIGURE M-2. Meeting Agenda

If the agenda is distributed in advance of the meeting, it should be accompanied by a <u>memo</u> or <u>e-mail</u> informing people of the meeting. The cover memo should include the following information:

- The purpose of the meeting
- The date and place
- The meeting start and stop times
- The names of the people invited
- Instructions on how to prepare

Figure M-3 on page 342 shows a cover memo to accompany an agenda.

Subject: Planning Meeting
Date: Tues, 06 May 2003 13:30:12 EST
From: Susan McLaughlin <smclaughlin@millenniumsoftware.com>
To: **New Products Advertising Managers; Equipment Sales**
                 Representatives; Customer Service Staff; Service Managers
Attachments: 〣 Sales Meeting Agenda.doc (29 KB)

Purpose of the Meeting

The purpose of this meeting is to get your ideas for the upcoming introduction and sales campaign for our new software.

Date, Time, and Location

Date: May 12, 2003
Time: 9:30 a.m.–11:00 a.m.
Place: Conference Room E (go to the ground floor, take a right off the
       elevator, third door on the left)

Attendees

The groups addressed above

Meeting Preparation

Everyone should be prepared to offer suggestions on the following items:

- Sales features of the new software
- Techniques for selling software
- Customer profile for potential business
- FAQs--questions customers may ask
- Anticipated service needs

Agenda

Please see the attached document.

FIGURE M–3. E-mail to Accompany an Agenda

Assign the Minute-Taking.   Delegate the minute-taking to someone other than the leader. The minute-taker should record major decisions made and tasks assigned. To avoid misunderstandings, the
minute-taker must record each assignment, the person responsible for
it, and the date on which it is due.

For a standing committee, it is best to rotate the responsibility of
taking minutes. See also <u>minutes of meetings</u> and <u>note-taking</u>.

Conducting the Meeting

Assign someone to write on a flip chart or use a computer to record information that needs to be viewed by everyone present.

During the meeting, keep to your agenda; however, allow room for
differing views, and foster an environment in which participants listen

respectfully to one another. (See also listening.) Create a productive environment.

- Consider the feelings, thoughts, ideas, and needs of others—do not let your own agenda blind you to other points of view.
- Help other participants feel valued and respected by listening to them and responding to what they say.
- Respond positively to the comments of others as best you can.
- Consider ways of doing things that are different from your own, particularly those from other cultures. See also global communication.

Deal with Conflict.   Despite your best efforts, conflict is inevitable. However, conflict is potentially valuable; when managed positively, it can stimulate creative thinking by challenging complacency and showing ways to achieve goals more efficiently or economically. See collaborative writing.

Members of any group are likely to vary greatly in their personalities and attitudes, and you may encounter people who approach meetings differently. Consider the following tactics for handling the interruptive, negative, rambling, overly quiet, and territorial personality types.

- The *interruptive person* rarely lets anyone finish a sentence and intimidates the group's quieter members. Tell that person in a firm but nonhostile tone to let the others finish in the interest of getting everyone's input. By addressing the issue directly, you signal to the group the importance of putting common goals first.
- The *negative person* has difficulty accepting change and often considers a new idea or project from a negative point of view. Such negativity, if left unchecked, can demoralize the group and deflate enthusiasm for new ideas. If the negative person brings up a valid point, ask for the group's suggestions to remedy the issue being raised. If the negative person's reactions are not valid or are outside the agenda, state the necessity of staying focused on the agenda and perhaps recommend a separate meeting to address those issues.
- The *rambling person* cannot collect his or her thoughts quickly enough to verbalize them succinctly. Restate or clarify this person's ideas. Try to strike a balance between providing your own interpretation and drawing out the person's intended meaning.
- The *quiet person* may be timid or may just be deep in thought. Ask for this person's thoughts, being careful not to embarrass the person. In some cases, you can have a quiet person jot down his or her thoughts and give them to you later.
- The *territorial person* fiercely defends his or her group against real or perceived threats and may refuse to cooperate with members of

other departments, companies, and so on. Point out that although such concerns may be valid, everyone is working toward the same overall goal and that goal takes precedence.

Close the Meeting.   To close the meeting, review all decisions and assignments. Paraphrase each to help the group focus on what they have agreed to do and to confirm the accuracy of the minutes. Now is the time to raise questions and clarify any misunderstandings. Set a date by which everyone at the meeting can expect to receive copies of the minutes. Finally, thank everyone for participating, and close the meeting on a positive note.

Writer's Checklist: Planning and Conducting Meetings

☑ Develop a purpose statement for the meeting to focus your planning.

☑ Invite only those essential to fulfilling the purpose of the meeting.

☑ Select a time and place convenient to all those attending.

☑ Create an agenda and distribute it a day or two before the meeting.

☑ Assign someone to take meeting minutes.

☑ Ensure that the minutes record key decisions; assignments; due dates; and the date, time, and location of the follow-up meeting.

☑ Follow the agenda to keep everyone focused.

☑ Respect the views of others and how they are expressed.

☑ Review strategies in this entry for dealing with conflict and with attendees whose style of expression may prevent getting everyone's best thinking.

☑ Close the meeting by reviewing key decisions and assignments.

memos

Memos—paper and electronic—are used for many of the types of workplace writing described in entries throughout this book. Memos are internal documents that, for example, announce policies, disseminate information, delegate responsibilities, instruct employees, and report results. They provide a record of decisions made and actions taken. They also can play a key role in the management of many organizations because managers use memos to inform and motivate employees.

Writing Memos

Keep in mind that many of the principles discussed in **correspondence** and **e-mail** apply to writing memos. To produce an effective memo, outline it first, even if you simply jot down points to be covered and then order them logically. With careful preparation, your memos will be both concise and adequately developed. Adequate development of your thoughts is crucial to the memo's clarity, as the following example indicates.

ABRUPT   Be more careful on the loading dock.

DEVELOPED   To prevent accidents on the loading dock, follow
                these procedures:
                1. Check . . .
                2. Load only . . .
                3. Replace . . .

Although the abrupt version is concise, it is not as clear and specific as the developed revision. Do not assume your reader will know what you mean. Readers who are pressed for time may misinterpret a vague memo. See also <u>conciseness/wordiness</u>.

Openings.   Although **methods of development** vary, a memo normally begins with a statement of its main point. Consider the following example:

- Because of recent hacker attacks on our Web site, I recommend that we no longer post the e-mail addresses of our laboratory employees.

When your reader is not familiar with the subject or with the background of a problem, provide an introductory paragraph before stating the main point of the memo. Doing so is especially important in memos that will serve as records of crucial information. Generally, longer or complex subjects benefit most from more thorough <u>introductions</u>. However, even when you are writing a short memo about a familiar subject, remind readers of the context. In the following example, words that provide context are shown in italics.

- *As Maria recommended,* I reviewed the office reorganization plan. I like most of the features; however, . . .

Do not state the main point first when (1) the reader is likely to be highly skeptical or (2) you are disagreeing with persons in positions of higher authority. In such cases, a more persuasive tactic is to state the problem first, then present the specific points supporting your final recommendation. See also <u>persuasion</u> and <u>refusal letters</u>.

Writing Style and Tone.   Whether your memo is formal or informal depends entirely on your <u>readers</u> and your <u>purpose</u>. A message to a coworker who is also a friend is likely to be informal, while an internal

M

proposal to several readers or to someone two or three levels higher in your organization is likely to be more formal. Consider the following versions of a statement:

TO AN EQUAL   I can't go along with the plan because I think it poses serious logistical problems. First, . . .

TO A SUPERIOR   The logistics of moving the department may pose serious problems. First, . . .

A memo that gives instructions to a subordinate should also be relatively formal, impersonal, and direct, unless you are trying to reassure or praise. When writing to subordinates, remember that *managing* does not mean *dictating*. If you are too formal, sprinkling your writing with fancy words, you may seem stuffy and pompous. In fact, <u>affectation</u> may both irritate and baffle readers, cause a loss of time, and produce costly errors. Consider the unintended secondary messages the following notice conveys:

- It has been decided that the office will be open the day after Thanksgiving.

"It has been decided" not only sounds impersonal but also communicates an authoritarian, management-versus-employee tone. The passive <u>voice</u> also suggests that the decision-maker does not want to say "I have decided" and thus be identified. One solution is to remove the first part of the sentence.

- The office will be open the day after Thanksgiving.

The best solution would be to suggest both that there is a good reason for the decision and that employees are privy to (if not a part of) the decision-making process.

- Because we must meet the December 15 deadline to be eligible for the government contract, the office will be open the day after Thanksgiving.

By subordinating the bad news (the need to work on that day), the writer focuses on the reasoning behind the decision to work. Employees may not necessarily like the message, but they will at least understand that the decision is not arbitrary and is tied to an important deadline.

Lists and Headings.   Lists can give impact to important points by making it easier for your reader to quickly grasp information. Be careful, however, not to overuse lists. A memo that consists almost entirely of lists is difficult to understand because it forces the reader to connect the separate and disjointed items on the page. Further, lists lose their effectiveness when they are overused.

Headings are another attention-getting device, particularly in long memos. They divide material into manageable segments, call attention to main topics, and signal a shift in topic. Readers can scan the headings and read only the section or sections appropriate to their needs.

Closings. A memo closing can accomplish many important tasks, such as building positive relationships with readers, encouraging colleagues and employees, and letting recipients know what you will do or what you expect of them.

- I will discuss the problem with the marketing consultant and let you know by Monday what we are able to change.

Although routine statements are sometimes unavoidable ("Thanks again for your help"), make your closing work for you by providing specific prompts to which the reader can respond. See also conclusions.

- If you would like further information, such as a copy of the questionnaire we used, please e-mail me at delgado@prn.com.

Format and Design

Memos vary greatly in format and customs. Although there is no single standard, Figure M–4 on page 348 shows a typical 8½-by-11-inch format with a printed company name.

Subject Lines. Subject lines announce the topic; because they also aid filing and later retrieval, they must be specific and accurate.

VAGUE	Subject: Tuition Reimbursement
VAGUE	Subject: Time-Management Seminar
SPECIFIC	Subject: Tuition Reimbursement for Time-Management Seminar

Capitalize all major words in a subject line, except articles, prepositions, and conjunctions with fewer than five letters unless they are the first or last words. Remember that the subject line should not substitute for an opening that provides context for the message.

Signature. The final step is signing or initialing a memo, a practice that lets readers know that you approve of its contents. Where you sign or initial the memo depends on the practice of your organization. Figure M–4 shows a typical placement of initials.

PROFESSIONAL PUBLISHING SERVICES
MEMORANDUM

TO: Barbara Smith, Publications Manager
FROM: Hannah Kaufman, Vice President   *HK*
DATE: April 14, 2003
SUBJECT: Schedule for ACM Electronics Brochure

ACM Electronics has asked us to prepare a comprehensive brochure for its
Milwaukee office by August 8, 2003. We have worked with electronics
firms in the past, so this job should be relatively easy to prepare. My guess
is that the job will take nearly two months. Ted Harris has requested time
and cost estimates for the project. Fred Moore in production will prepare
the cost estimates, and I would like you to prepare a tentative schedule for
the project.

Additional Personnel

In preparing the schedule, check the status of the following:
    1. Production schedule for all staff writers
    2. Available freelance writers
    3. Dependable graphic designers
Ordinarily, we would not need to depend on outside personnel; however,
because our bid for the *Wall Street Journal* special project is still under
consideration, we could be short of staff in June and July. Further, we have
to consider vacations that have already been approved.

Time Estimates

Please give me time estimates by April 18. A successful job done on time
will give us a good chance to obtain the contract to do ACM Electronics'
annual report for its stockholders' meeting this fall.

I know your staff can do the job.

cc: Ted Harris, President
     Fred Moore, Production Editor

FIGURE M–4. Typical Memo Format

Writer's Checklist: Preparing Printed and Electronic Memos

- ☑ Develop and organize your points before writing a draft.
- ☑ Open with the main point (provide a background if appropriate) and close purposefully.
- ☑ Ensure that the style and tone are appropriate for the sender and recipients.
- ☑ Use lists and headings strategically.
- ☑ Use the appropriate format for your organization.
- ☑ Prepare a useful subject line that clearly announces the topic.

metaphors (*see* figures of speech)

methods of development

A logical method of development satisfies the <u>readers'</u> need for shape and structure in a document, whether it is an <u>e-mail</u>, a <u>report</u>, or a Web page. It helps you as a writer move smoothly and logically from the <u>introduction</u> to a <u>conclusion</u>.

Choose the method of development that best suits your subject, your readers, and your <u>purpose</u>. Following are the most common methods, each of which is discussed in further detail in its own entry.

- <u>Cause-and-effect method of development</u> begins with either the cause or the effect of an event. This approach can also be used to develop a <u>report</u> that offers a solution to a problem, beginning with the problem and moving on to the solution.

- <u>Chronological method of development</u> emphasizes the time element of a sequence. A <u>trouble report</u> that traces events as they occurred in time, for example, could use this method of development.

- <u>Comparison method of development</u> is useful when writing about a new topic that is in many ways similar to another topic that is more familiar to the readers.

- <u>Definition method of development</u> extends definitions with additional details, examples, comparisons, or other explanatory devices. See also <u>defining terms</u>.

- <u>Division-and-classification method of development</u> is useful for describing physical objects or structures with component parts. *Division* could be used to describe a physical object, such as the

M

parts of a fax machine; *classification* could be used to organize individual components, such as a listing of Web sites grouped by topic.

- General and specific method of development proceeds either from general information to specific details or from specific information to a general conclusion.

- Order-of-importance method of development presents a sequence that reflects the relative importance of each detail. The information can be presented in either decreasing order of importance, as in a proposal that begins with the most important point, or increasing order of importance, as in a presentation that ends with the most important point.

- Sequential method of development emphasizes the order of elements and is particularly useful when writing step-by-step instructions.

- Spatial method of development describes the physical appearance of an object or area (such as a room) from top to bottom, inside to outside, front to back, and so on.

Methods of development often overlap — rarely does a writer rely on only one method. Nevertheless, you should select one primary method of development appropriate to your writing task and base your outline on it, and then subordinate any other methods to it. (See outlining.) For example, in describing the organization of a company, you could use elements from three methods of development. You could divide the larger topic — the company — into departments (division and classification), arrange the departments by their order of importance within the company (order of importance), and present their operations in sequence succession (sequential).

minutes of meetings

Organizations and committees keep official records of their meetings; such records are known as *minutes*. If you attend many business meetings, you may be asked to write and distribute the minutes of a meeting. For advice on conducting meetings, see meetings.

Because minutes are often used to settle disputes, they must be accurate, complete, and clear. When approved, minutes of meetings are official and can be used as evidence in legal proceedings.

Keep your minutes brief and to the point. Except for recording motions, which must be transcribed word for word, summarize what occurs and paraphrase discussions. To keep the minutes concise, follow a set format and use headings for each major point discussed. See also note-taking.

Avoid abstractions and generalities; always be specific. Refer to everyone in the same way—a lack of consistency in titles or names may suggest a deference to one person at the expense of another. Avoid adjectives and adverbs that suggest good or bad qualities, as in "Mr. Sturgess's *capable* assistant read the *comprehensive* report to the subcommittee." Minutes should be as objective and impartial as possible.

If a member of the committee is to follow up on something and report back to the committee at its next meeting, clearly state the person's name and the responsibility he or she has accepted. An example of minutes is shown in Figure M–5 below.

NORTH TAMPA MEDICAL CENTER

Minutes of the Monthly Meeting
Medical Audit Committee

DATE: July 25, 2003

PRESENT: G. Miller (Chair), C. Bloom, J. Dades, K. Gilley,
             D. Ingoglia (Secretary), S. Ramirez
ABSENT: D. Rowan, C. Tsien, C. Voronski, R. Fautier, R. Wolf

Dr. Gail Miller called the meeting to order at 12:45 p.m. Dr. David Ingoglia made a motion that the June 2, 2003, minutes be approved as distributed. The motion was seconded and passed.

The committee discussed and took action on the following topics.

(1) TOPIC: Meeting Time

Discussion: The most convenient time for the committee to meet.
Action taken: The committee decided to meet on the fourth Tuesday of every month, at 12:30 p.m.

FIGURE M–5. Minutes of a Meeting

Writer's Checklist: Preparing Minutes of Meetings

Include the following in meeting minutes:

- ☑ The name of the group or committee holding the meeting
- ☑ The topic of the meeting

Writer's Checklist: Preparing Minutes of Meetings (continued)

☑ The kind of meeting (a regular meeting or a special meeting called to discuss a specific subject or problem)

☑ The number of members present and, for committees or boards of ten or fewer members, their names

☑ The place, time, and date of the meeting

☑ A statement that the chair and the secretary were present or the names of any substitutes

☑ A statement that the minutes of the previous meeting were approved or revised

☑ A list of any reports that were read and approved

☑ All the main motions that were made, with statements as to whether they were carried, defeated, or tabled (vote postponed), and the names of those who made and seconded the motions (motions that were withdrawn are not mentioned)

☑ A full description of resolutions that were adopted and a simple statement of any that were rejected

☑ A record of all ballots with the number of votes cast for and against resolutions

☑ The time the meeting was adjourned (officially ended) and the place, time, and date of the next meeting

☑ The recording secretary's signature and typed name and, if desired, the signature of the chairperson

mixed constructions

A *mixed construction* is a sentence in which the elements do not sensibly fit together. The problem may be a **grammar** error, a **logic error**, or both, as in the following example:

- I have a degree in accounting along with 12 years of experience sets a foundation for a strong background in analyzing problems and assessing solutions. [The **verb** *sets* has no logical subject.]

The following revision puts the elements together in a **sentence construction** (a *complex sentence*) that is both grammatically and logically complete.

- I have a degree in accounting and 12 years of experience, which provide me with a strong background in analyzing problems and assessing solutions.

modifiers

Modifiers are words, phrases, or clauses that expand, limit, or make more precise the meaning of other elements in a sentence. Although we can create sentences without modifiers, we often need the detail and clarification they provide.

WITHOUT MODIFIERS Production decreased.

WITH MODIFIERS *Automobile* production decreased *rapidly*.

Most modifiers function as <u>adjectives</u> or <u>adverbs</u>. Adjectives describe qualities or impose boundaries on the words they modify (*loud* machinery, *ten* automobiles, *this* printer, *an* animal). An adverb modifies an adjective, another adverb, a <u>verb</u>, or an entire <u>clause</u>.

- Under test conditions, the brake pad showed *much* less wear than it did under actual conditions.
  [The adverb *much* modifies the adjective *less*.]

- The redesigned brake pad lasted *much* longer.
  [The adverb *much* modifies another adverb, *longer*.]

- The wrecking ball hit the wall of the building *hard*.
  [The adverb *hard* modifies the verb *hit*.]

- *Surprisingly,* the machine failed even after all the durability and performance tests it had passed.
  [The adverb *surprisingly* modifies an entire clause: *the machine failed*.]

Adverbs are <u>intensifiers</u> when they increase the impact of adjectives (*very* fine, *too* high) or adverbs (*rather* quickly, *very* slowly). Be cautious using intensifiers; their overuse can lead to exaggeration and hence to inaccuracies.

Stacked (Jammed) Modifiers

Stacked modifiers are strings of modifiers preceding nouns that make writing unclear or difficult to read.

- Your *staffing-level authorization reassessment* plan should result in a major improvement.

The noun *plan* is preceded by three long modifiers, a string that slows the reader down and makes the sentence awkward. Stacked modifiers are often the result of an overuse of <u>jargon</u> or <u>buzzwords</u>. See how breaking up the stacked modifiers makes the example easier to read.

M

- Your plan for reassessing the staffing-level authorizations should result in a major improvement.

Misplaced Modifiers

A modifier is misplaced when it modifies the wrong word or phrase. A misplaced modifier can cause <u>ambiguity</u>.

- We *almost* lost all of the parts.
  [The parts were *almost* lost but were not.]

- We lost *almost* all of the parts. [Most of the parts were in fact lost.]

To avoid ambiguity, place modifiers as close as possible to the words they are intended to modify. Likewise, place phrases near the words they modify. Note the two meanings possible when the phrase is shifted in the following sentences:

- The equipment *without the accessories* sold the best.
  [Different types of equipment were available, some with and some without accessories.]

- The equipment sold the best *without the accessories*.
  [One type of equipment was available, and the accessories were optional.]

Place clauses as close as possible to the words they modify.

REMOTE	We sent the brochure to four local firms *that had four-color art.*
CLOSE	We sent the brochure *that had four-color art* to four local firms.

Squinting Modifiers

A modifier squints when it can be interpreted as modifying either of two sentence elements simultaneously, thereby confusing readers about which is intended. See also <u>dangling modifiers</u>.

- We agreed *on the next day* to make the adjustments.
  [Did they agree *to make the adjustments on the next day?* Or *on the next day,* did they agree to make the adjustments?]

A squinting modifier can sometimes be corrected simply by changing its position, but often it is better to rewrite the sentence.

- We agreed that *on the next day* we would make the adjustments.
  [The adjustments were to be made on the next day.]

- *On the next day,* we agreed that we would make the adjustments.
  [The agreement was made on the next day.]

mood

The grammatical term *mood* refers to the verb functions that indicate whether the <u>verb</u> is intended to make a statement, ask a question, give a command, or express a hypothetical possibility.

The *indicative mood* states a fact, gives an opinion, or asks a question.

- The setting *is* correct.

- *Is* the setting correct?

The *imperative mood* expresses a command, suggestion, request, or entreaty. In the imperative mood, the implied subject *you* is not expressed. (*Install* the system today.)

The *subjunctive mood* expresses something that is contrary to fact, conditional, hypothetical, or purely imaginative; it can also express a wish, a doubt, or a possibility. In the subjunctive mood, *were* is used instead of *was* in clauses that speculate about the present or future, and

---

ESL TIPS FOR DETERMINING MOOD

In written and especially in spoken English, there is an increasing tendency to use the indicative mood where the subjunctive traditionally has been used. Note the differences between traditional and contemporary usage in the following examples.

Traditional use of the subjunctive mood

- I wish he *were* here now.

- If I *were* going to the conference, I would room with him.

- I requested that she *show* up on time.

Contemporary (informal) use of the indicative mood

- I wish he *was* here now.

- If I *was* going to the conference, I would room with him.

- I requested that she *shows* up on time.

As a nonnative speaker of English, you are faced with a choice: Do you use the subjunctive and, consequently, in some circles sound sophisticated, intellectual, or even weird? Or do you use the indicative and in other circles sound uneducated? The answer might be to master both uses and be able to move freely between the different circles. In formal business and technical writing, however, it is best to use the more traditional expressions.

M

the base form (*be*) is used following certain verbs, such as *propose, request,* or *insist.* (See also progressive <u>tense</u>.)

- If we *were* to close the sale today, we would meet our monthly quota.
- The senior partner insisted that she [I, you, we, they] *be* in charge of the project.

The most common use of the subjunctive mood is to express clearly that the writer considers a condition to be contrary to fact. If the condition is not considered to be contrary to fact, use the indicative mood.

SUBJUNCTIVE   If I *were* president of the firm, I would change several hiring policies.

INDICATIVE   Although I *am* president of the firm, I don't feel that I control every aspect of its policies.

Ms. / Miss / Mrs.

Ms. is widely used in business and public life to address or refer to a woman, especially if her marital status is either unknown or irrelevant to the context. More traditionally, *Miss* is used to refer to an unmarried woman, and *Mrs.* is used to refer to a married woman. Some women may indicate a preference for *Ms., Miss,* or *Mrs.,* which you should honor. If a woman has an academic or professional title, use the appropriate form of address (*Doctor, Professor, Captain*) instead of *Ms., Miss,* or *Mrs.*

mutual / common

Common is used when two or more persons (or things) share something or possess it jointly.

- We have a *common* desire to make the program succeed.
- The fore and aft guidance assemblies have a *common* power source.

Mutual may also mean "shared" (*mutual* friend, of *mutual* benefit), but it usually implies something given and received reciprocally and is used with reference to only two persons or parties.

- Melek mistrusts Roth, and I am afraid the mistrust is *mutual.* [Roth also mistrusts Melek.]

N

narration

Narration is the presentation of a series of events in a prescribed (usually **chronological**) sequence. Much narrative writing explains how something happened: a laboratory study, a site visit, an accident, the decisions in an important meeting. See also **trip reports** and **trouble reports**.

Effective narration rests on two key writing techniques: the careful, accurate sequencing of events and a consistent **point of view** on the part of the narrator. Narrative sequence and essential shifts in the sequence are signaled in three ways: chronology (clock and calendar time), transitional words pertaining to time (*before, after, next, first, while, then*), and verb tenses that indicate whether something has happened (past **tense**) or is underway (present tense). The point of view indicates the writer's relation to the information being narrated as reflected in the use of **person**. Narration usually expresses a first- or third-person point of view. First-person narration indicates that the writer is a participant, and third-person narration indicates that the writer is writing about what happened to someone or something else.

The narrative shown in Figure N–1 on page 358 reconstructs the final hours of the flight of a small aircraft that attempted to land at the airport in Hailey, Idaho. Instead, the plane crashed into the side of a mountain, killing the pilot and the copilot. Because of the disastrous outcome of the flight, the investigators needed to "tell the story" in detail so that any lessons learned could be made available to other fliers. To do that, they had to recount the events as closely as possible. Thus, the chronology of the events is specified throughout the narrative in local (Hailey, Idaho) time.

To tell the story, the narrator used the third-person point of view throughout, except for the first-person reports from witnesses. The words of each witness were quoted for their bearing on what happened, from that person's vantage point. Investigators sequenced the events as precisely as possible from the beginning of the flight until the crash site was located by referencing verified clock times and approximating those that could not be verified. The verb tenses throughout indicate a past action: *departed, canceled, cleared, called, reported.*

History of the Flight*

At 0613 m.s.t.[1] on January 3, 1983, N805C, a Canadair Challenger owned and operated by the A.E. Staley Company departed Decatur, Illinois, on a flight to Friedman Memorial Airport, Hailey, Idaho. The route of the flight was via Capitol, Illinois, Omaha, Nebraska, Scotts Bluff, Nebraska, Riverton, Wyoming, Idaho Falls, Idaho, direct 43°30′ north latitude, 114°17′ west longitude.

The en route portion of the flight was uneventful, and about 35 nmi east of Idaho Falls N805C was cleared by the Salt Lake City, Utah, Air Route Traffic Control Center (ARTCCs) to descend from 39,000 feet to 22,000 feet. N805C descended to 22,000 feet and the flightcrew then requested a descent to 17,000 feet. About 35 nmi east of Sun Valley Airport, after being cleared, N805C descended to 17,000 feet. About 0901, N805C's flightcrew canceled their flight plan and, shortly thereafter, changed the transponder from 1311, the assigned discrete code, to 1200, the visual flight rules (VFR) code. At 0901:07, the data analysis reduction tool (DART) radar data showed a 1311 beacon code at 17,000 feet about 11 nmi east of the Sun Valley Airport. At 0901:37, the DART radar data showed a 1200 VFR transponder beacon code with no altitude readout about 2 nmi west of the 1311 beacon code that was recorded at 0901:07.

At 0904:10, DART radar data recorded a 1200 code target at 13,500 feet almost directly over the Sun Valley Airport. According to an employee of the airport's fixed base operator, N805C's flightcrew called on the airport's UNICOM[2] frequency and requested a landing advisory and asked if a food order had been placed. The flightcrew then stated that there would "be a quick turn," and placed a fuel request. This was the last transmission heard from N805C. The employee said that she provided the latest altimeter setting to N805C, and "since we did not have the cloud conditions in the area, I was glad when other pilots were able to give reports as they saw things from the air."

The flightcrew of Cessna Citation, N13BT, which had landed at Sun Valley about 0903, also heard N805C report "over the field." According to N13BT's pilot, N805C reported over the field "sometime during our final approach or landing." According to the pilot, the weather at the airport when he landed "was 800 (feet) overcast with 10 miles visibility. The tops of the overcast fog bank was about 6,800 (feet) m.s.l." He said that the overcast was "solid northwest up the valley. Visibility appeared lower (to the) northwest."

About 0908, Trans Western Flight 1301, a Convair 580, landed at Sun Valley Airport. Flight 1301 had descended through a hole in the overcast about 15 nmi southwest of Bellevue, Idaho, which is about 3 nmi southeast of the Sun Valley

---

*National Transportation Safety Board, "Aircraft Accident Report: A. E. Staley Manufacturing Co., Inc., Canadair Challenger CL-600, N805C, Hailey, Idaho, January 3, 1983." National Technical Information Service, Springfield, Va., 1983.

[1]All times, unless otherwise noted, are mountain standard time based on the 24-hour clock.

[2]UNICOM. The non-government air/ground radio communications facility which may provide airport advisory information at certain airports. The Sun Valley UNICOM did not record, nor was it required to record or log, the time of radio communications.

FIGURE N–1. Narration

Airport. The first officer said that he gave position reports to the Sun Valley UNI-COM when the flight was 15 nmi from the airport, when it was 10 nmi from the airport over Bellevue turning on final approach for runway 31, and when it was 1 mile from the runway. The first officer said that he could see the visual approach slope indicator (VASI) lights for runway 30 during the landing approach. The captain and the first officer said that they neither saw N805C nor heard radio transmissions from N805C.

About 0900, a man who was driving his truck north on the highway between Bellevue and Hailey, Idaho, saw a twin engine, cream-colored jet, break through the clouds when he was about 2.5 miles north of Bellevue. He saw that the landing gear was down but he did not see any lights on the airplane. When the airplane appeared, "it was about 300 to 500 yards from the west hills adjacent to the airport and about 1,000 feet from the valley floor." The airplane was in a noseup attitude. The witness said that after the airplane descended below the clouds "and (the pilot) saw how close to the hills he was, he then started a sharp right turn." The airplane disappeared from his view into "low hanging clouds" over the northwest side of the hangar at the airport.

Between 0900 and 0930, another man, who was in the yard of his home in northeast Hailey, saw a jet airplane east of his home. The airplane was "white or silver with a blue tint." (N805C was painted white with blue and gold stripes along the length of the fuselage and tops of the wings.) The airplane was below the clouds, and he had "a good view of the airplane for about 10 to 15 seconds." He said that the airplane had a noseup attitude and "the wings were rocking up and down about 20°." The witness said that the clouds obscured all but the lower peaks of the mountains to the east and that after he lost sight of the airplane he thought it was "odd that the aircraft was under cloud cover."

Shortly after 0900, a woman who was located in an apartment in southeast Hailey, heard a jet airplane fly over "in a northerly direction." She thought that this was "odd because jets don't go over us heading north from the airport. The engines sounded very loud. . . ."

A fourth witness said that, between 0900 and 0940, she heard a jet airplane overfly her house in northeast Hailey. The woman was in the living room of her house when she heard the airplane and thought that "it must be low because of the loudness of the (engine) noise," and that "the sound of the jet did not trail off as they do as they fly farther away from you. The sound stopped less than 30 seconds from the time I first heard it." At the time, the clouds were resting on and hiding the tops of the mountains to the east.

About 1030, the chief pilot of the A. E. Staley Manufacturing Company, who was to board N805C at Sun Valley, arrived at the airport. Since the airplane was overdue, he instituted inquiries to several nearby airports to determine where the airplane had landed. At 1300, he asked air traffic control to make a full communications search. At 1400, after being told that the airplane had not been found, he requested an air search. While waiting for search and rescue teams to arrive, the chief pilot rented an airplane and at about 1700, found the accident site. The impact site, elevation about 6,510 feet, was about 2.2 nmi north of Sun Valley Airport at coordinates 43°32′50″ N latitude, 114°17′35″ W longitude.

FIGURE N-1. Narration *(continued)*

Important but secondary information was either mentioned in a footnote or, like the color of the plane, inserted in parentheses. Information about the color of the plane was important to the narration because it verified a sighting by an eyewitness; otherwise, it would have been unnecessary.

Although narration often exists in combination with other <u>forms of discourse</u>, once a narrative is underway, it should not be interrupted by lengthy explanations or analyses. Explain only what is necessary for <u>readers</u> to follow the events.

nature

Nature, when used to mean "kind" or "sort," is vague. Avoid this usage in your writing. Say exactly what you mean.

- The ~~nature of~~ the contract caused the problem. *(exclusionary clause in)*

needless to say

Although the phrase *needless to say* sometimes occurs in speech and writing, you can either eliminate the phrase or comment on a statement to follow with a more descriptive word.

- ~~Needless to say,~~ departmental cutbacks have meant decreased efficiency. *(Understandably,)*

neo- (*see* prefixes)

new words

New words, usages, or expressions (also called *neologisms*) continually find their way into the language from a variety of sources. Some come from technology (*download*), scientific research (*in vitro*), and brand or trade names (*Xerox*). Others are formed by blending existing words (*netiquette* from Internet plus etiquette), <u>acronyms</u> (*spool* from simultaneous peripheral operations online). Some words have come from other languages, such as *skiing* (Norwegian) and *whiskey* (Gaelic).

Business and technology are responsible for many new words. Some are necessary and unavoidable because they communicate new concepts concisely and accurately. However, it is best to avoid creating a new word if an existing word will do. If you do use a new word or expression you believe a reader may not know, define it the first time you use it. See also **affectation**, **buzzwords**, **defining terms**, and **word choice**.

newsletter articles

If your organization publishes a **newsletter**, you may be asked to contribute an article on a subject in your area of expertise. In fact, an article is a good way to promote your work or your department.

Before you begin to write, consider the traditional *who, what, where, when,* and *why* of journalism (*Who* did it? *What* was done? *Where* was it done? *When* was it done? *Why* was it done?) and then add *how,* which may be of as much interest to your colleagues as any of the five *w*'s.

Next, determine whether the company has an official policy or position on your subject. If so, adhere to it as you prepare your article. If there is no company policy, determine as nearly as you can management's attitude toward your subject.

Gather several fairly recent issues of the newsletter and study the **style** and **tone** of the writing and the approach used for various kinds of subjects. Understand those perspectives before you begin to work on your own article. Ask yourself the following questions about your subject: What is its significance to the organization? What is its significance to my coworkers? The answers to those questions should help you establish the style, tone, and approach for your article and also heavily influence your conclusion.

Research for a newsletter article frequently consists of **interviewing for information**. Interview everyone concerned with your subject. Get all available information and all points of view. Be sure to give maximum credit to the maximum number of people.

Writer's Checklist: Newsletter Articles

Writing a newsletter article requires a more journalistic approach than writing a **report**. Because newsletters are usually not required reading, you do not have a captive audience. Use the following tips to achieve **emphasis**.

☑ Write an intriguing **title** to catch the **readers'** attention; **rhetorical questions** often work well.

Writer's Checklist: Newsletter Articles (continued)

☑ Include eye-catching **photographs** or **visuals** that entice your audience to read the lead **paragraph** of your **introduction**. See also **layout and design**.

☑ Fashion a lead, or first paragraph, that will encourage further reading. The first paragraph generally makes the **transition** from the title to the substantive body of the article.

☑ Offer a well-developed presentation of your subject to hold the readers' interest all the way to the end of the article.

☑ Write a **conclusion** that emphasizes the significance of your subject to your audience and stresses the points you want your readers to retain.

In preparing your newsletter article, you may find it helpful to follow the steps listed in the Checklist of the Writing Process. Figure N–2 is an article written for a newsletter.

newsletters

Newsletters are publications that are designed to inform and to create and sustain interest and membership in an organization. They can also be used to sell products and services. There are two main types of newsletters: organizational newsletters and subscription newsletters.

Organizational newsletters are sent to employees or members of an association to keep them informed about issues regarding their company or group, such as the development of new products or policies, or the accomplishments of individuals or teams. Stories included in organizational newsletters can also be created to call members to take a specific action. For example, a health club's newsletter could detail new equipment to be installed and explain how to use it. If you are asked to contribute an article to a newsletter, see **newsletter articles**.

Subscription newsletters are designed to attract and build a readership interested in buying specific products or services or in learning more about investing or financial matters. Subscribers are buying information, and they expect a certain level of value for their money. For example, a person with experience in the stock market could create a financial newsletter and charge subscribers a monthly fee for the investing advice in that newsletter; a person who collects movie memorabilia could create an online newsletter that includes stories about ways to find and sell rare movie posters.

Before you begin to develop a newsletter, decide on its specific purpose and the specific **readers** you will be targeting; then make sure

Ken Cook Co.

Partners in Product Documentation

www.kencook.com

June 2000

NEW PARTS MANUAL CONCEPT FOR LIFE FITNESS

Vincent A. Hebein, Manager of Logistics and Product Manager of Customer Support Services, of Life Fitness had a problem. He needed to create a catalog for Customer Support Services that would raise the bar for product support catalogs in the fitness and sports accessories industry.

Ken Cook Co. had the answer: Instead of using confusing line and exploded view drawings to illustrate replacement parts for Life Fitness equipment, why not use a "see through" photograph to illustrate the parts placement on the machine? Warren Metzger, Senior Technical Writer, and Dave Duecker and Pete Zielinski, Documentation Design Specialists, of Ken Cook Co. put the process together.

After some experimentation, we found a way of "peeling off" enough of the outside covers to visualize the components inside. You need a photograph of the machine with the outside covers on and another one with the outside covers off without moving the camera. Peter used Adobe® Photoshop® to bring both images together and reveal just the right amount of inside detail to get the desired effect. Dave used QuarkXPress™ to design the page layout and index the closeup photograph of each component to the illustration of each machine prepared by Peter. All pages and the cover were designed to complement the Life Fitness product catalog released earlier this year.

Metzger directed the photography at the new Customer

Support Services facility in Franklin Park, Illinois. Tracy Pecs of Customer Support Services chose the replacement parts we needed to illustrate and made sure that we had the correct parts available to photograph. Aaron Sztuk of Life Fitness co-ordinated the effort of removing the shrouds and covers of each machine.

"This catalog is really a showcase," according to Jeff Mayer, Sales and Marketing Executive of Ken Cook Co. "Many applications lend themselves to this process from automobiles to outboards," said Mayer. Brett Kehoe, Account Executive, and Beth Bergeson, Coordinator, at Ken Cook Co. helped Mayer manage the project and keep it on course.

Life Fitness, a division of Brunswick Corporation, is the global leader in designing and manufacturing a full line of reliable, high-quality fitness equipment for commercial and consumer use. Its cardiovascular and strength-training products, including the renowned Lifecycle exercise bike, are used in health, fitness and wellness facilities as

well as in homes worldwide. The company is headquartered near Chicago and distributes its equipment in more than 80 countries.

Life Fitness is the largest commercial fitness equipment provider in the world with the No. 1 market share in this segment. The Lifecycle exercise bike is currently the No. 1 bike in commercial facilities and has set the standard for efficient, easy-to-use workout programs since its introduction 30 years ago in 1968. Likewise, Life Fitness treadmills, introduced in 1991, are the No. 1 treadmills in commercial facilities. The company is also a leading manufacturer of cross-trainers, stairclimbers and strength-training equipment.

FEATURES	
New Parts Manual Concept for Life Fitness	1
KEN COOK CO. increases production with new equipment	2
KEN COOK CO. seeks ISO 9000 certification	3

FIGURE N–2. Newsletter Article

the newsletter's appearance and editorial choices create a sense of identification among the readership as shown in Figure N–3 on page 364. See also **persuasion** and **promotional writing**.

You will need to acquire a *mailing list* (names and addresses of your readers), and you will need to decide on the most strategic way to get the newsletter to the readers, whether through interoffice mail, the post office, or online. Because it can be time-consuming and technically problematic to send out hundreds or thousands of online newsletters by yourself, you may also need to subscribe to a list-hosting service.

As you develop and edit articles, you need to <u>research</u> the topics and interview relevant sources. As you research trade journals, business and technology magazines, newspapers, or the <u>Internet</u>, find specific angles for the articles that will appeal to your select audience. Attempt

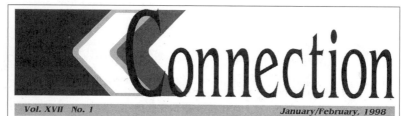

Connection

Vol. XVII No. 1   January/February, 1998

Accounts Grow To Include Literature Management Services

Festo Corporation, Hauppauge, NY; and Motorola-Core Technologies, Consumer Systems Group, Austin, TX; have contracted Ken Cook Co. to provide comprehensive Literature Management and Order Fulfillment services. The expansion of services to these manufacturers follows an industry-wide trend to subcontract non-core competencies to an outside source.

Festo Corporation

An electronic information sharing environment built between Festo Corporation and Ken Cook Co. provides seamless document management for this manufacturer of pneumatic products and components. Now this 15-year partnership has grown to include total Literature Management and Order Fulfillment services.

According to Bruce Radtke, KCC Account Representative, 175 pallets of Festo materials have been received, bar coded and bulk cartoned at the Ken Cook warehouse. Another 100 bins are available for literature picking and distribution. "We accept and fulfill fax orders directly from Festo sales regions," he says. "Up to 50 line items per order are processed."

Complete graphics, printing and binding services are also provided to maintain inventory levels. Ken Cook processes and tracks all orders electronically and provides monthly reports of shipping activity and inventory in a PC format accessible to the Festo Advertising Department.

"Ken Cook and Festo have a long-standing business partnership," says Rich Day, Director of Marketing at Festo. "Over the years they have demonstrated their credibility, and professional and

technical expertise through quality products and services. They understand our business, our customers and our needs."

"Realizing Festo is an ISO 9001 certified company, we understand that quality, customer satisfaction and timeliness of shipments are extremely important concerns," adds Bruce. "We fully expect to meet or exceed Festo requirements and expectations in this area. We believe that the expertise gained by managing similar programs with other clients will enable us to fulfill their quality assurance requirements."

Motorola-Core Technologies, Consumer Systems Group

Anticipating significant growth in product sales, the Motorola-Core Technologies, Consumer Systems Group (MCTG) contracted Ken Cook Co. as its sole source for Literature Management and Order Fulfillment. "The Group wants to be

continued on page 2

In This Issue:

KEN COOK CO.

FIGURE N–3. Company Newsletter (Front Page)

to provide content that they won't read elsewhere. Other strategies include interviewing and profiling customers or your employees. Be aware that your fact-checking needs to be meticulous. Unlike the general readership of a newspaper, newsletter readers are often specialists in their fields. See also <u>interviewing for information</u>.

As shown in Figure N–3, a newsletter's format should be simple and consistent, yet visually appealing to your readership. Use the active <u>voice</u> and a conversational <u>tone</u>. Boldface names of customers or association members included in your articles to identify them and bring attention to their contributions. Use subheadings and bullets to break up the text and make the newsletter easy to read. Keep your sentences simple and paragraphs short. See <u>conciseness/wordiness</u> and <u>layout and design</u>.

Using word-processing or desktop publishing software, create newspaper columns and one or two <u>visuals</u> per page that complement the text. On the front page, identify the organization, include the date and volume and issue numbers, and include a contents box. Depending on the length and quantity of the text and <u>photographs</u> or other visuals to be included, newsletters are usually $8\frac{1}{2}'' \times 11''$ or $17'' \times 22''$ pages folded in half. If your budget is generous and the scale of your project is substantial, you may want to work with a professional printer to produce your newsletter.

nominalizations

N

A *nominalization* is a weak verb (*make, do, conduct, perform*) combined with a noun, when the verb form of the noun would communicate the same idea more effectively and concisely.

- The quality assurance team will ~~perform an evaluation of~~ the new software.

  evaluate

If you use nominalizations just to make your writing sound more formal, the result will be <u>affectation</u>. You may occasionally have a legitimate use for a nominalization. For example, you might use a nominalization to slow down the <u>pace</u> of your writing. See also <u>technical writing style</u> and <u>conciseness/wordiness</u>.

none

None can be considered either a singular or a plural <u>pronoun</u>, depending on the context. (See also <u>agreement</u>.)

- *None* of the material *has* been ordered.
  [Always use a singular <u>verb</u> with a singular <u>noun</u>, in this case, "material."]

- *None* of the clients *has* been called yet.
  [Use a singular verb even with a plural noun (*clients*) if the intended emphasis is on the idea of *not one*.]

- *None* of the clients *have* been called yet.
  [Use a plural verb if you intend *none* to refer to all clients.]

For <u>emphasis</u>, substitute *no one* or *not one* for *none* and use a singular verb.

- We paid the retail price for three of the machines, ~~none~~ of which was worth the money.

not one (inserted above "none")

nor / or

Nor always follows *neither* in sentences with continuing negation. (They will *neither* support *nor* approve the plan.) Likewise, *or* follows *either* in sentences. (The firm will accept *either* a short-term *or* a long-term loan.)

Two or more singular subjects joined by *or* or *nor* usually take a singular <u>verb</u>. However, when one subject is singular and one is plural, the verb agrees with the subject nearer to it. See also <u>conjunctions</u>.

> SINGULAR   Neither the *architect* nor the *client was* happy with the design.

> PLURAL   Neither the *architect* nor the *clients were* happy with the design.

> SINGULAR   Neither the *architects* nor the *client was* happy with the design.

note-taking

The purpose of note-taking is to summarize and record information you extract during <u>research</u>. (For taking notes at a meeting, see <u>minutes of meetings</u>.) The challenge in taking notes is to condense another writer's thoughts into your own words without distorting the original thinking or plagiarizing. As you extract information, let your knowledge of the audience and the **purpose** of your writing guide you. Resist the temptation to copy your source word for word as you take notes; instead, paraphrase the author's idea or concept. You must do more than just change a few words in the original passage; otherwise, you will be guilty of **plagiarism**. See also **paraphrasing**.

On occasion, when your source concisely sums up a great deal of information or points to a trend important to your subject, you are justified in directly quoting the source and incorporating it into your document. As a general rule, you will rarely need to quote anything longer than a paragraph. If you decide to use a direct quote, enclose the material in **quotation marks** in your notes. In your finished writing, provide the sources of your **quotations**. See also **documenting sources**.

When taking notes on abstract ideas, as opposed to factual data, do not sacrifice clarity for brevity—notes expressing concepts can lose their meaning if they are too brief. The critical test is whether a week later you understand the note and can recall the significant ideas of the passage. Consider the information in the following paragraph:

> Long before the existence of bacteria was suspected, techniques were in use for combating their influence in, for instance, the decomposition of meat. Salt and heat were known to be effective, and these do in fact kill bacteria or prevent them from multiplying. Salt acts by the osmotic effect of extracting water from the bacterial cell fluid. Bacteria are less easily destroyed by osmotic action than are animal cells because their cell walls are constructed in a totally different way, which makes them very much less permeable.

The paragraph says essentially three things:

1. Before the discovery of bacteria, salt and heat were used to combat the effects of bacteria.
2. Salt kills bacteria by extracting water from their cells by osmosis, hence its use in curing meat.
3. Bacteria are less affected by the osmotic effect of salt than are animal cells, because bacterial cell walls are less permeable.

If your readers' needs and your objective involved tracing the origin of the bacterial theory of disease, you might want to note that salt was traditionally used to kill bacteria long before people realized what caused meat to spoil. It might not be necessary to your topic to say anything about the relative permeability of bacterial cell walls. Figure N–4 on page 368 is an example of a first note taken from the paragraph on the use of salt as a preservative; Figure N–5 on page 368 is an example of a subsequent note.

Although you should record notes in a way you find efficient, noth-

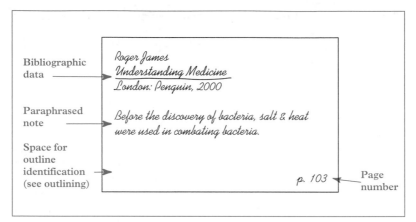

FIGURE N–4. First Note from a Source

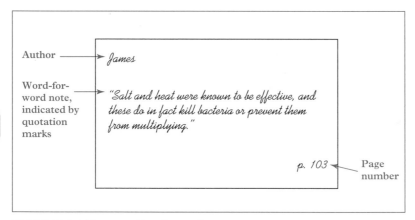

FIGURE N–5. Subsequent Note from a Source

ing replaces 3″ × 5″ index cards (as shown in Figures N–4 and N–5) for some tasks. They are especially useful for <u>outlining</u> a complex project.

Writer's Checklist: Taking Notes

Ask the following questions:

- ☑ What information do you need to fulfill your purpose? How much do your <u>readers</u> know about your subject? What are their needs?
- ☑ Record only the most important ideas and concepts. Be sure to record all vital names, dates, and definitions.

Writer's Checklist: Taking Notes (continued)

☑   When in doubt about whether to take a note, consider the difficulty of finding the source again should you want it later.

☑   Use a quotation when a source summarizes a great deal of information or points to an important trend.

☑   Ensure proper credit: Record the author, title, publisher, place and date of publication, and page number. (On subsequent notes from the same source, you will need to include only the author and the page number.)

☑   Use your own shorthand and record notes in a way that you find efficient, whether in an electronic document or on note cards.

☑   Photocopy and highlight passages that you intend to quote.

☑   Print out key sections from online documents or download the information to a notes file: Be sure to copy the full URL and your retrieval date for Web sites.

☑   Check your notes for accuracy against the original material before moving on to another source.

nouns

DIRECTORY

N

A *noun* names a person, place, thing, concept, action, or quality.

Types of Nouns

The two basic types of nouns are proper nouns and common nouns. *Proper nouns,* which are capitalized, name specific people, places, and things (H. G. Wells, Boston, United Nations, Nobel Prize). See also capitalization.

Common nouns, which are not capitalized unless they begin sentences, name general classes or categories of persons, places, things, concepts, actions, and qualities (writer, city, organization, award). Common nouns include collective nouns, abstract nouns, concrete nouns, count nouns, and mass nouns.

Collective nouns are common nouns that indicate a group or collection. They are plural in meaning but singular in form (audience, jury, brigade, staff, committee). (See the subsection Collective Nouns below for advice on using singular or plural forms with collective nouns.)

Abstract nouns are common nouns that refer to things that cannot be discerned by the five senses (loyalty, pride, valor, peace, devotion).

Concrete nouns are common nouns used to identify those things that can be discerned by the five senses (paper, keyboard, glue, nail, grease).

Count nouns are concrete nouns that identify things that can be separated into countable units (desks, envelopes, printers, pencils, books).

Mass nouns are concrete nouns that identify things that are a mass rather than individual units and that cannot be easily separated into countable units (water, air, electricity, oil, cement).

Noun Usage

Nouns function as subjects of <u>verbs</u>, direct and indirect objects of verbs and <u>prepositions</u>, subjective and objective <u>complements</u>, or <u>appositives</u>.

- The *metal* failed during the test. [subject]
- The bricklayer cemented the *blocks* efficiently. [direct object of a verb]
- The state presented our *department* a safety award. [indirect object]
- The event occurred within the *year*. [object of a preposition]
- A dynamo is a *generator*. [subjective complement]
- The regional manager was appointed *chairperson*. [objective complement]
- George Thomas, the *treasurer*, gave his report last. [appositive]

Words normally used as nouns can also be used as <u>adjectives</u> and <u>adverbs</u>.

- It is *company* policy. [adjective]
- He went *home*. [adverb]

Collective Nouns

When a collective noun refers to a group as a whole, it takes a singular verb and pronoun.

- The staff *was* divided on the issue and could not reach *its* decision until May 15.

When a collective noun refers to individuals within a group, it takes a plural verb and pronoun.

● The staff *returned* to *their* offices after the conference.

A better way to emphasize the individuals on the staff would be to use the phrase *the staff members.*

● The staff members *returned* to *their* offices after the conference.

Treat organization names and titles as singular.

● The LRM Corporation *has* grown 30 percent in the last three years; *it* will move to a new facility in January.

Some collective nouns regularly take singular verbs (*crowd*); others do not (*people*).

● The crowd *was* growing impatient.

● Many people *were* able to watch the space shuttle land safely.

Forming Plurals

Most nouns form the plural by adding -*s*.

● *Dolphins* are capable of communication.

Nouns ending in -*ch*, -*s*, -*sh*, -*x*, and -*z* form the plural by adding -*es*.

● search/searches, glass/glasses, wish/wishes, six/sixes, buzz/buzzes

Nouns that end in a consonant plus -*y* form the plural by changing the -*y* to -*ies*.

● delivery/deliveries

Some nouns ending in -*o* add -*es* to form the plural, but others add only -*s*.

● tomato/tomatoes, dynamo/dynamos

Some nouns ending in -*f* or -*fe* add -*s* to form the plural; others change the -*f* or -*fe* to -*ves*.

● cliff/cliffs, fife/fifes, hoof/hooves, knife/knives

Some nouns require an internal change to form the plural.

● woman/women, man/men, mouse/mice, goose/geese

Some nouns do not change in the plural form.

● Many *fish*, several *deer*, fifty *sheep*

Hyphenated and open compound nouns form the plural in the main word.

- sons-in-law, high schools, editors-in-chief

Compound nouns written as one word add *s* to the end.

- Use seven *tablespoonfuls* of ground coffee.

If you are unsure of the proper usage, check a dictionary. See **possessive case** for a discussion of how nouns form possessives.

number (grammar)

Number is the grammatical property of **nouns**, **pronouns**, and **verbs** that signifies whether one thing (singular) or more than one (plural) is being referred to. (See also **agreement**.) Nouns normally form the plural by simply adding *s* or *es* to their singular forms.

- *Partners* in successful *businesses* are not always personal friends.

Some nouns require an internal change to form the plural.

- woman/women, man/men, goose/geese, mouse/mice

All pronouns except *you* change internally to form the plural.

- I/we, he/they, she/they, it/they

By adding an *s* or *es,* most verbs show the singular of the third **person**, present **tense**, indicative **mood**.

- he *stands,* she *works,* it *goes*

The verb *be* normally changes form to indicate the plural.

SINGULAR   I *am* ready to begin work.

PLURAL   We *are* ready to begin work.

numbers

DIRECTORY

The standards for using numbers vary; however, unless you are following an organizational or professional style guide, observe the following guidelines.

Numerals or Words

Write numbers from zero to ten as words and numbers above ten as numerals. Spell out approximate numbers.

- We've had over *a thousand* requests this month.

In most writing, do not spell out ordinal numbers, which express degree or sequence (42nd), unless they are single words (tenth, sixteenth). When several numbers appear in the same sentence or paragraph, write them the same way, regardless of other rules and guidelines.

- The company owned *150* trucks, employed *271* people, and rented *7* warehouses.

Spell out numbers that begin a sentence, even if they would otherwise be written as figures.

- *One hundred and fifty* people attended the meeting.

If spelling out such a number seems awkward, rewrite the sentence so that the number does not appear at the beginning.

- The meeting was attended by *150* people.

Plurals

Indicate the plural of numerals by adding -*s* (7s, the late 1990s). Form the plural of a written number (like any **noun**) by adding -*s* or -*es* or by dropping -*y* and adding -*ies* (elevens, sixes, twenties). See also **apostrophes**.

Measurements

Express units of measurement as numerals (3 miles, 45 cubic feet, 9 meters). When numbers run together in the same phrase, write one as a numeral and the other as a word.

- The order was for ~~12~~ *twelve* 6-foot tables.

Generally give percentages as numerals and write out the word *percent*, except when the number is in a **table**. (Approximately *85 percent* of the land has been sold.)

Fractions

Express fractions as numerals when they are written with whole numbers (27½ inches, 4¼ miles). Spell out fractions when they are expressed without a whole number (one-fourth, seven-eighths). Always write decimal numbers as numerals (5.21 meters).

Money

In general, use numerals to express exact or approximate amounts of money.

- We need to charge $28.95 per unit.
- The new system costs $60,000.

Use words to express indefinite amounts of money.

- The printing system may cost several thousand dollars.

Use numbers and words for rounded amounts of money over one million dollars.

- The contract is worth $6.8 million dollars.

Use numbers for more complex or exact amounts.

- The corporation paid $2,452,500 in taxes last year.

For amounts under a dollar, ordinarily use numerals and the word *cents,* unless other figures that require dollar signs appear in the same sentence.

- The pens cost 50 cents each.
- The business-card holders cost $10.49 each, the pens cost $.50 each, and the pencil-cup holders cost $6.49 each.

Time

Express hours and minutes as numerals when *a.m.* or *p.m.* follows (11:30 a.m., 7:30 p.m.). Spell out time that is not followed by *a.m.* or *p.m.* (four o'clock, eleven o'clock).

Dates

The year and the day of the month should be written as numerals. In the United States, dates are usually written in the month-day-year sequence, in which the year may or may not be followed by a comma.

- *August 26, 2025,* is the payoff date for the loan.

Use the strictly numerical form for <u>dates</u> (8/26/25) in informal writing only and never in <u>international correspondence</u>, where dates are often written in the day-month-year sequence.

* *26 August 2025* is the payoff date for the loan.

In the day-month-year sequence there is no comma following the year.

Addresses

Spell out numbered streets from one to ten unless space is at a premium (East Tenth Street). Write building numbers as numerals. The only exception is the building number *one* (4862 East Monument Street; One East Monument Street). Write highway numbers as numerals (U.S. 70, Ohio 271, I-94).

ESL TIPS FOR PUNCTUATING NUMBERS

The rules for punctuating numbers in U.S. English are summarized as follows.
    A comma separates numbers with five or more digits into groups of three, starting from the right.

* 57,890 cubic feet
* $187,291
* 5,289,112,001 atoms

In numbers with four digits, the comma is optional.

* 1,902 cases *or* 1902 cases

Do not use a comma in years, house numbers, zip codes, and page numbers.

* The Boeing 777 was first flown commercially by United Airlines in June *1995*.
* Autotech Industries is located at *92401* East Alameda Drive in Los Angeles.
* The zip code is *91601*.
* The citation is located on page *1204*.

Use a period to represent the decimal point.

* Their stock values increased at a monthly rate of *4.2* percent.
* The jackpot for last week's lottery was *$3,742,097.43*.

See also <u>global communication</u> and <u>global graphics</u>.

N

Documents

In manuscripts, page numbers are written as numerals, but chapter and volume numbers may appear as numerals or words.

- Page 37, Chapter 2 or Chapter Two, Volume 1 or Volume One

Express figure and table numbers as numerals (Figure 4, Table 3).

Do not follow a word representing a number with a numeral in parentheses that represents the same number. Doing so is redundant.

- Send five ~~(5)~~ copies of the report.

N

O

objective (*see* **purpose**)

objects

The three kinds of objects are direct objects, indirect objects, and objects of **prepositions**. All objects are **nouns** or noun equivalents: **pronouns**, **verbals**, and noun **phrases** and **clauses**. See also **complements**.

Direct Objects

A *direct object* answers the question "What?" or "Whom?" about a verb and its subject.

- We sent a *full report.*
  [*Full report,* the direct object, answers the question, "We sent what?"]

- Bill telephoned the *client.*
  [*Client,* the direct object, answers the question, "Bill telephoned whom?"]

A **verb** whose meaning is completed by a direct object is called a *transitive verb.*

Indirect Objects

An *indirect object* is a noun or noun equivalent that occurs with a direct object after certain kinds of transitive verbs, such as *give, wish, cause,* and *tell.* The indirect object answers the question "To whom or what?" or "For whom or what?" The indirect object always precedes the direct object.

- We sent the *general manager* a full report.
  [*Report* is the direct object; the indirect object, *general manager,* answers the question, "We sent a full report *to whom?*"]

- The purchasing department bought *Sheila* a new printer.
  [*Printer* is the direct object; the indirect object, *Sheila,* answers the question, "The purchasing department bought a new printer *for whom?*"]

Object of a Preposition

The *object of a preposition* is a noun or pronoun that is introduced by a preposition, forming a prepositional phrase.

- At the *meeting,* the district managers approved the contract.
  [*Meeting* is the object, and *at the meeting* is the prepositional phrase.]

observance / observation

The meaning of *observance* as "keeping or complying with a duty, custom, or law" is sometimes confused with *observation,* which is the "act of noticing or recording something."

- The ~~observation~~ *observance* of Veterans Day as a paid holiday varies from one organization to another.

OK / okay

The expression *okay* (also spelled *OK*) is common in informal writing, but it should be avoided in more formal documents, such as reports.

- Mr. Sturgess ~~gave his okay to~~ *approved* the project.

on / onto / upon

On is normally used as a preposition meaning "attached to" or "located at." (Install the phone *on* the wall.) *Onto* implies movement to a position on or movement up and on. (The commuters surged *onto* the platform.)

Similarly, *on* stresses a position of rest (A book lay *on* the table), and *upon* emphasizes movement (She put a book *upon* the table).

one

When used as an indefinite **pronoun,** *one* may help you avoid repeating a **noun.** (We need a new plan, not an old *one.*) *One* is often redundant in phrases in which it restates the noun, and it may take the proper emphasis away from the **adjective.**

- The training program was not ~~a unique one.~~ *unique.*

One can also be used in place of a noun or personal pronoun in a statement. (*One* cannot ignore *one's* physical condition.) Using *one* in that way is formal and impersonal; in any but the most formal writing, you are better advised to address your reader directly and personally as *you.* (*You* cannot ignore *your* physical condition.) See also **point of view.**

one of those . . . who

A dependent **clause** beginning with *who* or *that* and preceded by *one of those* takes a plural **verb.**

- She is *one of those* managers *who are* concerned about their writing.

- This is *one of those* policies *that make* no sense when you examine them closely.

In those two examples, *who* and *that* refer to plural antecedents (*managers* and *policies*) and thus take plural verbs (*are* and *make*). See also **agreement.**

If the phrase *one of those* is preceded by *the only,* however, the verb should be singular.

- She is *the only one of those* managers *who is* concerned about her writing.
  [The verb is singular because its subject, *who,* refers to a singular antecedent, *one.* If the sentence were reversed, it would read, "Of those managers, she is *the only one who is* concerned about her writing."]

- This is *the only one of those* policies *that makes* no sense when you examine it closely.
  [If the sentence were reversed, it would read, "Of those policies, this is *the only one that makes* no sense when you examine it closely."]

O

only

In writing, the word *only* should be placed immediately before the word or phrase it modifies.

- We ~~only~~ lack financial backing. _{*only*}

(*only* inserted before "lack")

Be careful with the placement of *only* because it can change the meaning of a sentence.

- *Only* he said that he was tired.
  [He alone said that he was tired.]
- He *only* said that he was tired.
  [He actually was not tired, although he said he was.]
- He said *only* that he was tired.
  [He said nothing except that he was tired.]
- He said that he was *only* tired.
  [He said that he was nothing except tired.]

openings (*see* introductions)

oral presentations (*see* presentations)

order-of-importance method of development

The order-of-importance **method of development** is a particularly effective and common organizing strategy. This method can use one of two ordering strategies—decreasing order, which is often best for written documents, and increasing order, which is especially effective for oral **presentations**.

Decreasing Order

Decreasing order arranges the writer's major points in descending order of importance. It begins with the most important fact or point, then moves to the next most important, and so on, ending with the least important. This order is especially appropriate for a **memo** or other **correspondence** addressed to a busy decision-maker, who may be able to reach a decision after considering only the most important points. In a **report** addressed to various **readers**, some of whom may be interested in only the major points and others who may need all the information, decreasing order may be ideal for your **purpose**. The advantages of decreasing order are that (1) it gets the reader's attention immediately by presenting the most important point first, (2) it makes a strong initial impression, and (3) it ensures that even the most hurried reader will not miss the most important point. The example shown in Figure O–1 uses decreasing order of importance.

Increasing Order

Increasing order begins from the least important point or fact, then progresses to the next more important, and builds finally to the most important or strongest point.

Memorandum

To: Tawana Shaw, Director, Human Resources Department
From: Frank W. Nemitz, Chief, Product Marketing   *FWN*
Date: November 13, 2003
Subject: Selection of Manager of the Technical Writing Department

Top-Ranked Candidate
The most qualified candidate for Manager of the Technical Writing Department is Michelle Bryant, who is currently acting manager of the department. In her 12 years in the department, Ms. Bryant has gained wide experience in all facets of its operations. She has maintained a consistently high production record and has demonstrated the skills and knowledge required for the supervisory duties she is now handling in an acting capacity. Another consideration is that she has continually been rated "outstanding" in all categories in her job-performance appraisals. However, her supervisory experience is limited to her present three-month tenure as acting manager of the department, and she lacks the college degree required by the job description.

Second-Ranked Candidate
Michael Bastick, Graphics Coordinator, my second choice, also has strong potential for the position. An able administrator, he has been with the company for seven years. Further, he is enrolled in a management-training course at Metro State University's downtown campus. He is ranked second because he lacks supervisory experience and because his most recent work as a technical writer has been limited.

Third-Ranked Candidate
Jane Fine, my third-ranked candidate, has shown herself to be an exceptionally skilled writer in her three years with the Public Relations Department. Despite her obvious potential, she does not yet have the breadth of experience in technical writing that is required to manage the Technical Writing Department. Jane Fine also lacks on-the-job supervisory experience.

FIGURE O–1.   Decreasing-Order-of-Importance Method of Development

Increasing order of importance is effective for writing in which (1) you want to save your strongest points until the end or (2) you need to build the ideas point by point to an important conclusion. Many oral presentations benefit especially from increasing order because it leaves the audience with the strongest points freshest in their minds. The disadvantage of increasing order, especially for written documents, is that it begins weakly, and the reader or listener may become impatient or distracted before reaching your main point. In the example given in Figure O–2, the writer begins with the least productive source of applicants and builds up to the most productive source.

INTEROFFICE MEMORANDUM

To: Sun-Hee Kim, Vice President, Operations
From: Harry Matthews, Human Resources Department   *HM*
Date: May 19, 2003
Subject: Recruiting Qualified Electronics Technicians

As our company continues to expand, and with the planned opening of the Lakeland Facility late next year, we need to increase and refocus our recruiting program to keep our company staffed with qualified electronics technicians. Over the past three years, we have relied on our in-house internship program and on local and regional technical school graduates to fill these positions.

Technical School Recruitment
Although our in-house internship program provides a qualified pool of employees, technical school enrollments in the area have in the past provided candidates who are already trained. Each year, however, fewer technical school graduates are being produced, and even the most vigorous Career Day recruiting has yielded disappointing results.

Military Veteran Recruitment
In the past, we relied heavily on the recruitment of skilled veterans from all branches of the military. This source of qualified applicants all but disappeared when the military offered attractive re-enlistment bonuses for skilled technicians in uniform. As a result, we need to become aggressive in our attempts to reach this group through advertising. I would like to meet with you soon to discuss the details of a more dynamic recruiting program for skilled technicians leaving the military.

I am certain that with the right recruitment campaign, we can find the skilled employees essential to our expanding role in electronics products and consulting.

FIGURE O–2. Increasing-Order-of-Importance Method of Development

organization

Organization is essential to the success of any writing project, from a formal report to a Web design, or to an effective presentation. Good organization is achieved by outlining, using a logical and appropriate method of development that suits your subject, your readers, and your purpose.

During the organization stage of the writing process, you must consider a layout and design that will be helpful to your reader and a format appropriate to your subject and purpose. If you intend to include visuals with your writing, a good time to think about them is when you have completed your outline, especially if they need to be prepared by someone else while you are writing and revising the draft. See also "Five Steps to Successful Writing."

organizational charts

An *organizational chart* shows how the various components of an organization are related to one another. This type of visual is useful when you want to give your readers an overview of an organization or to display the lines of authority within it, as in Figure O–3.

The title of each organizational component (office, section, division) is placed in a separate box. The boxes are then linked to a central authority, as shown in Figure O–3. If your readers need the information, include the name of the person and position title in each box.

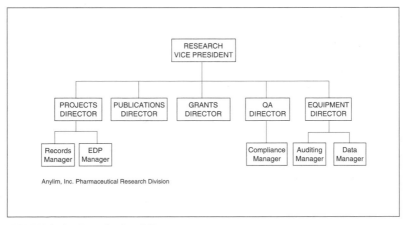

Anylim, Inc. Pharmaceutical Research Division

FIGURE O–3. Organizational Chart

When you incorporate an organizational chart into a document, place it as close as possible to the text that refers to it.

orient / orientate

Avoid *orientate* as a long variant of *orient,* meaning "to accustom somebody or yourself to a new situation or set of surroundings." See also **affectation**.

- She will need a few weeks to ~~orientate~~ *orient* herself to our system.

outlining

An *outline* is the skeleton of the document you are going to write; at the least, it should list the main topics of your subject and subtopics in a logical **method of development**.

Advantages of Outlining

An outline provides structure to your writing by ensuring that it has a beginning (**introduction**), a middle (main body), and an end (**conclusion**). Using an outline offers many other benefits.

- Larger and more complex subjects are easier to handle by breaking them into manageable parts in the outline.
- **Logic errors** are much easier to detect and correct in an outline than in a draft.
- Parts of an outline are easily moved around so you can select the most effective arrangement of your ideas.
- Like a road map, an outline indicates a starting point and keeps you moving logically so you do not lose your way before you arrive at your conclusion.
- Creating a good outline frees you from concerns of **organization** while you are **writing a draft**.
- An outline enables you to provide **coherence** and **transition** so that one part flows smoothly into the next without omitting important details.

Types of Outlines

Two types of outlines are most common: short topic outlines and lengthy sentence outlines. A *topic outline* consists of short phrases arranged in the order of your primary method of development. A topic outline is especially useful for short documents such as letters, e-mails, or memos. See also correspondence.

For a large writing project, create a topic outline first, and then use it as a basis for creating a sentence outline. A *sentence outline* summarizes each idea in a complete sentence that may become the topic sentence for a paragraph in the rough draft. If most of your notes can be shaped into topic sentences for paragraphs in your rough draft, you can be relatively sure that your document will be well organized. See also note-taking and research.

The less certain you are about your writing ability or about your subject, the more important it is to prepare a detailed outline.

Creating an Outline

When you are outlining large and complex subjects with many pieces of information, the first step is to group related notes into categories. Sort the notes by major and minor division headings. Use an appropriate method of development to arrange items and label them with roman numerals. For example, the major divisions for this discussion of outlining could be as follows:

   I. Advantages of outlining
   II. Types of outlines
   III. Creating an outline

The second step is to establish your minor points by deciding on the minor divisions within each major division. Arrange your minor points using a method of development under their major division and label them with capital letters.

   II. Types of outlines
      A. Topic outlines   Division and Classification
      B. Sentence outlines
   III. Creating an outline
      A. Establish major and minor divisions.
      B. Sort notes by major and minor divisions.   Sequential
      C. Complete the sentence outline.

You will often need more than two levels of headings. If your subject is complicated, you may need three or four levels of headings to better organize all of your ideas in proper relationship to one another. In that event, use the following numbering scheme:

I. First-level heading
  A. Second-level heading
    1. Third-level heading
      a. Fourth-level heading

The third step is to mark each of your notes with the appropriate roman numeral and capital letter. Organize the notes logically within each minor heading, and mark each with the appropriate, sequential arabic number. As you do, make sure your organization is logical and your headings have parallel structure. For example, all the second-level headings under "III. Creating an outline" are complete sentences in the active <u>voice</u>.

Treat <u>visuals</u> as an integral part of your outline, and plan approximately where each should appear. At each place, either include a rough sketch of the visual or write "illustration of . . ." As with other information in an outline, feel free to move or delete visuals. Planning your graphics requirements from the beginning ensures their harmonious integration throughout all versions of the draft to the finished document.

The outline samples shown earlier use a combination of numbers and letters to differentiate the various levels of information. You could also use a decimal numbering system, such as the following, for your outline.

1. FIRST-LEVEL HEADING
  1.1 Second-level heading
  1.2 Second-level heading
    1.2.1 Third-level heading
    1.2.2 Third-level heading
      1.2.2.1 Fourth-level heading
      1.2.2.2 Fourth-level heading
  1.3 Second-level heading
2. FIRST-LEVEL HEADING

This system should not go beyond the fourth level because the numbers get too cumbersome beyond that point. In many documents, the decimal numbering system is carried over from the outline to the final version of the document for ease of cross-referencing sections.

Create your draft by converting your notes into complete sentences and <u>paragraphs.</u> If you have a complete sentence outline, the most difficult part of the writing job is over. However, whether you have a topic or a sentence outline, remember that an outline is not set in stone; it may need to change as you write the draft, but it should always be your point of departure and return.

DIGITAL TIP **CREATING AN OUTLINE**

Using the outline feature of your word-processing software permits you to format your outline automatically; fill in, rearrange, and update your outline; experiment with the organization and scope of information; rearrange sections and subsections easily; and create alphanumeric or decimal numbering outlining styles. For more on this topic, see <*www.bedfordstmartins .com/alred*> and select *Digital Tips*.

outside [of]

In the phrase *outside of,* the word *of* is redundant.

- Place the rack outside ~~of~~ the incubator.

In addition, do not use *outside of* to mean "aside from" or "except for."

Except for
- ~~Outside of~~ his frequent absences, Jim has a good work record.

over [with]

In the expression *over with,* the word *with* is redundant; such words as *completed* or *finished* often better express the thought.

- You may enter the test chamber when the experiment is over ~~with~~.

completed.
- You may enter the test chamber when the experiment is over ~~with.~~

O

P

pace

Pace is the speed at which you present ideas to the reader. Your goal should be to achieve a pace that fits your <u>readers</u>, <u>purpose</u>, and <u>subject</u>. The more knowledgeable the reader is about the subject, the faster your pace can be. But be careful, though, not to lose control of the pace. In the first version of the following passage, facts are piled on top of each other at a rapid pace. In the second version, even though its length is no greater, the same facts are presented in two, more easily assimilated sentences. In addition, the second version achieves a different and more desirable <u>emphasis</u>.

> **RAPID** The hospital's generator produces 110 volts at 60 hertz and is powered by a 90-horsepower engine. It is designed to operate under normal conditions of temperature and humidity, for use under emergency conditions, and may be phased with other units of the same type to produce additional power when needed.

> **CONTROLLED** The hospital's generator, which is powered by a 90-horsepower engine, produces 110 volts at 60 hertz under normal conditions of temperature and humidity. Designed especially for use under emergency conditions, this generator may be phased with other units of the same type to produce additional power when needed.

paragraphs

A *paragraph* performs three functions: (1) it develops the unit of thought stated in the topic sentence; (2) it provides a logical break in the material; and (3) it creates a visual break on the page, which signals a new topic.

388

Topic Sentence

A *topic sentence* states the paragraph's main idea; the rest of the paragraph supports and develops that statement with carefully related details. The topic sentence is often the first sentence because it tells the reader what the paragraph is about.

- *The arithmetic of searching for oil is stark.* For all the scientific methods of detection, the only way the oil driller can actually know for sure that there is oil in the ground is to drill a well. The average cost of drilling an oil well is over $300,000 and drilling a single well may cost over $8,000,000. And once the well is drilled, the odds against its containing any oil at all are 8 to 1!

On rare occasions, the topic sentence logically falls in the middle of a paragraph.

- . . . [It] is time to insist that science does not progress by carefully designed steps called "experiments," each of which has a well-defined beginning and end. *Science is a continuous and often a disorderly and accidental process.* We shall not do the young psychologist any favor if we agree to reconstruct our practices to fit the pattern demanded by current scientific methodology.
  —B. F. Skinner, "A Case History in Scientific Method"

The topic sentence is usually most effective early in the paragraph, but a paragraph can lead up to the topic sentence, which is sometimes done to achieve <u>emphasis</u>.

- Energy does far more than simply make our daily lives more comfortable and convenient. Suppose you wanted to stop—and reverse—the economic progress of this nation. What would be the surest and quickest way to do it? Find a way to cut off the nation's oil resources! . . . The economy would plummet into the abyss of national economic ruin. *Our economy, in short, is energy-based.*
  — *The Baker World* (Los Angeles: Baker Oil Tools)

Paragraph Length

Paragraph length should aid the reader's understanding of ideas. A series of short, undeveloped paragraphs can indicate poor <u>organization</u> and sacrifice unity by breaking a single idea into several pieces. A series of long paragraphs, however, can fail to provide the reader with manageable subdivisions of thought. A paragraph should be just long enough to deal adequately with the subject of its topic sentence. A new paragraph should begin whenever the subject changes significantly.

Occasionally, a one-sentence paragraph is acceptable if it is used as a transition between larger paragraphs or in letters and memos in which one-sentence openings and closings are appropriate. See also **introductions** and **conclusions**.

Writing Paragraphs

Careful paragraphing reflects the writer's logical thinking and organization and helps the reader follow the writer's thoughts. A good working outline makes it easy to group ideas into appropriate paragraphs. (See also **outlining**.) Notice how the following partial topic outline plots the course of the subsequent paragraphs:

TOPIC OUTLINE (PARTIAL)

I. Advantages of Chicago as location for new facility
  A. Transport infrastructure
      1. Rail
      2. Air
      3. Truck
      4. Sea (except in winter)
  B. Labor supply
      1. Engineering and scientific personnel
          a. Many similar companies in the area
          b. Several major universities
      2. Technical and manufacturing personnel
          a. Existing programs in community colleges
          b. Possible special programs designed for us

RESULTING PARAGRAPHS

Probably the greatest advantage of Chicago as a location for our new facility is its excellent transport infrastructure. The city is served by three major railroads. Both domestic and international air cargo service is available at O'Hare International Airport; Midway Airport's convenient location adds flexibility for domestic air cargo service. Chicago is a major hub of the trucking industry, and most of the nation's large freight carriers have terminals there. Finally, except in the winter months when the Great Lakes are frozen, Chicago is a seaport, accessible through the St. Lawrence Seaway.

Chicago's second advantage is its abundant labor force. An ample supply of engineering and scientific staff is assured not only by the presence of many companies engaged in activities similar to ours but also by the presence of several major universities in the metropolitan area. Similarly, technicians and manufacturing personnel are in abundant supply. The colleges in the Chicago City

College system, as well as half a dozen other two-year colleges in the outlying areas, produce graduates with associate's degrees in a wide variety of technical specialties appropriate to our needs. Moreover, three of the outlying colleges have expressed an interest in developing off-campus courses attuned specifically to our requirements.

Paragraph Unity and Coherence

A good paragraph has <u>unity</u> and <u>coherence,</u> as well as adequate development. *Unity* is singleness of purpose, based on a topic sentence that states the core idea of the paragraph. When every sentence in the paragraph develops the core idea, the paragraph has unity.

Coherence is holding to one point of view, one attitude, one tense; it is the joining of sentences into a logical pattern. A careful choice of transitional words ties ideas together and thus contributes to coherence in a paragraph as shown in the following example. Notice how the boldfaced words provide <u>transition</u> by tying together the ideas in the paragraph.

TOPIC SENTENCE *Over the past several months, I have heard complaints about the Merit Award Program.* **Specifically,** many employees feel that this program should be linked to annual **salary increases.** They believe that **salary increases** would provide a much better incentive than the current $500 to $700 cash awards for exceptional service. **In addition,** these **employees believe** that their supervisors consider the cash awards a satisfactory alternative to salary increases. Although I believe few supervisors follow this practice, the fact that the **employees believe** that the practice is widespread justifies a reevaluation of the Merit Award Program.

Simple enumeration (*first, second, then, next,* and so on) also provides effective transition within paragraphs. Notice how the boldfaced words and phrases give coherence to the following paragraph.

- Most adjustable office chairs have nylon tubes that hold metal spindle rods. To keep the chair operational, lubricate the spindle rods occasionally. **First,** loosen the set screw in the adjustable bell. **Then,** lift the chair from the base. **Next,** apply the lubricant to the spindle rod and the nylon washer. **When you have finished,** replace the chair and tighten the set screw.

parallel structure

Parallel sentence structure requires that sentence elements that are alike in function be alike in construction as well. It achieves an economy of words, clarifies meaning, expresses the equality of its ideas, and achieves **emphasis**. Parallel structure assists **readers** because it allows them to anticipate the meaning of a sentence element on the basis of its parallel construction.

Parallel structure can be achieved with words, **phrases**, or **clauses**.

- If you want to earn a satisfactory grade in the training program, you must be *punctual, courteous,* and *conscientious.* [parallel words]

- If you want to earn a satisfactory grade in the training program, you must recognize the importance *of punctuality, of courtesy,* and *of conscientiousness.* [parallel phrases]

- If you want to earn a satisfactory grade in the training program, *you must arrive punctually, you must behave courteously,* and *you must study conscientiously.* [parallel clauses]

Correlative **conjunctions** (*either . . . or, neither . . . nor, not only . . . but also*) should always use parallel structure. Both parts of the pairs should be followed immediately by the same grammatical form: two similar words, two similar phrases, or two similar clauses.

- Viruses carry either *DNA* or *RNA,* never both. [parallel words]

- Clearly, neither *serological tests* nor *virus isolation studies* alone would have been adequate. [parallel phrases]

- Either *we must increase our production efficiency* or *we must decrease our production goals.* [parallel clauses]

To make a parallel construction clear and effective, it is often best to repeat an **article**, a **pronoun**, a helping **verb**, a **preposition**, a subordinating conjunction, or the mark of an infinitive (*to*).

- The association has *a* mission statement and *a* code of ethics.

- The driver *must* check the gauge regularly and *must* act quickly when the indicator falls below the red line.

Parallel structure is especially important in creating **lists**, outlines, **tables of contents**, and **headings**, because it lets readers know the relative value of each item in a table of contents and each heading in the body of a document. See also **outlining**.

Faulty Parallelism

Faulty parallelism results when joined elements are intended to serve equal grammatical functions but do not have equal grammatical form. Faulty parallelism sometimes occurs because a writer tries to compare items that are not comparable.

NOT PARALLEL
> The company offers special college training to help nonexempt employees move into professional careers like engineering management, software development, *service technicians,* and *sales trainees.*
> [Notice that occupations—*engineering management* and *software development*—are being compared to people—*service technicians* and *sales trainees.*]

To avoid faulty parallelism, make certain that each element in a series is similar in form and structure to all others in the same series.

PARALLEL
> The company offers special college training to help nonexempt employees move into professional careers like *engineering management, software development, technical services,* and *sales.*

paraphrasing

Paraphrasing is restating or rewriting in your own words the essential ideas of another writer. Because the paraphrase does not quote the source word for word, **quotation marks** are not necessary. However, paraphrased material should be credited because the *ideas* are taken from someone else. The following example is an original passage explaining the concept of object blur. The paraphrased version restates the essential information of the passage in a form appropriate for a **report**.

ORIGINAL
> One of the major visual cues used by pilots in maintaining precision ground reference during low-level flight is that of object blur. We are acquainted with the object-blur phenomenon experienced when driving an automobile. Objects in the foreground appear to be rushing toward us, while objects in the background appear to recede slightly.

PARAPHRASED   Object blur refers to the phenomenon by which observers in a moving vehicle report that foreground objects appear to rush at them, while background objects appear to recede slightly.

Strive to put the original ideas into your words without distorting them. See also ethics in writing, note-taking, plagiarism, and quotations.

parentheses

Parentheses are used to enclose explanatory or digressive words, phrases, or sentences. The material in parentheses often clarifies a sentence or passage without altering its meaning. Parenthetical information may not be essential to a sentence — in fact, parentheses de-emphasize the enclosed material — but it may be interesting or helpful to some readers. Parenthetical material applies to the word or phrase immediately preceding it.

- She severely bruised her shin (or *tibia*) in the accident.

Parenthetical material does not affect the punctuation of a sentence. A comma or period should appear outside the closing parenthesis following a parenthetical word, phrase, or clause.

- She severely bruised her shin (or *tibia*), and he tore the cartilage in his knee (or *meniscus*).

Note that when parentheses within a sentence contain an independent clause, the ending punctuation for the clause is omitted. However, such constructions are better avoided with subordination.

- The scans showed little damage ~~(the attending physician was pleased)~~ *, which pleased the attending physician,* but later tests revealed extensive bruising.

When a complete sentence within parentheses stands independently, the ending punctuation goes inside the final parenthesis.

- Most of our regional managers report increases of 15 to 30 percent. (The only important exceptions are the Denver and Houston offices.)

Parentheses also are used to enclose numerals or letters that indicate sequence.

- The following sections deal with (1) preparation, (2) research, (3) organization, (4) writing, and (5) revision.

Do not follow spelled-out <u>numbers</u> with numerals in parentheses representing the same numbers.

- Send five ~~(5)~~ copies of the report.

Use <u>brackets</u> to set off a parenthetical item that is already within parentheses.

- We should be sure to give Emanuel Foose (and his brother Emilio [1812–1882]) credit for his part in founding the institute.

See also <u>documenting sources</u> and <u>quotations</u>.

parts of speech

The term *parts of speech* describes the class of words to which a particular word belongs, according to its function in a sentence (naming, asserting, describing, joining, acting, modifying, exclaiming). Many words can function as more than one part of speech. See also <u>functional shift</u>.

PART OF SPEECH	FUNCTION
<u>noun</u>, <u>pronoun</u>	naming/referring
<u>verb</u>	asserting
<u>adjective</u>, <u>adverb</u>	describing/modifying
<u>conjunction</u>, <u>preposition</u>	joining/linking
<u>interjection</u>	exclaiming

P

party

In legal language, *party* refers to an individual, a group, or an organization. (The injured *party* sued my client.) The term is inappropriate in all but legal writing; when you are referring to a person, use the word *person*.

- *person*
- The ~~party~~ whose file you requested is here now.

Party is, of course, appropriate when it refers to a group. (Jim arranged a tour of the facility for the members of our *party*.)

passive voice (*see* **voice**)

per

When *per* is used to mean "for each," "by means of," "through," or "on account of," it is appropriate (*per* annum, *per* capita, *per* diem, *per* head). When used to mean "according to" (*per* your request, *per* your order), the expression is jargon and should be avoided.

we discussed,
- As ~~per our discussion,~~ I will send revised instructions.
     ^

percent / percentage

Percent is normally used instead of the symbol % (only 15 *percent*), except in tables, where space is at a premium. *Percentage,* which is never used with numbers, indicates a general size (only a small *percentage*).

periods

A *period* usually indicates the end of a declarative or imperative sentence. Periods also link when used as leaders (for example, in a <u>table of contents</u>) and indicate omissions when used as <u>ellipses</u>. Periods may also end questions that are actually polite requests and questions to which an affirmative response is assumed. (Will you please send me the financial statement.)

Periods in Quotations

Use a <u>comma,</u> not a period, after a declarative sentence that is quoted in the context of another sentence.

- "There is every chance of success," she stated.

A period is conventionally placed inside <u>quotation marks</u>. See also <u>quotations</u>.

- He stated clearly, "My vote is yes."

Periods with Parentheses

If a sentence ends with a parenthesis, the period should follow the parenthesis.

- The institute was founded by Harry Denman (1902–1972).

If a whole sentence (beginning with an initial capital letter) is enclosed in **parentheses,** the period (or other end mark) should be placed inside the final parenthesis.

- The project director listed the problems her staff faced. (This was the third time she had complained to the board.)

Other Uses of Periods

Use periods after initials in names (Wilma T. Grant, J. P. Morgan). Use periods as decimal points with **numbers** (27.3 degrees Celsius, $540.26, 6.9 percent). Use periods to indicate **abbreviations** (Ms., Dr., Inc.). When a sentence ends with an abbreviation that ends with a period, do not add another period. (Please meet me at 3:30 a.m.) Use periods following the numerals in a numbered list.

- 1. Enter your name.
  2. Enter your address.
  3. Enter your telephone number.

Period Faults

The incorrect use of a period is sometimes referred to as a *period fault.* When a period is inserted prematurely, the result is a **sentence frag-ment.**

> FRAGMENT After a long day at the office during which we fin-ished the quarterly report. We left hurriedly for home.
>
> SENTENCE After a long day at the office, during which we fin-ished the quarterly report, we left hurriedly for home.

When two independent clauses are joined without any punctuation, the result is a *fused* or *run-on* sentence. Adding a period between the clauses is one way to correct a run-on sentence.

> RUN-ON Bill was late for ten days in a row Ms. Sturgess had to fire him.
>
> CORRECT Bill was late for ten days in a row. Ms. Sturgess had to fire him.

Other options are to add a comma and a coordinating **conjunction** (*and, but, for, or, nor, so, yet*) between the clauses, to add a **semicolon,** or to add a semicolon with a conjunctive **adverb** (*therefore, however*). See also **sentence construction.**

person (grammar)

Person refers to the form of a personal <u>pronoun</u> that indicates whether the pronoun represents the speaker, the person spoken to, or the person or thing spoken about. A pronoun representing the speaker is in the *first* person. (*I* could not find the answer in the manual.) A pronoun that represents the person or people spoken to is in the *second* person. (*You* will be a good manager.) A pronoun that represents the person or people spoken about is in the *third* person. (*They* received the news quietly.) The following list shows first-, second-, and third-person pronouns. See also <u>case</u>, <u>number</u>, and <u>one</u>.

PERSON	SINGULAR	PLURAL
First	I, me, my, mine	we, us, our, ours
Second	you, your, yours	you, your, yours
Third	he, him, his, she, her, hers, it, its	they, them, their, theirs

personal / personnel

Personal is an <u>adjective</u> meaning "of or pertaining to an individual person" (a *personal* problem). *Personnel* is a <u>noun</u> meaning a "group of people engaged in a common job" (military *personnel*). Be careful not to use *personnel* when the word you need is *persons, people,* or a more descriptive word.

- The remaining ~~personnel~~ *employees* will be moved next Thursday.
     ∧

persons / people

The word *persons* is used to refer to a specific category or number of people, often in legal or official contexts. (Admittance is limited to *persons* age 18 and over.) In all other contexts, use *people*. (We need more qualified *people* to fill the vacant positions.)

persuasion

Persuasive writing attempts to convince the <u>reader</u> to adopt the writer's point of view or take a particular action. Much workplace writing uses persuasion to reinforce ideas that readers already have, to convince readers to change their current ideas, or to lobby for a particular suggestion or policy. You may find yourself pleading for safer working conditions, justifying the expense of a new program, or writing a **proposal** for a large purchase. See also **audience** and **purpose**.

In persuasive writing, the way you present your ideas is as important as the ideas themselves. You must support your appeal with logic and a sound presentation of facts, statistics, and examples. Avoid ambiguity: Do not wander from your main point and, above all, never make false claims. You should also acknowledge any real or potentially conflicting opinions; doing so allows you to anticipate and overcome objections and builds your credibility. See also **ethics in writing** and **logic errors**.

The memo shown in Figure P–1 was written by an MIS administrator to persuade the engineering sales staff to accept and participate

---

Memo

TO: Engineering Sales Staff
FROM: Bernadine Kovak, MIS Administrator *BK*
DATE: April 7, 2003
SUBJECT: Plans for Changeover to NRT/R4 System

As you all know, our workload has jumped by 30 percent in the past month. It has increased because our customer base and resulting technical support services have grown dramatically. This growth is a result, in part, of our recent merger with Datacom.

This growth has meant that we have all experienced the difficulty of providing our customers with up-to-date technical information when they need it. In the next few months, we anticipate that the workload will increase another 20 percent. Even a staff as experienced as ours cannot handle such a workload without help.

To cope with this expansion, we will install in the next month the NRT/R4 server and QCS enterprise software with Web-based applications and global sales and service network. This system will speed processing dramatically and give us access to all relevant company-wide databases. It should enable us to access the information both we and our customers need.

The new system, unfortunately, will cause some disruption at first. We will need to transfer many of our legacy programs and software applications to the new format. And all of us need to learn to navigate in the R4 and QCS environments. However, once we have made these adjustments, I believe we will welcome the changes.

I would like to put your knowledge and experience to work in getting the new system into operation. Let's meet in my office to discuss the improvements on Friday, April 11, at 1:00 p.m. I will have details of the plan to discuss with you. I'm also eager to get your comments, suggestions, and— most of all—your cooperation.

P

FIGURE P–1. Persuasive Memo

in a change to a new computer system. Notice that not everything in this memo is presented in a positive light. Change brings disruption, and the writer acknowledges that fact.

A writer also gains credibility, and thus persuasiveness, through the readers' impressions of the document's appearance. For this reason, consideration of <u>layout and design</u> is important, especially in documents such as <u>résumés</u>.

phenomenon / phenomena

A *phenomenon* is an observable thing, fact, or occurrence (a natural *phenomenon*). Its plural form is *phenomena*.

photographs

Photographs are the best way to show the surface of an object, record an event, or demonstrate the development of a phenomenon over a period of time. They are not always the best type of illustration, however. Photographs cannot depict the internal workings of a mechanism or show the below-the-surface details of objects or structures (see Figure P-2). Such details are better represented in <u>drawings</u>.

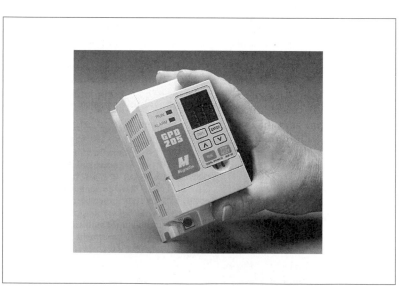

FIGURE P-2. Photo (of Control Device). Photo courtesy of Ken Cook Company.

If you take a photograph yourself, stand close enough to the subject so it fills the picture frame. To get precise and clear photographs, choose camera angles carefully. Select important details and the camera angles that will record them. To show relative size, place a familiar object— such as a ruler or a person—near the subject being photographed. For example, the photograph in Figure P–2 on page 400 shows a control device being held by a human hand to illustrate its relative size and features.

Treat photographs as you do other visuals. Give the photograph a figure number, callouts (labels) to identify key features in the photograph, and any other important information. Position the figure number and caption so the reader can view them and the photograph from the same orientation.

phrases

Below the level of the sentence, words are combined into two groups: clauses and phrases. *Phrases* are based on nouns, nonfinite verb forms, or verb combinations without subjects.

P

- She encouraged her staff *by her calm confidence.* [phrase]

A phrase may function as an adjective, an adverb, a noun, or a verb.

- The subjects *on the agenda* were all discussed. [adjective]

- We discussed the project *with great enthusiasm.* [adverb]

- *Working hard* is her way of life. [noun]

- The chief engineer *should have been notified.* [verb]

Even though phrases function as adjectives, adverbs, nouns, or verbs, they are normally named for the kind of word around which they are constructed: preposition, participle, infinitive, gerund, verb, or noun. A phrase that begins with a preposition is a *prepositional phrase,* a phrase that begins with a participle is a *participial phrase,* and so on. For typical verb phrases and prepositional phrases that give speakers of <u>English as a second language</u> trouble, see <u>idioms</u>.

Prepositional Phrases

A *preposition* is a word that shows relationship and combines with a noun or <u>pronoun</u> (its <u>object</u>) to form a modifying phrase. A prepositional phrase, then, consists of a preposition plus its object and the object's modifiers.

- *After the meeting,* the district managers adjourned *to the executive dining room.*

Prepositional phrases, because they normally modify nouns or verbs, usually function as adverbs or adjectives. A prepositional phrase may function as an adverb of motion. (Turn the dial four degrees *to the left.*) A prepositional phrase may function as an adverb of manner. (Answer customers' questions *in a courteous fashion.*) A prepositional phrase may function as an adverb of place and may appear in different places in the sentence.

- *In home and office computer systems,* security is essential.
- Security is essential *in home and office computer systems.*

A prepositional phrase may function as an adjective. When functioning as adjectives, prepositional phrases follow the nouns they modify.

- Garbage *with a high protein content* can be processed into animal food.

Be careful when you use prepositional phrases because separating a prepositional phrase from the noun it modifies can cause <u>ambiguity</u>.

| AMBIGUOUS | *The woman* standing by the security guard *in the gray suit* is our division manager. |
| CLEAR | *The woman in the gray suit* who is standing by the security guard is our division manager. |

Watch as well for the overuse of prepositional phrases where <u>modifiers</u> would be more economical.

| OVERUSED | The man *with gray hair in the blue suit with pinstripes* is the former president *of the company.* |
| ECONOMICAL | The *gray-haired* man in the blue *pinstripe* suit is the former *company* president. |

Participial Phrases

A *participle* is any form of a verb that is used as an adjective. A participial phrase consists of a participle plus its object and its modifiers.

- The division *having the largest number of patents* will work with NASA.

The relationship between a participial phrase and the rest of the sentence must be clear to the reader. For that reason, every sentence containing a participial phrase must have a noun or pronoun that the participial phrase modifies; if it does not, the result is a dangling participial phrase.

Dangling Participial Phrases. A dangling participial phrase occurs when the noun or pronoun that the participial phrase is meant to modify is not stated but only implied in the sentence. See also <u>dangling modifiers</u>.

DANGLING	*Being unhappy with the job,* his efficiency suffered. [His efficiency was not unhappy with the job; what the participial phrase really modifies — *he* — is not stated but merely implied.]
CORRECT	*Being unhappy with the job,* he grew less efficient. [Now what that participial phrase modifies — *he* — is explicitly stated.]

Misplaced Participial Phrases. A participial phrase is misplaced when it is too far from the noun or pronoun it is meant to modify and so appears to modify something else. Such an error can make the writer look ridiculous. See also <u>modifiers</u>.

MISPLACED	We saw a large warehouse *driving down the highway.*
CORRECT	*Driving down the highway,* we saw a large warehouse.

Infinitive Phrases

An *infinitive* is the bare form of a verb (*go, run, talk*) without the restrictions imposed by <u>person</u> and <u>number</u>. An infinitive is generally preceded by the word *to* (which is usually a preposition but in this use is called the *sign,* or *mark,* of the infinitive). An infinitive phrase consists of the word *to* plus an infinitive and any objects or modifiers.

- *To succeed in this field,* you must be willing *to assume responsibility.*

Do not confuse a prepositional phrase beginning with *to* with an infinitive phrase. In an infinitive phrase, *to* is followed by a verb; in a prepositional phrase, *to* is followed by a noun or a pronoun.

> **PREPOSITIONAL PHRASE**   We went *to the building site.*
>
> **INFINITIVE PHRASE**   Our firm tries *to provide a comprehensive training program.*

The implied subject of an introductory infinitive phrase should be the same as the subject of the sentence. If it is not, the phrase is a dangling modifier. In the following example, the implied subject of the infinitive is *you* or *one,* not *practice.*

- To learn a new language, ~~practice is needed.~~ *you must practice.*

Gerund Phrases

A gerund is a <u>verbal</u> ending in *-ing* that is used as a noun. A gerund phrase consists of a gerund plus any objects or modifiers and always functions as a noun.

> **SUBJECT**   *Writing a technical manual* is a difficult task.
>
> **DIRECT OBJECT**   She liked *designing the system.*

Verb Phrases

A verb phrase consists of a main verb and its helping verb.

- He *is* [helping verb] *working* [main verb] hard this summer.

Words can appear between the helping verb and the main verb of a verb phrase. (He *is **always** working.*) The main verb is always the last verb in a verb phrase.

- You *will have filed* your tax return on time if you begin now.
- You *are* not *filing* your tax return too early if you begin now.

Questions often begin with a verb phrase. (*Will* he *audit* their account soon?) The adverb *not* may be appended to a helping verb in a verb phrase. (He *did not work* today.)

Noun Phrases

A noun phrase consists of a noun and its modifiers. (Have *the two new employees* fill out *these forms.*)

plagiarism

Plagiarism is the use of someone else's unique ideas without acknowledgment, or the use of someone else's exact words without <u>quotation marks</u> and appropriate credit. Plagiarism is considered to be the theft of someone else's creative and intellectual property and is not accepted in business, science, journalism, academia, and other fields. For detailed guidance on quoting correctly, see <u>quotations</u>.

You may, however, quote or paraphrase the words and ideas of another if you document your source. (See also <u>paraphrasing</u> and <u>documenting sources</u>.) Although you do not enclose paraphrased ideas or materials in quotation marks, you must document their sources. Paraphrasing a passage without citing the source is permissible only when the information paraphrased is common knowledge in a field. *Common knowledge* refers to information on a topic widely known and readily available in handbooks, manuals, atlases, and other references. If you intend to publish, reproduce, or distribute material that includes quotations from published works, you may need to obtain written permission from the <u>copyright</u> holder to do so.

In the workplace, employees often borrow from in-house manuals, reports, and other company documents. Using such boilerplate information is neither plagiarism nor a violation of copyright. See also <u>ethics in writing</u>.

point of view

Point of view is the writer's relation to the information presented, as reflected in the use of grammatical <u>person</u>. The writer usually expresses point of view in first-, second-, or third-person personal <u>pronouns</u>. Use of the first person indicates that the writer is a participant or observer. Use of second and third person indicates that the writer is giving directions, <u>instructions</u>, or advice or writing about other people or something impersonal.

FIRST PERSON	*I* scrolled down to find the settings option.
SECOND PERSON	Scroll down to find the settings option and double-click. [*You* is understood.]
THIRD PERSON	*He* scrolled down to find the settings option.

Consider the following sentence, revised from an impersonal to a more personal point of view. Although the meaning of the sentence does not change, the revision indicates that people are involved in the communication.

- *I regret*   *we cannot accept*
  ~~It is regrettable~~ that the equipment shipped on the 12th ~~is un-acceptable~~.

Many people think they should avoid the pronoun *I* in their technical writing. Such practice, however, leads to awkward sentences with people referring to themselves in the third person as *one* or as *the writer* instead of as *I*.

- *I*
  ~~One~~ can only conclude that the absorption rate is too fast.

- *I believe*
  ~~The writer believes~~ that this project will be completed by the end of June.

However, do not use the personal point of view when an impersonal point of view would be more appropriate or more effective because you need to emphasize the subject matter over the writer or the reader.

- *The evidence suggests*
  ~~One can only conclude~~ that the absorption rate is too fast.

In the following example, it does not help to personalize the situation; in fact, the impersonal version may be more tactful.

> PERSONAL   I received objections to my proposal from several of your managers.
>
> IMPERSONAL   Several managers have raised objections to the proposal.

Whether you adopt a personal or an impersonal point of view depends on the **purpose** and the **readers** of the document. For example, in an informal **e-mail** to an associate, you would most likely adopt a personal point of view. However, in a **report** to a large group, you would probably emphasize the subject by using an impersonal point of view.

In **correspondence** on company stationery, use of the pronoun *we* may be interpreted as reflecting company policy, whereas *I* clearly reflects personal opinion. Which pronoun to use should be decided according to whether the matter discussed in the letter is a corporate or an individual concern.

- *I* understand your frustration with the price increase, but *we* must now include the import tax.

> **ESL TIPS FOR STATING AN OPINION**
>
> In some cultures, stating an opinion in writing is considered impolite or un-
> necessary, but in the United States, readers expect to see a writer's opinion
> stated clearly and explicitly. The opinion should be followed by specific ex-
> amples to help the reader understand the writer's point of view.

positive writing

Presenting positive information as though it were negative is confusing
to readers.

> NEGATIVE   If the error does *not* involve data transmission, the
>   backup function will *not* be used.

In this sentence, the reader must reverse two negatives to understand
the exception that is being stated. (See also double negatives.) The following sentence presents the exception in a positive and straightforward manner.

> POSITIVE   The backup function is used only when the error in-
>   volves data transmission.

Negative facts or conclusions should be stated negatively; stating a negative fact or conclusion positively is deceptive because it can mislead
the reader. See also ethics in writing.

> DECEPTIVE   In the first quarter of this year, employee exposure to
>   airborne lead was within 10 percent of acceptable
>   state health standards.

> ACCURATE   In the first quarter of this year, employee exposure to
>   airborne lead was 10 percent below acceptable state
>   health standards.

Even if what you are saying is negative, do not use more negative words
than necessary.

> NEGATIVE   We are withholding your shipment until we receive
>   your payment.

> POSITIVE   We will forward your shipment as soon as we receive
>   your payment.

See also "you" viewpoint.

possessive case

A **noun** or **pronoun** is in the *possessive case* when it represents a person, place, or thing that possesses something. Possession is generally expressed with an **apostrophe** and an *s* (the *report's* title), with a prepositional **phrase** using *of* (the title *of the report*), or with the possessive form of a pronoun (*our* report).

General Guidelines

Singular nouns usually show the possessive case with *'s*, and plural nouns that end in -*s* show the possessive case with only an apostrophe.

SINGULAR	PLURAL
a *manager's* office	the *managers'* reports
an *employee's* job satisfaction	the *employees'* paychecks
the *company's* stock value	the *companies'* joint project

Singular nouns that end in -*s* generally form the possessive with *'s* or (when the word that follows begins with an *s* or *sh* sound) with only an apostrophe.

- the *witness's* testimony

- the *witness'* statement [*s* sound follows *witness*]

However, singular nouns of one syllable that end with -*s* form the possessive by adding *'s*.

- the *bus's* schedule

Plural forms of such nouns use only the apostrophe to show the possessive.

- the *witnesses'* reports, the *busses'* schedules

Plural nouns that do not end in -*s* show the possessive with *'s*.

- *children's* clothing, *women's* resources, *men's* room

Proper nouns and ancient names (*Moses, Ramses, Xerxes*) that end in consecutive -*s* or -*z* sounds form the possessive by adding only an apostrophe.

- Jesús Castillo was assigned to the new Global Systems Division. *Jesús'* responsibilities will include expanding our European operations.

Apostrophes are not used in some cases for possessive nouns that function as **adjectives** (*taxpayers* meeting, *carpenters* union, *consumers* group).

Compound Nouns

Compound nouns form the possessive with 's following the final letter.

- the *pipeline's* diameter, the *editor-in-chief's* desk

Plurals of some compound expressions are often best expressed with a prepositional phrase (presentations *of the editors-in-chief*).

Coordinate Nouns

Coordinate nouns show joint possession with 's following the last noun.

- *Michelson and Morely's* famous experiment on the velocity of light was completed in 1887.

Coordinate nouns show individual possession with 's following each noun.

- The difference between *Barker's* and *Washburne's* test results was statistically insignificant.

Possessive Pronouns

The possessive form of a pronoun (*its, whose, his, her, our, your, their*) is also used to show possession and does not require an apostrophe. (Even good systems have *their* flaws.) Only the possessive form of a pronoun should be used with a gerund (a noun formed from an -*ing* verb).

- The safety officer insisted on *our* wearing protective clothing. [*Wearing* is the gerund.]

Possessive pronouns are also used to replace nouns. (The responsibility was *theirs*.)

Indefinite Pronouns

Some indefinite pronouns (*all, any, each, few, most, none, some*) form the possessive case with the preposition *of*.

- Both desks were stored in the warehouse, but water ruined the surface *of each*.

Other indefinite pronouns (*everyone, someone, anyone, no one*), however, use 's.

- *Everyone's* contribution is welcome.

See also its/it's and case (grammar).

practicable / practical

Practicable means that something is possible or feasible. *Practical* means that something is both possible and useful.

- The program is *practical*, but, considering the company's recent financial problems, is it *practicable?*

Practical can also refer to a person, implying common sense and a tendency toward what is useful or workable.

prefixes

A *prefix* is a letter or group of letters placed in front of a root word that changes the meaning of the root word. When a prefix ends with a vowel and the root word begins with a vowel, the prefix is often separated from the root word with a <u>hyphen</u> (re-enter, pro-active, anti-inflammatory). Some words with the double vowel are written without a hyphen (cooperate) and others with or without a hyphen (re-elect or reelect).

Prefixes, such as *neo-* (derived from a Greek word meaning "new"), are usually hyphenated when used with a proper <u>noun</u> (for example, neo-Darwinism, neo-Keynesianism). Such prefixes are not normally hyphenated when used with common nouns, unless the base word begins with the same vowel (neonatal, neo-orthodoxy).

A hyphen may be necessary to clarify the meaning of a prefix; for example, *reform* means "correct" or "improve," and *re-form* means "change the shape of." When in doubt, check a current <u>dictionary</u>.

P

preparation

The preparation stage of the writing process is essential. By determining your <u>readers'</u> needs, your primary <u>purpose</u>, and your <u>scope</u> of coverage, you understand the information you will need to gather during <u>research</u>. See also "Five Steps to Successful Writing."

Writer's Checklist: Preparing to Write

☑ Determine who your readers are and learn certain key facts about them — their knowledge, attitudes, and needs relative to your subject.

☑ Determine the document's primary purpose: What exactly do you want your readers to know, believe, or be able to do when they have finished reading your document?

Writer's Checklist: Preparing to Write (continued)

☑   Establish the scope of your document — the type and amount of detail you must include — by considering any external constraints, such as word limits for **trade journal articles** or the space limitations of **Web design**, and by understanding your purpose and readers' needs.

☑   Consider **selecting the medium** appropriate for your message — **e-mail**, **report**, **memo**, and so on.

prepositions

A *preposition* is a word that links a **noun** or **pronoun** (the preposition's **object**) to another sentence element by expressing such relationships as direction (*to, into, across, toward*), location (*at, in, on, under, over, beside, among, by, between, through*), time (*before, after, during, until, since*), or position (*for, against, with*). Together, the preposition, its object, and the object's **modifiers** form a prepositional phrase that acts as a modifier.

- Answer customers' questions *in a courteous manner.*
  [The prepositional phrase *in a courteous manner* modifies the verb *answer.*]

See **phrases** for advice on using prepositional phrases and, for typical prepositional phrases that give speakers of **English as a second language** trouble, see **idioms**.

Preposition Functions

The object of a preposition (the word or phrase following the preposition) is always in the objective **case**. When the object is a compound noun, both nouns should be in the objective case. For example, the phrase "between you and *me*" is frequently and incorrectly written as "between you and *I.*" *Me* is the objective form of the pronoun, and *I* is the subjective form.

Many words that function as prepositions also function as **adverbs**. If the word takes an object and functions as a connective, it is a preposition; if it has no object and functions as a modifier, it is an adverb.

PREPOSITIONS	The manager sat *behind* the desk *in* her office.
ADVERBS	The customer lagged *behind;* then he came *in* and sat down.

Certain verbs, adverbs, and adjectives are used with certain prepositions (interested *in,* aware *of,* equated *with,* adhere *to,* capable *of,* object *to,* infer *from*). See **idioms** for a more detailed list of such usages. See also **English as a second language**.

Prepositions at the End of a Sentence

A preposition at the end of a sentence can be an indication that the sentence is awkwardly constructed.

- *She was at the*
  The branch office ~~is where she was at~~.
  ^

However, if a preposition falls naturally at the end of a sentence, leave it there. (I don't remember which file name I saved it *under.*)

Prepositions in Titles

When a preposition appears in a <u>title</u>, it is capitalized only if it is the first word in the title or if it has more than four letters. See also <u>capitalization</u>.

- The article "New Concerns About Distance Education" was reviewed recently in the newspaper column "In My Opinion."

Preposition Errors

Do not use redundant prepositions, such as "off *of,*" "in back *of,*" "inside *of,*" and "at *about.*" See also <u>conciseness/wordiness</u>.

> EXACT   The client arrived at ~~about~~ four o'clock.
>
> APPROXIMATE The client arrived ~~at~~ about four o'clock.

Avoid unnecessarily adding the preposition *up* to verbs.

- *to*
  Call ~~up and~~ see if he is in his office.
  ^

Do not omit necessary prepositions.

- *to*
  He was oblivious and not distracted by the view from his office window.
  ^

presentations

The steps required to prepare an effective presentation parallel the steps you follow to write a document. As with writing a document, you must determine your **purpose** and analyze your **audience**. You must gather the facts that will support your point of view and proposal and logically organize that information. Presentations do, however, differ from written documents in a number of important ways. They are intended for listeners, not readers. Because you are speaking, your manner of delivery, the way you organize the material, and your supporting **visuals** require as much attention as your content.

Determining Your Purpose

Every presentation is given for a purpose, even if it is only to share information. To determine the primary purpose of your presentation, use the following question as a guide: What do I want the audience to *know,* to *believe,* and to *do* when I have finished the presentation? Based on the answer to that question, write a purpose statement that answers the questions *What?* and *Why?*.

- The purpose of my presentation is to convince my company's chief information officer of the need to improve the appearance, content, and customer use of our company's Web site [what] so that she will be persuaded to allocate additional funds for site-development work in the next fiscal year [why].

Analyzing Your Audience

Once you have determined the desired end result of the presentation, you need to analyze your audience so that you can tailor your presentation to their needs. (See **readers**.) Ask yourself these questions about your audience:

- What is your audience's level of experience or knowledge about your topic?
- What is the general educational level and age of your audience?
- What is your audience's attitude toward the topic you are speaking about, and—based on that attitude—what concerns, fears, or objections might your audience have?
- Do any subgroups in the audience have different concerns or needs?
- What questions might your audience ask about this topic?

Gathering Information

Once you have focused the presentation, you need to find the facts that support your point of view or the action you propose. As you gather information, keep in mind that you should give the audience only the facts necessary to accomplish your goals; too much will overwhelm them and too little will not adequately inform your listeners or support your recommendations. For detailed guidance about gathering information, see **research**, **Internet research**, and **library research**.

Structuring the Presentation

When structuring the presentation, focus on your audience. Listeners are freshest at the outset and refocus their attention near the end. Take advantage of that pattern. Give your audience a brief overview of your presentation at the beginning, use the body to develop your ideas, and end with a summary of what you covered and, if appropriate, a call to action. See also **methods of development**.

The Introduction.   Include in the **introduction** an opening that focuses your audience's attention, such as in the following examples.

- You have to write an important report, but you'd like to incorporate lengthy sections of an old report into your new one. The problem is that you don't have an electronic version of the old report. You will have to rekey many pages. You groan because that seems an incredible waste of time. Have I got a solution for you! [Definition of a problem]

- As many as 50 million Americans have high blood pressure. [An attention-getting statement]

- Would you be interested in a full-sized computer keyboard that is waterproof and noiseless, and can be rolled up like a rubber mat? [A rhetorical question]

- As I sat at my computer one morning, deleting my eighth spam message of the day, I decided that it was time to take action to eliminate this time-waster. [A personal experience]

- According to researchers at the Massachusetts Institute of Technology, "Garlic and its cousin, the onion, confer major health benefits — including fighting cancer, infections, and heart disease." [An appropriate quotation]

Following your opening, use the introduction to set the stage for your audience by providing an overview of the presentation, which can include general or background information that will be needed to understand any more detailed information in the body of your presentation. It can also show how you have organized the material.

- This presentation analyzes three different scanner models for us to consider purchasing. Based on a comparison of all three, I will recommend the one I believe best meets our needs. To do so, I'll discuss the following five points:
  1. Why we need a scanner [the problem]
  2. The basics of scanner technology [general information]
  3. The criteria I used to compare the three scanner models [comparison]
  4. The scanner models I compared and why [possible solutions]
  5. The scanner I propose we buy [proposed solution]

The Body.   If applicable, present the evidence that will persuade the audience to agree with your conclusions and act on them. (See **persuasion**.) If there is a problem, demonstrate that it exists and offer a solution or range of possible solutions. For example, if your introduction stated that the problem is low profits, high costs, outdated technology, or high employee absenteeism, you could use the following approach.

P

1. Prove your point.
   a. Marshal the facts and data you need.
   b. Present the information using easy-to-understand visuals.
2. Offer solutions.
   a. "Increase profits by lowering production costs."
   b. "Cut overhead to reduce costs or abolish specific programs or product lines."
   c. "Replace outdated technology or upgrade existing technology."
   d. "Offer employees more flexibility in their work schedules or other incentives."
3. Anticipate questions ("How much will it cost?") and objections ("We're too busy now — when would we have time to learn the new software?") and incorporate the answers into your presentation.

The Closing.   Fulfill the goals of your presentation in the closing. If your purpose is to motivate the listeners to take action, ask them to do what you want them to do; if your purpose is to get your audience to think about something, summarize what you want them to think about. Many presenters make the mistake of not actually closing—they simply quit talking, shuffle papers, and then walk away.

Because your closing is what your audience is most likely to remember, it is the time to be strong and persuasive. Consider the following possible closing.

- Based on all the data, I believe that the Worthington scanner best suits our needs. It produces 3,000 units a month more than its closest competitor and creates electronic files we can use on our Web site and on paper. The Worthington is also compatible with our current computer network and includes staff training at our site. Although the initial cost is higher than that for the other two models, the additional capabilities, longer life-cycle for replacement parts, and lower maintenance costs make it a better value.

    I recommend we allocate the funds necessary for this scanner by the 15th of this month in order to be well prepared for next quarter's customer presentations.

The closing brings the presentation full circle and asks the audience to fulfill the purpose of the presentation—exactly what a closing should do. See also conclusions.

Transitions.   Planned transitions should appear between the introduction and the body, between points in the body, and between the body and the closing. Transitions are simply a sentence or two to let the audience know that you are moving from one topic to the next. They also prevent a choppy presentation and provide the audience with assurance that you know where you are going and how to get there.

- Before getting into the specifics of each scanner I compared, I'd like to demonstrate how scanners work in general. That information will provide you with the background you'll need to compare the differences among the scanners and their capabilities discussed in this presentation.

It is also a good idea to pause for a moment after you have delivered a transition between topics to let your listeners shift gears with you. Remember, they do not know your plan.

Using Visuals

Well-planned visuals not only add interest and emphasis to your presentation, they also clarify and simplify your message because they communicate clearly, quickly, and vividly. Charts, graphs, and illustrations greatly increase audience understanding and retention of information, especially for complex issues and technical information that could otherwise be misunderstood or overlooked. A bar graph, pie chart, diagram, or concise summary of key points can eliminate misunderstanding and save many words.

You can create and present the visual components of your presentation by using a variety of media—flip charts, whiteboard or chalkboard, overhead transparencies, slides, or computer presentation software.

Flip Charts.   Flip charts are ideal for smaller groups in a conference room or classroom and are also ideal for **brainstorming** with your audience.

Whiteboard or Chalkboard.   The whiteboard or chalkboard common to classrooms is convenient for creating sketches and for jotting notes during your presentation. If your presentation requires extensive notes or complex drawings, create them before the presentation to minimize audience restlessness.

Overhead Transparencies.   With transparencies you can create a series of overlays to explain a complex device or system, adding (or removing) the overlays one at a time. You can also lay a sheet of paper over a list of items on a transparency, uncovering one item at a time as you discuss it, to focus audience attention on each point in the sequence.

Presentation Software.   Presentation software, such as Power Point, Corel Presentations, and Freelance Graphics, lets you create your presentation on your computer. You can develop charts and graphs with data from spreadsheet software or locate visuals on the Web, and then import those files into your presentation. This software also offers standard templates and other features that help you design effective visuals and integrated text. Enhancements include a selection of typefaces, highlighting devices, background textures and colors, and clip-art images. Avoid using too many enhancements, which may distract viewers from your message. Images can also be printed out for use as overhead transparencies or handouts.

Rehearse your presentation using your electronic slides, and practice your transitions from slide to slide. Also practice loading your presentation and anticipate any technical difficulties that might arise. Should you encounter a technical snag during the presentation, stay calm and give yourself time to solve the problem. If you cannot solve the problem, move on without the technology. As a backup, carry a printout of your electronic presentation. Carry an extra copy of your presentation on diskette for backup. Figure P–3 shows slides for a brief presentation.

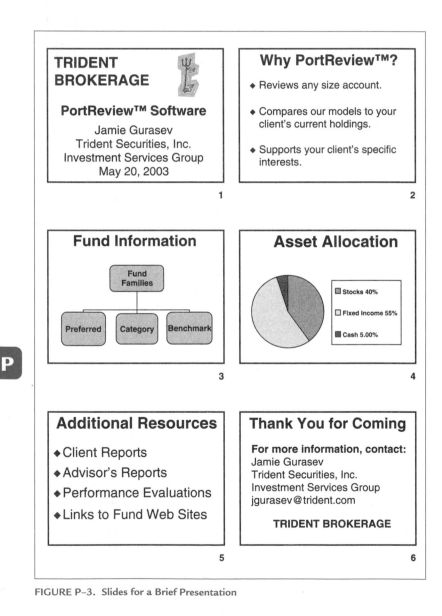

FIGURE P–3. Slides for a Brief Presentation

Writer's Checklist: Using Visuals in a Presentation

☑ Use text sparingly. Use bulleted or numbered <u>lists</u>, keeping them parallel in content and grammatical form. Use numbers if the sequence is important and bullets if it is not. See <u>parallel structure</u>.

☑ Limit the number of bulleted or numbered items to 5 or 6 per visual. Each visual should contain no more than 40 to 45 words. Any more will clutter the visual and force you to use a smaller type size that could impair the audience's ability to read it.

☑ Make your visuals consistent in type style, size, and spacing.

☑ Use a type size visible to members of the audience in the back of the room. Type should be boldface and no smaller than 30 points. For headings, 45- or 50-point type works even better.

☑ Use graphs and charts to show data trends. Use only one or two illustrations per visual to avoid clutter and confusion.

☑ Make the contrast between your text and the background sharp. Use light backgrounds with dark lettering and avoid textured or decorated backgrounds.

☑ Use no more than 12 visuals per presentation. Any more will tax the audience's concentration.

☑ Match your delivery of the content to your visuals. Do not put one set of words or images on the screen and talk about the previous visual or, even worse, the next one.

☑ Do not read the text on your visual word for word. Your audience can read the visuals; they look to you to cover the salient points in detail.

Delivering a Presentation

Once you have outlined and drafted your presentation and prepared your visual, you are ready to practice your presentation and delivery techniques.

Practice. Familiarize yourself with the sequence of the material — major topics, notes, and visuals — in your outline. Once you feel comfortable with the content, you are ready to practice the presentation itself.

PRACTICE ON YOUR FEET AND OUT LOUD. Try to practice in the room where you will give the presentation. Practicing on-site helps you get the feel of the room: the acoustics, the lighting, the arrangement of the chairs, the position of electrical outlets and switches, and so forth. Practice out loud to make clear exactly how long your presentation will take, highlight problems such as awkward transitions, and help eliminate verbal tics, such as "um," "you know," and "like."

PRACTICE WITH YOUR VISUALS AND TEXT.   Integrate your visuals into your practice sessions to help your presentation go more smoothly. Operate the equipment (computer, slide projector, or overhead projector) until you are comfortable with it. Decide if you want to use a remote control or to have someone else advance your slides. Even if things go wrong, being prepared and practiced will give you the confidence and poise to go on.

Delivery Techniques That Work.   Your delivery is both audible and visual. In addition to your words and message, your nonverbal communication affects your audience. To make an impression on your listeners and keep their attention, you must be animated. Your words will have more staying power when they are delivered with physical and vocal animation. If you want listeners to share your point of view, show enthusiasm for your topic. The most common delivery techniques include making eye contact; using movement and gestures; and varying voice inflection, pace, and projection.

EYE CONTACT.   The best way to establish rapport with your audience is through eye contact. In a large audience, directly address those people who seem most responsive to you in different parts of the room. Doing that helps you establish rapport with your listeners by holding their attention and gives you important visual cues that let you know how your message is being received. Are people engaged and actively listening? Or are they looking around or staring at the floor? Such cues tell you that you may need to speed up or slow down the pace of your presentation.

MOVEMENT.   Animate the presentation with physical movement. Take a step or two to one side after you have been talking for a minute or so. That type of movement is most effective at transitional points in your presentation between major topics or after pauses or emphases. Too much movement, however, can be distracting, so try not to pace.

Another way to integrate movement into your presentation is to walk to the screen and point to the visual as you discuss it. Touch the screen with the pointer and then turn back to the audience before beginning to speak (remember the three *t*'s: touch, turn, and talk).

GESTURES.   Gestures both animate your presentation and help communicate your message. Most people gesture naturally when they talk; nervousness, however, can inhibit gesturing during a presentation. Keep one hand free and use that hand to gesture.

VOICE. Your voice can be an effective tool in communicating your sincerity, enthusiasm, and command of your topic. Use it to your advantage to project your credibility. *Vocal inflection* is the rise and fall of your voice at different times, such as the way your voice naturally rises at the end of a question ("You want it *when?*"). A conversational delivery and eye contact promote the feeling among members of the audience that you are addressing them directly. Use vocal inflection to highlight differences between key and subordinate points in your presentation.

PACE. Be aware of the speed at which you deliver your presentation. If you speak too fast, your words will run together, making it difficult for your audience to follow. If you speak too slowly, your listeners will become impatient and distracted.

PROJECTION. Most speakers think they are projecting more loudly than they are. Remember that your presentation is ineffective for anyone in the audience who cannot hear you. If listeners must strain to hear you, they may give up trying to listen. Correct projection problems by practicing out loud with someone listening from the back of the room.

Presentation Anxiety. Everyone experiences nervousness before a presentation. Survey after survey reveals that for most people dread of public speaking ranks among their top five fears. Instead of letting fear inhibit you, focus on channeling your nervous energy into a helpful stimulant. The best way to master anxiety is to know your topic thoroughly—knowing what you are going to say and how you are going to say it will help you gain confidence and reduce anxiety as you become immersed in your subject.

P

Writer's Checklist: Preparing for and Delivering a Presentation

- ☑ Practice your presentation with visuals; practice in front of listeners, if possible.
- ☑ Visit the location of the presentation ahead of time to familiarize yourself with the surroundings.
- ☑ Prepare a set of notes that will trigger your memory during the presentation.
- ☑ Make as much eye contact as possible with your audience to establish rapport and maximize opportunities for audience feedback.

☑ Animate your delivery by integrating movement, gestures, and vocal inflection into your presentation. However, keep your movements and speech patterns natural.

☑ Speak loudly and slowly enough to be heard and understood.

☑ Do not read the text on your visuals word for word; explain the salient points in detail.

For information and tips on communicating with cross-cultural audiences, see **global graphics** and **international correspondence**.

press releases

Companies and organizations write *press releases* (or news releases) to announce new products and services, new policies, special events (such as branch openings, company anniversaries, mergers, and grand openings), management changes, and sponsorship of social-action and cultural programs. The purpose of a press release is both to inform the public about the company and its products and services and to promote a favorable image. Even the announcement of an unfavorable event, such as the closing of a division, should try to present the company in a good light. Large corporations and institutions usually have their own public-relations staffs or use outside agencies. However, if you work for a small company without public-relations resources, you may be called on to write a press release.

The press release should be clear, concise, and written with particular attention to the five *w*'s: *who, what, where, when,* and *why.* Write the release in decreasing **order-of-importance method of development**. Put all critical information in the first paragraph, information of the next level of importance in the second paragraph, and so on. Editors must make your release fit the space they have available; if they must cut your release, they will delete the last paragraph first, and then the next to last, and so on. Make sure your facts are accurate and be careful to define any unfamiliar terms.

Releases are usually sent to local newspapers, television and radio stations, and other special groups, such as trade publications or professional associations. Send the release to a specific person whenever possible. Otherwise, try to address the news release to a particular editor, such as to the business editor or the technology editor. See also **newsletter articles**.

Begin with the place and date of the announcement, as shown in Figure P–4.

Bentley Plastics
3535 Michigan Avenue
Chicago, IL 60653
E-mail: bentleypr@bentley.org
Web: http://www.bentleyplas.org

FOR IMMEDIATE RELEASE

For More Information Call:
Marjorie Kohls
(312) 712-1946

CHICAGO, ILLINOIS, MAY 20, 2003 . . . Mark Williams has joined Bentley Plastics as vice president–marketing. He will direct the Illinois and Indiana district sales offices and coordinate overseas distribution through the company's Singapore office. Williams will travel extensively to the Far East while developing marketing channels for Bentley.

Formerly, Williams was director of services with International Marketing Associates, a consulting group in New York. While at IMA, he developed a computer-based marketing center that linked textile firms in the United States, Finland, and Great Britain. A graduate of the Columbia University Graduate School of Business, Williams is the author of *Marketing Dynamics,* an informal examination of psychological appeals to the buying public. The book has been used in marketing classrooms at several universities.

Bentley Plastics, headquartered in Chicago, produces polymer tubing and coils. The company's main plants are in Skokie, Illinois, and Gary, Indiana. Since 1978 Bentley has also been producing tubing for French distribution through the firm of Jourdan and Sons, Paris.

-30-

FIGURE P–4. Press Release

Writer's Checklist: Preparing Press Releases

☑ Print the release on company letterhead and double- or triple-space for easy reading and editing.

☑ Leave the top third of the page blank for editors to mark any necessary headings on the copy.

☑ Give the name and phone number of someone in the company who can give further information.

☑ If the release is longer than one page, type *More* at the bottom of each page except the last.

☑ Type -30- or *End* or ### to indicate where the press release ends.

WEB LINK WRITING PRESS RELEASES

Press-Release-Writing.com, a service of Accurate Online Solutions, offers a Web site that includes tips, samples, and resources for writing press releases. See *<www.bedfordstmartins.com/alred>* and select *Links for Technical Writing.*

---

principal / principle

Principal, meaning an "amount of money on which interest is earned or paid" or a "chief official in a school or court proceeding," is sometimes confused with *principle,* which means "basic truth or belief."

• The bank will pay 6.5 percent on the *principal.*

• He sent a letter to the *principal* of the high school.

• She is a person of unwavering *principles.*

Principal is also an adjective, meaning "main" or "primary." (My *principal* objection is that it will be too expensive.)

process explanation

Many kinds of technical writing explain a process, an operation, or a procedure. A process explanation describes the steps that a mechanism or system uses to accomplish a certain result, such as the steps necessary to design a product. In your opening, present a brief overview of the process or let readers know why it is important for them to become familiar with the process you are explaining. Be sure to define terms that readers might not understand and provide visuals to clarify the process. See also defining terms, instructions, and introductions.

In describing a process, transitional words and phrases create unity within paragraphs, and headings often mark the transition from one step to the next. Illustrations, like the one used in Figure P–5 on page 426, can help convey your message.

progress and activity reports

Progress and activity reports document ongoing activities: Progress reports are often used to report on major projects, whereas activity reports focus on the work and achievements of individual employees.

Progress Reports

A *progress report* provides information about a project—its status, whether it is on schedule and within budget, and so on. Progress reports are often submitted by a contracting company to a client company. They are used mainly for projects that involve many steps over a period of time and are issued at regular intervals to state what has been done and what remains to be done. Progress reports help keep projects running smoothly by helping managers assign work, adjust schedules, allocate budgets, and schedule supplies and equipment. All progress reports dealing with a particular project should have the same format.

The introduction to the first progress report should identify the project, any materials needed, and the project's completion date. Subsequent reports summarize the progress to date; include the status of schedules and costs; list the steps that remain to be taken; and conclude with recommendations about changes in the schedule, materials, and so on. Figure P–6 on page 427 shows the initial progress report submitted by a construction company to a client.

Activity Reports

Within an organization, employees often submit activity reports on the progress of ongoing projects. Managers may combine the activity reports (also called *status reports*) of several individuals or teams into larger activity reports and, in turn, submit those larger reports to their own managers.

Because the activity report is issued periodically (usually monthly) and contains material familiar to its readers, it normally needs no

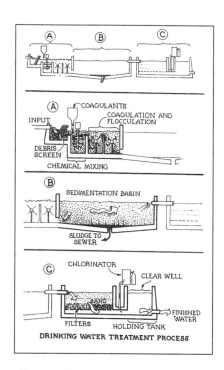

DRINKING WATER TREATMENT PROCESS

Drinking Water Treatment Process

After it has been transported from its source to a local water system, most surface water must be processed in a treatment plant before it can be used. Some groundwater, on the other hand, is considered chemically and biologically pure enough to pass directly from a well into the distribution system that carries it to the home.

Although there are innumerable variations, surface water is usually treated as follows: First, it enters a storage lagoon where a chemical, usually copper sulfate, is added to control algae growth. From there, water passes through one or more screens that remove large debris. Next, a coagulant, such as alum, is mixed into the water to encourage the settling of suspended particles. The water flows slowly through one or more sedimentation basins so that larger particles settle to the bottom and can be removed. Water then passes through a filtration basin partially filled with sand and gravel where yet more suspended particles are removed.

At that point in the process, the Safe Drinking Water Act has mandated an additional step for communities using surface water. . . . Water is to be filtered through activated carbon to remove any microscopic organic material and chemicals that have escaped the other processes.

FIGURE P–5. Process Explanation (with Illustration)

Hobard Construction Company
9032 Salem Avenue
Lubbock, TX 79409

www.hobardcc.com
(808) 769-0832
Fax: (808) 769-5327

August 18, 2003

Walter M. Wazuski
County Administrator
109 Grand Avenue
Manchester, NH 03103

Dear Mr. Wazuski:

The renovation of the County Courthouse is progressing on schedule and within budget. Although the cost of certain materials is higher than our original bid indicated, we expect to complete the project without exceeding the estimated costs because the speed with which the project is being completed will reduce overall labor expenses.

Costs
Materials used to date have cost $78,600, and labor costs have been $193,000 (including some subcontracted plumbing). Our estimate for the remainder of the materials is $59,000; remaining labor costs should not exceed $64,000.

Work Completed
As of August 15, we had finished the installation of the circuit-breaker panels and meters, of level-one service outlets, and of all subfloor wiring. The upgrading of the courtroom, the upgrading of the records-storage room, and the replacement of the air-conditioning units are in the preliminary stages.

Work Schedule
We have scheduled the upgrading of the courtroom to take place from August 25 to October 3, the upgrading of the records-storage room from October 6 to November 12, and the replacement of the air-conditioning units from November 14 to December 17. We see no difficulty in having the job finished by the scheduled date of December 23.

Sincerely yours,

Tran Nuguélen

Tran Nuguélen
ntran@hobardcc.com

FIGURE P–6. Progress Report

introduction or conclusion, although it may need a brief opening to provide context. Although format varies from company to company, the following sections are typical: Current Projects, Current Problems, Plans for the Next Period, and Current Staffing Level (for managers).

The activity report shown in Figure P–7 was submitted by a manager of application programs (Wayne Tribinski) who supervises 11 employees. The reader of the report (Kathryn Hunter) is Tribinski's manager and the Director of Engineering.

pronoun reference

A pronoun should refer clearly to a specific antecedent. Avoid vague and uncertain references.

- We got the account after we wrote the proposal. *, which was a big one,* ~~It was a big one.~~

For coherence, place pronouns as close as possible to their antecedents — the likelihood of an ambiguous reference increases with distance. See also ambiguity.

- The office building next to City Hall *, praised for its architectural design, is* ~~was praised for its architectural design.~~

Avoid general and hidden references. A general (or broad) reference or one that has no real antecedent is a problem that often occurs when the word *this* is used by itself.

- He deals with personnel problems in his work. This *experience* helps him in his personal life.

A hidden reference is one that has only an implied antecedent.

- A high-lipid, low-carbohydrate diet is "ketogenic" because it favors *the* ~~their~~ formation *of ketone bodies* .

Do not repeat an antecedent in parentheses following the pronoun. If you feel you must identify the pronoun's antecedent in that way, rewrite the sentence.

AWKWARD   The senior partner first met Bob Evans when he (Evans) was a trainee.

IMPROVED   Bob Evans was a trainee when the senior partner first met him.

INTEROFFICE MEMO

Date: June 5, 2003
To: Kathryn Hunter, Director of Engineering
From: Wayne Tribinski, Manager, Applications Programs   *W T*
Subject: Activity Report for May 2003

We are dealing with the following projects and problems, as of May 31.

Projects

1. For the *Software Training Mailing Campaign,* we anticipate producing a set of labels for mailing software training information to customers by June 10.
2. The *Search Project* is on hold until the PL/I training has been completed, probably by the end of June.
3. The project to provide a database for the *Information Management System* has been expanded in scope to provide a database for all training activities. We are rescheduling the project to take the new scope into account.

Problems

The *Information Management System* has been delayed. The original schedule was based on the assumption that a systems analyst who was familiar with the system would work on this project. Instead, the project was assigned to a newly hired systems analyst who was inexperienced and required much more learning time than expected.

Bill Michaels, whose activity report is attached, is correcting a problem in the *CNG Software.* This correction may take a week.

Plans for Next Month

- Complete the *Software Training Mailing Campaign.*
- Resume the *Search Project.*
- Restart the project to provide a database on information management with a schedule that reflects its new scope.
- Write a report to justify the addition of two software engineers to my department.
- Congratulate publicly the recipients of Meritorious Achievement Awards: Bill Thomasson and Nancy O'Rourke.

Current Staffing Level

Current staff: 11
Open requisitions: 0

Attachment

FIGURE P–7. Activity Report

For advice on avoiding pronoun reference problems with gender, see **biased language**. See also **agreement** and **pronouns**.

pronouns

DIRECTORY
Case   431
Gender   432
Number   432
Person   433

A *pronoun* is a word that is used as a substitute for a **noun** (the noun for which a pronoun substitutes is called the *antecedent*). Using pronouns in place of nouns relieves the monotony of repeating the same noun over and over. See also **pronoun reference**.

Personal pronouns refer to the person or people speaking (*I, me, my, mine; we, us, our, ours*); the person or people spoken to (*you, your, yours*); or the person, people, or thing(s) spoken of (*he, him, his; she, her, hers; it, its; they, them, their, theirs*). See also **person** and **point of view**.

- If *their* figures are correct, *ours* must be in error.

Demonstrative pronouns (*this, these, that, those*) indicate or point out the thing being referred to.

- *This* is my desk. *These* are my coworkers. *That* will be a difficult job. *Those* are incorrect figures.

Relative pronouns (*who, whom, which, that*) perform a dual function: (1) They take the place of nouns and (2) they connect and establish the relationship between a dependent **clause** and its main clause.

- The department manager decided *who* would be hired.

Interrogative pronouns (*who, whom, what, which*) are used only to ask questions.

- *What* is the trouble?

Indefinite pronouns specify a class or group of persons or things rather than a particular person or thing (*all, another, any, anyone, anything, both, each, either, everybody, few, many, most, much, neither, nobody, none, several, some, such*).

- Not *everyone* liked the new procedures; *some* even refused to follow them.

A *reflexive pronoun*, which always ends with the suffix *self* or *selves*, indicates that the subject of the sentence acts upon itself. See also <u>sentence construction</u>.

- The electrician accidentally shocked *herself.*

The reflexive pronouns are *myself, yourself, himself, herself, itself, oneself, ourselves, yourselves,* and *themselves. Myself* is not a substitute for *I* or *me* as a personal pronoun.

- Victor and ~~myself~~ *I* completed the report on time.

- The assignment was given to Ingrid and ~~myself~~ *me*.

Intensive pronouns are identical in form to the reflexive pronouns, but they perform a different function: to emphasize their antecedents.

- I *myself* asked the same question.

Reciprocal pronouns (*one another, each other*) indicate the relationship of one item to another. *Each other* is commonly used when referring to two persons or things and *one another* when referring to more than two.

- Salih and Kara work well with *each other.*

- The crew members work well with *one another.*

Case

Pronouns have forms to show the subjective, objective, and possessive cases, as shown in the entry <u>case</u>. A pronoun that is used as the subject of a clause or sentence is in the subjective case (*I, we, he, she, it, you, they, who*). The subjective case is also used when the pronoun follows a linking <u>verb</u>.

- *She* is my boss.

- My boss is *she.*

A pronoun that is used as the object of a verb or preposition is in the objective case (*me, us, him, her, it, you, them, whom*).

- Ms. Davis hired Tom and *me.* [object of verb]

- Between *you* and *me*, she's wrong. [object of preposition]

A pronoun that is used to express ownership is in the possessive case (*my, mine, our, ours, his, hers, its, your, yours, their, theirs, whose*).

- He took *his* notes with him on the business trip.

- We took *our* notes with us on the business trip.

A pronoun appositive takes the case of its antecedents.

- Two systems analysts, Joe and *I*, were selected to represent the company.
  [*Joe and I* is in apposition to the subject, *systems analysts*, and must therefore be in the subjective case.]

- The systems analysts selected two members—Joe and *me*.
  [*Joe and me* is in apposition to *two members*, which is the object of the verb, *selected*, and therefore must be in the objective case.]

If you have difficulty determining the case of a compound pronoun, try using the pronoun singly.

- In his letter, Eldon mentioned *him* and *me*.
  In his letter, Eldon mentioned *him*.
  In his letter, Eldon mentioned *me*.

- *They* and *we* must discuss the terms of the merger.
  They must discuss the terms of the merger.
  We must discuss the terms of the merger.

When a pronoun modifies a noun, try it without the noun to determine its case.

- [*We/Us*] pilots fly our own planes.
  We fly our own planes.
  [You would not write, "*Us* fly our own planes."]

- He addressed his remarks directly to [*we/us*] technicians.
  He addressed his remarks directly to *us*.
  [You would not write, "He addressed his remarks directly to *we*."]

Gender

A pronoun must agree in gender with its antecedent. A problem sometimes occurs because the masculine pronoun has traditionally been used to refer to both sexes. To avoid the sexual bias implied in such usage, use *he or she* or the plural form of the pronoun, *they*.

- ~~Each~~ may stay or go as ~~he chooses.~~
  All   they choose.

If you use the plural form of the pronoun, be sure to change the indefinite pronoun *each* to its plural form, *all*. See also **biased language**.

Number

Number is a frequent problem with only a few indefinite pronouns (*each, either, neither,* and those ending with *-body* or *-one,* such as *any-*

body, anyone, everybody, everyone, nobody, no one, somebody, someone) that are normally singular and so require singular verbs and are referred to by singular pronouns.

- As *each member arrives* for the meeting, please hand *him or her* a copy of the confidential report. *Everyone* must return the copy before *he or she* leaves. *Everybody* on the committee *understands* that *neither* of our major competitors *is* aware of the new process we have developed.

Person

Third-person personal pronouns usually have antecedents.

- Gina presented the report to the members of the board of directors. *She* [Gina] first summarized *it* [the report] for *them* [the directors] and then asked for questions.

First- and second-person personal pronouns do not normally require antecedents.

- *I* like my job.
- *You* were there at the time.
- *We* all worked hard on the project.

ESL TIPS FOR USING POSSESSIVE PRONOUNS

In many languages, possessive pronouns agree in number and gender with the nouns they modify. In English, however, possessive pronouns agree in number and gender with their antecedents. Check your writing carefully for agreement between a possessive pronoun and the word, phrase, or clause that it refers to.

- The *woman* brought *her* brother a cup of soup.
- *Robert* sent *his* mother flowers on Mother's Day.

P

proofreaders' marks

Publishers have established symbols called *proofreaders' marks*, that writers and editors use to communicate in the production of publications. Familiarity with those symbols makes it easy for you to communicate your changes to others. Figure P–8 on page 434 lists standard proofreaders' marks.

Mark in Margin	Instruction	Mark on Manuscript	Corrected Type
	Delete	the ~~lawyer's~~ Bible	the Bible
	Insert	the bible	the lawyer's bible
stet	Let stand	the ~~lawyer's~~ bible	the lawyer's bible
cap	Capitalize	the bible	the Bible
lc	Make lowercase	the Law	the law
ital	Italicize	the lawyer's bible	the *lawyer's* bible
tr	Transpose	the bible lawyer's	the lawyer's bible
C	Close space	the Bi ble	the Bible
sp	Spell out	2 bibles	two bibles
#	Insert space	the Bible	the Bible
¶	Start paragraph	The lawyer's . . .	The lawyer's . . .
run in	No paragraph	. . . marks. / Below is a . . .	. . . marks. Below is a . . .
sc	Set in small capitals	the bible	the BIBLE
rom	Set in roman type	the *bible*	the bible
bf	Set in boldface	the bible	the **bible**
lf	Set in lightface	the **bible**	the bible
⊙	Insert period	The lawyers have their own bible	The lawyers have their own bible.
⌄	Insert comma	However we cannot . . .	However, we cannot . . .
= / =	Insert hyphens	half and half	half-and-half
⊙	Insert colon	We need the following	We need the following:
⌃	Insert semicolon	Use the law don't . . .	Use the law; don't . . .
⌄	Insert apostrophe	John's law book	John's law book
⌄/⌄	Insert quotation marks	The law is law.	The "law" is law.
(/)/	Insert parentheses	John's law book	John's (law) book
[/]/	Insert brackets	(John Martin 1920–1962 went . . .)	(John Martin [1920–1962] went . . .)
⌐N	Insert en dash	1920 1962	1920–1962
⌐M	Insert em dash	Our goal victory	Our goal—victory
⌄	Insert superior type	3 = 9	$3^2 = 9$
⌃2	Insert inferior type	HSO₄	H_2SO_4

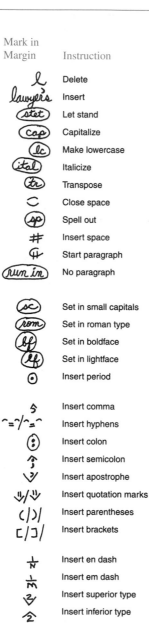
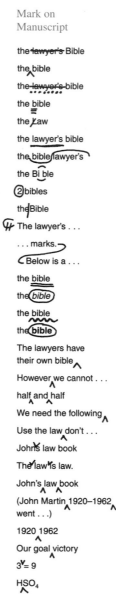

FIGURE P–8. Proofreaders' Marks

P

proofreading

Computer grammar checkers and spell checkers, while a help to proof-reading, can make writers overconfident. If a typographical error results in a legitimate English word (for example, *coarse* instead of *course*), the spell checker will not flag the misspelling. Therefore, you still must proofread your work carefully—both on your monitor and on paper. You may find some of the tactics discussed in <u>revision</u> useful when proofreading; in fact, you may find passages during proofreading that will require further revision.

Whether the material you proofread is your own writing or that of someone else, consider proofreading in several stages. Although you need to tailor the stages to the specific document and to your own problem areas, the following Writer's Checklist should provide a useful starting point for proofreading.

Writer's Checklist: Proofreading in Stages

FIRST-STAGE REVIEW

☑ Appropriate <u>format</u>, as for <u>reports</u> or <u>correspondence</u>

☑ Typographical consistency (<u>headings</u>, spacing, fonts)

SECOND-STAGE REVIEW

☑ Specific <u>grammar</u> and <u>usage</u> problems

☑ Appropriate <u>punctuation</u>

☑ Correct <u>abbreviations</u> and <u>capitalization</u>

☑ Correct <u>spelling</u> (especially names and places)

☑ Complete Web or <u>e-mail</u> addresses and the like

☑ Accurate figures in tables and lists

☑ Cut-and-paste errors; for example, a result of moved or deleted text and numbers

FINAL-STAGE REVIEW

☑ Final check of your goals: <u>readers'</u> needs and <u>purpose</u>

☑ Appearance of the document (see <u>layout and design</u>)

☑ Review by colleague for crucial documents

Use standard <u>proofreaders' marks,</u> especially for proofreading some-one else's document.

P

🖱 DIGITAL TIP **PROOFREADING FOR FORMAT CONSISTENCY**

An effective way to check formatting, spacing, and typographical consistency as well as general appearance of documents is to view the whole page on-screen in your word-processing program. You may also wish to view multiple pages or "tile" separate documents side by side for comparison. For more on this topic, see <*www.bedfordstmartins.com/alred*> and select *Digital Tips*.

---

proposals

P

A *proposal* is a document written to persuade <u>readers</u> to follow a plan or course of action that you are proposing. (See also <u>persuasion</u>.) You may need to send a proposal to a colleague within your organization (an internal proposal) or to a potential customer or client outside the organization (an external proposal). Often, you will collaborate with others in preparing a proposal.

Strategies

Regardless of the type of proposal you will write, you must begin by considering the audience and purpose, the project management, and the proposal structure.

Audience and Purpose.   A proposal offers a plan to fill a need, and readers will evaluate your plan based on how well you answer their questions about what you are proposing to do, how you plan to do it, when you plan to do it, and how much it will cost. Because proposals often require more than one level of approval, take all your readers into account as you answer their questions. Consider especially their levels of technical knowledge of the subject. For example, if your primary reader is an expert on your subject but his or her supervisor who must also approve the proposal is not, provide an <u>executive summary</u> written in nontechnical language. You might also include a <u>glossary</u> of terms used in the body of the proposal or an <u>appendix</u> that explains highly detailed information in nontechnical language. If your primary reader is not an expert but his or her supervisor is, write the proposal with the nonexpert in mind and include an appendix that contains the technical details.

Often the complicated task of writing a persuasive proposal can be simplified by composing a concise statement of exactly the problem or opportunity that your proposal is designed to address. Determining the problem that your proposal addresses helps you and your reader understand the value, scope, and limitations of your proposed solution. Doing so will also enable you to construct a clear statement of the proposal's <u>purpose</u>.

Project Management.   Proposal writers are often faced with writing high-quality, persuasive proposals under tight organizational deadlines. Breaking the task down into manageable parts is the key to accomplishing your goal. For details about working with others on writing projects, see <u>collaborative writing</u>.

Proposal Structure.   Proposals vary widely in length, formality, and structure. A short or medium-length proposal typically consists of an <u>introduction</u>, a body, and a <u>conclusion</u>.

P

INTRODUCTION.   The *introduction* should state the purpose and scope of your proposal, as well as the problem you propose to solve and your solution to it. It should also indicate the dates on which you propose to begin and complete work, any special benefits of your proposed approach, and the cost of the project. For a sales (external) proposal, refer to any previous positive associations your organization has had with the potential client or customer.

BODY.   The *body* should offer the details of your solution to the problem and explain (1) the product or service you are offering, (2) how the job will be done, (3) how you will perform the work and any special materials you may use, (4) a schedule when each phase of the project will be completed, and (5) a breakdown of project costs.

CONCLUSION.   The *conclusion* should persuasively resell your proposal by emphasizing the benefits of your solution, product, or service over any competing ideas. Also include details about the time period during which the proposal is valid. Effective conclusions show confidence in your solution, your appreciation for the opportunity to submit the proposal, and your willingness to provide further information as well as encouraging your reader to act on your proposal.

Longer proposals expand the basic introduction, body, and conclusion into distinct sections. The number of sections varies by company or organization. The list that follows shows the variety of sections that may be found in longer, more formal proposals. Each of these sections are described on the following pages. Which components should be included depends on the reader, the purpose, and the contents of the proposal. For more information on many of the following components, see <u>formal reports</u>.

FRONT MATTER
Copy of Request for proposals (if applicable)
Cover letter
Title page
Table of contents
List of figures

BODY OF PROPOSAL
Executive summary
Introduction
Background or Problem statement
Product description
Detailed solutions (rationale)
Cost analysis
Delivery schedule
Staffing
Site preparation
Training (if needed)
Statement of responsibilities
Description of vendor
Organizational sales pitch
Authorization request and deadline
Conclusion

BACK MATTER
Appendixes
Bibliography
Glossary

FRONT MATTER

- *Copy of the Request for Proposals (RFP).* Include a copy of the RFP with a formal proposal.
- *Cover Letter or Letter of Transmittal.* In the <u>cover letter</u>, express your appreciation for the opportunity to submit your proposal, any assistance the customer has provided, and any previous positive associations. Then summarize the proposal's recommendations and express confidence that they will satisfy the customer's needs.
- *Title Page.* Include the <u>title</u> of the proposal, the date, the organization to which it is being submitted, and your company name.
- *Table of Contents.* Include a <u>table of contents</u> in longer proposals to guide readers to important headings, which should be listed according to beginning page numbers.
- *List of Figures.* If your proposal has six or more figures, include a list of figures to help guide your reader to relevant material.

BODY

- *Executive Summary.* Briefly summarize the proposal's highlights in persuasive, nontechnical language in this crucial section often read by decision-makers.
- *Introduction.* Explain the reasons for the proposal, emphasize reader benefits, and, when appropriate, discuss your understanding of the problem.
- *Background or Problem Statement.* Describe the problem or opportunity your proposal addresses. To make your proposal more persuasive, your problem statement should illustrate how your proposal will benefit your client's organization.
- *Product Description.* If your proposal offers products as well as services, include a general description of the products and any technical specifications.
- *Detailed Solutions (Rationale).* Explain in a detailed section that will be read by technical specialists exactly how you plan to do what you are proposing.
- *Cost Analysis.* Itemize the estimated costs of all the products and services you are offering.
- *Delivery Schedule.* Outline how you will accomplish the work and show a timetable for each phase of the project.
- *Staffing.* Describe the expertise (education, experience, and certifications) of key personnel who will work on the project and include their <u>résumés</u>.

- *Site Preparation.* If your recommendations include modifying the customer's physical facilities, include a site-preparation description that details the required modifications.

- *Training Requirements.* If the products and services you are proposing require training the customer's employees, specify the required training and its cost.

- *Statement of Responsibilities.* To prevent misunderstandings about what you and your customer's responsibilities will be, state those responsibilities.

- *Organizational Sales Pitch.* Describe your company, its history, and its present position in the industry. An organizational sales pitch is designed to sell your company and its general capability in the field. It promotes the company and concludes the proposal on an upbeat persuasive note.

- *Authorization Request and Deadline.* Close with a request for approval and a deadline that explains how long the proposed prices are valid.

- *Conclusion.* Include a persuasive <u>conclusion</u> that summarizes the proposal's key points and stresses your company's strong points.

BACK MATTER
- *Appendixes.* Provide material of interest to specialized readers — technical graphics, statistical analysis, charts, tables, or sources of further information — in an <u>appendix</u>.

- *Bibliography.* List all sources consulted to prepare the proposal. See <u>bibliographies</u> and <u>documenting sources</u>.

- *Glossary.* If your proposal contains terms that will be unfamiliar to your intended audience, list and define them in the <u>glossary</u>.

Internal Proposals

Two common types of internal proposals are often distinguished from each other by the frequency with which they are written and by the degree of change proposed.

Routine Internal Proposals. *Routine internal proposals* are common in most organizations. Usually in <u>memo</u> format, they typically include small spending requests, permission to hire new employees or increase salaries, and requests to attend conferences or purchase new equipment. In writing these routine proposals, use the introduction-body-conclusion format, and highlight any benefits to be realized.

Formal Internal Proposals.   *Formal internal proposals* are usually proposals to commit relatively large sums of money. These proposals have various names, but the most common designation is a *capital appropriations request* or a *capital appropriations proposal.*

The introduction to a formal internal proposal should persuasively establish that a problem exists that needs a solution and provide any background information your reader needs to help him or her make the decision in question. The introduction should also briefly describe any supporting evidence you are including, such as a feasibility study you may have conducted. See also **feasibility reports**.

The body of a formal internal proposal should offer a practical solution to the problem, compare possible alternatives, respond to possible objections, and describe all the equipment, property, or services you are proposing to purchase and their key benefits to the organization. You should include any justifications for each expenditure: the calculated return on investment or the internal rate of return; the volume of use over, for example, the next two years; the product life expectancy, technical support, or warranties; and the most economical purchasing option.

The conclusion should emphasize the benefits of what you are proposing and express your willingness to provide any further information that may be required.

Typical Internal Proposal.   Figures P–9 through P–11 show a typical internal proposal.

External Proposals

External proposals to be submitted to other companies or organizations could be either solicited or unsolicited.

Unsolicited and Solicited Proposals.   *Unsolicited proposals* are proposals submitted to a company without a prior request for a proposal. Companies often operate for years with a problem they have never recognized (unnecessarily high maintenance costs, for example, or poor inventory-control methods). You might prepare an unsolicited proposal if you were convinced that the potential customer could realize substantial benefits by adopting your solution to a problem. Of course, you would need to convince the customer of the need for what you are proposing and that your solution would be the best one. Many unsolicited proposals are preceded by an inquiry to determine potential interest. Once you have received a positive response, you would conduct a detailed study of the prospective customer's needs to determine whether you can be of help, and, if so, exactly how. You would then prepare your proposal on the basis of your study.

ABO, Inc.
InterOffice Memo

To: Joan Marlow, Director, Human Resources Division
From: Leslie Galusha, Chief, Employee   *LG*
          Benefits Department
Date: June 12, 2003
Subject: Employee Fitness and Health-Care Costs

Health-care and worker-compensation insurance costs at ABO, Inc., have risen 200 percent over the last five years. In 1998, costs were $960 per employee per year; in 2003, they have reached $1,560 per employee per year. This doubling of costs mirrors a national trend, with health-care costs anticipated to continue to rise at the same rate for the next 10 years. Controlling these escalating expenses will be essential. They are eating into ABO's profit margin because the company currently pays 80 percent of the costs for employee coverage.

Healthy employees bring direct financial benefits to companies in the form of lower employee insurance costs, lower absenteeism rates, and reduced turnover. Regular physical exercise promotes fit, healthy people by reducing the risk of coronary heart disease, diabetes, osteoporosis, hypertension, and stress-related problems. I propose that to promote regular, vigorous physical exercise for our employees, ABO implement a health-care program that focuses on employee fitness.

P

FIGURE P–9. Introduction for an Internal Proposal

Solicited proposals are proposals prepared in response to a request for goods or services. Procuring organizations that would like competing companies to bid for a job commonly issue a *Request for Proposals (RFP)* or an *Invitation for Bids (IFB)*.

An IFB is commonly issued by government agencies to solicit bids on clearly defined products or services. An IFB is restrictive, binding the bidder to produce an item that meets the exact requirements of the agency. The goods or services to be procured are defined in the IFB by references to performance standards stated in **specifications**. Bidders must be prepared to prove that their product will meet all requirements of the specifications.

Background
The U.S. Department of Health and Human Services recently estimated that health-care costs in the United States will triple by the year 2013. Corporate expenses for health care are rising at such a fast rate that, if unchecked, in eight years they will significantly erode corporate profits.

Researchers have found that people who do not participate in a regular and vigorous exercise program incur double the health-care costs and are hospitalized 30 percent more days than people who exercise regularly. Nonexercisers are also 41 percent more likely to submit medical claims over $5,000 at some point during their careers than are those who exercise regularly.

U.S. companies are recognizing this trend. Tenneco, Inc., for example, found that the average health-care claim for unfit men was $1,003 per illness compared with an average claim of $562 for those who exercised regularly. For women, the average claim for those who were unfit was $1,535, more than double the average claim of $639 for women who exercised. Additionally, Control Data Corporation found that nonexercisers cost the company an extra $115 a year in health-care expenses.

These figures are further supported by data from independent studies. A model created by the National Institutes of Health (NIH) estimates that the average white-collar company could save $466,000 annually in medical costs (per 1,000 employees) just by promoting wellness. NIH researchers estimated that for every $1 a firm invests in a health-care program, it saves up to $3.75 in health-care costs. Another NIH study of 667 insurance-company employees showed savings of $1.65 million over a five-year period. The same study also showed a 400-percent drop in absentee rates after the company implemented a company-wide fitness program.

Solution
The benefits of regular, vigorous physical activity for employees and companies are compelling. To achieve these benefits at Acme, I propose that we choose from one of two possible options: Build in-house fitness centers at our warehouse facilities, or offer employees several options for membership at a national fitness club.

FIGURE P–10. Body of an Internal Proposal

Conclusion and Recommendation

I recommend that ABO, Inc., participate in the corporate membership program at AeroFitness Clubs, Inc., by subsidizing employee memberships. By subsidizing memberships, ABO shows its commitment to the importance of a fit workforce. Club membership allows employees at all five ABO warehouses to participate in the program. The more employees who participate, the greater the long-term savings in ABO's health-care costs. Building and equipping fitness centers at all five warehouse sites would require an initial investment of nearly $2 million. These facilities would also occupy valuable floor space—on average, 4,000 square feet at each warehouse. Therefore, this option would be very costly.

Enrolling employees in the corporate program at AeroFitness would allow them to attend on a trial basis. Those interested in continuing could then join the club and pay half of the membership cost, less a 30-percent discount on $400 a year. The other half of the membership ($140) would be paid for by ABO. If an employee leaves the company, he or she would have the option of purchasing ABO's share of the membership. Employees not wishing to keep their membership could buy the membership back from ABO and sell it to another employee.

Implementing this program will help ABO, Inc., reduce its health-care costs while building stronger employee relations by offering employees a desirable benefit. If this proposal is adopted, I have some additional thoughts about publicizing the program to encourage employee participation. I look forward to discussing the details of this proposal with you and answering any questions you may have.

FIGURE P–11. Conclusion for an Internal Proposal

In contrast to an IFB, an RFP is flexible. It is negotiable, and it does not necessarily specify exactly what goods or services are required. Often an RFP will define a problem and allow those who respond to suggest possible solutions. For example, if a procuring agency wanted to develop a way to make existing offices accessible to wheelchairs, an RFP would normally be the means used to find the best method and select the most qualified vendor. In some instances, RFPs are presented in several stages: (1) development of a concept, (2) construction of a prototype or mockup, and (3) production of the device or plan selected.

The procuring organization generally publishes its IFB or RFP in trade journals, on its Web site, or in a specialized venue such as the

Commerce Business Daily <www.cdcbdweb.com> and the *Contracts Reference Guide* <www.hp.com.gov/federal>.

Sales Proposals. A persuasive *sales proposal* must demonstrate above all that the prospective customer's purchase of the seller's products or services *will solve a problem, improve operations,* or *offer other benefits*. Sales proposals vary greatly in size and sophistication—from several pages written by one person, to dozens of pages written collaboratively by several people, to hundreds of pages written by a team of professional proposal writers. A short sales proposal might bid for the construction of a single home, a moderate-length proposal might bid for the installation of a computer network, and a large proposal might bid for the construction of a multimillion-dollar water purification system. Always keep in mind that, once submitted, a sales proposal is a legally binding document that promises to offer goods or services within a specified time and for a specified price.

Your first task in writing a sales proposal is to find out exactly what your prospective customer needs. To do that, survey your potential customer's business, and determine whether your organization can satisfy the customer's needs. Before preparing a sales proposal, try to find out who your principal competitors are. Then compare your company's strengths with those of the competing firms, determine your advantages over your competitors, and emphasize those advantages in your proposal.

Typical Sales Proposal. Figures P–12 through P–20 show an example of a major sales proposal. A writer's checklist, Writing Persuasive Proposals, follows on page 456.

☀ WEB LINK **ANNOTATED SAMPLE SALES PROPOSAL**

For a complete and annotated version of this proposal, see <*www.bedford stmartins.com/alred*> and select *Model Documents Gallery*.

P

Grant and Research Proposals. Grant and research proposals are written to request the approval of, and usually funding for, particular projects. For example, a professor of education might submit a research or grant proposal to the Department of Education to request funding for research on the relationship of class size to educational performance. Many government and private agencies solicit research and grant proposals. The granting agencies usually have their own requirements for format and content of the proposals submitted to them, but the proposals must always be persuasive. Tailor your grant or research proposal to your audience carefully by explaining the project's goals, your plan for achieving those goals, and your qualifications to perform the project.

The Waters Corporation
17 North Waterloo Blvd.
Tampa, Florida 33607
Phone: (813) 919-1213   Fax: (813) 919-4411
http://www.waters.comp.com

September 2, 2003

Mr. John Yeung, General Manager
Cookson's Retail Stores, Inc.
101 Longuer Street
Savannah, Georgia 31499

Dear Mr. Yeung:

The Waters Corporation appreciates the opportunity to respond to Cookson's Request for Proposals dated July 25, 2003. We would like to thank Mr. Becklight, Director of your Management Information Systems Department, for his invaluable contributions to the study of your operations that we conducted before preparing our proposal.

It has been Waters's privilege to provide Cookson's with retail systems and equipment since your first store opened many years ago. Therefore, we have become very familiar with your requirements as they have evolved during the expansion you have experienced since that time. Waters's close working relationship with Cookson's has resulted in a clear understanding of Cookson's philosophy and needs.

Our proposal describes a Waters Interactive Terminal/Retail Processor System designed to meet Cookson's network and processing needs. It will provide all of your required capabilities, from the point-of-sale operational requirements at the store terminals to the host processor. The system uses the proven Retail III modular software, with its point-of-sale applications, and the superior Interactive Terminal, with its advanced capabilities and design. This system is easily installed without extensive customer reprogramming.

FIGURE P–12. Cover Letter for a Sales Proposal

Mr. J. Yeung
Page 2
September 2, 2003

The Waters Interactive Terminal/Retail Processor System, which is
compatible with much of Cookson's present equipment, not only will
answer your present requirements but will provide the flexibility to add
new features and products in the future. The system's unique hardware
modularity, efficient microprocessor design, and flexible programming
capability greatly reduce the risk of obsolescence.

Thank you for the opportunity to present this proposal. You may
be sure that we will use all the resources available to the Waters
Corporation to ensure the successful implementation of the new
system.

Sincerely yours,

Janet A. Curtain

Janet A. Curtain
Executive Account Manager
General Merchandise Systems
(*JCurtain@netcom.TF.com*)

Enclosure: Proposal

P

FIGURE P–12. Cover Letter for a Sales Proposal *(continued)*

EXECUTIVE SUMMARY

The Waters 319 Interactive Terminal/615 Retail Processor System will provide your management with the tools necessary to manage people and equipment more profitably with procedures that will yield more cost-effective business controls for Cookson's.

The equipment and applications proposed for Cookson's were selected through the combined effort of Waters and Cookson's Management Information Systems Director, Mr. Becklight. The architecture of the system will respond to your current requirements and allow for future expansion.

The features and hardware in the system were determined from data acquired through the comprehensive survey we conducted at your stores in February of this year. The total of 71 Interactive Terminals proposed to service your four store locations is based on the number of terminals currently in use and on the average number of transactions processed during normal and peak periods. The planned remodeling of all four stores was also considered, and the suggested terminal placement has been incorporated into the working floor plan. The proposed equipment configuration and software applications have been simulated to determine system performance based on the volumes and anticipated growth rates of the Cookson stores.

The information from the survey was also used in the cost justification, which was checked and verified by your controller, Mr. Deitering. The cost-effectiveness of the Waters Interactive Terminal/Retail Processor System is apparent. Expected savings, such as the projected 46-percent reduction in sales audit expenses, are realistic projections based on Waters's experience with other installations of this type.

P

FIGURE P-13. Executive Summary for a Sales Proposal

GENERAL SYSTEM DESCRIPTION

The point-of-sale system that Waters is proposing for Cookson's includes two primary Waters products. These are the 319 Interactive Terminal and the 615 Retail Processor.

Waters 319 Interactive Terminal

The primary component in the proposed retail system is the Interactive Terminal. It contains a full microprocessor, which gives it the flexibility that Cookson's has been looking for.

The 319 Interactive Terminal provides you with freedom in sequencing a transaction. You are not limited to a preset list of available steps or transactions. The terminal program can be adapted to provide unique transaction sets, each designed with a logical sequence of entry and processing to accomplish required tasks. In addition to sales transactions recorded on the selling floor, specialized transactions such as theater-ticket sales and payments can be designed for your customer-service area

The 319 Interactive Terminal also functions as a credit authorization device, either by using its own floor limits or by transmitting a credit inquiry to the 615 Retail Processor for authorization.

Data collection formats have been simplified so that transaction editing and formatting are much more easily accomplished. The IS manager has already been provided with documentation on these formats and has outlined all data-processing efforts that will be necessary to transmit the data to your current systems. These projections have been considered in the cost justification.

Waters 615 Retail Processor

The Waters 615 Retail Processor is a minicomputer system designed to support the Waters family of retail terminals. The . . .

FIGURE P-14. General Description of Products for a Sales Proposal

PAYROLL APPLICATION

Current Procedure

Your current system of reporting time requires each hourly employee to sign a time sheet; the time sheet is reviewed by the department manager and sent to the Payroll Department on Friday evening. Because the week ends on Saturday, the employee must show the scheduled hours for Saturday and not the actual hours; therefore, the department manager must adjust the reported hours on the time sheet for employees who do not report on the scheduled Saturday or who do not work the number of hours scheduled.

The Payroll Department employs a supervisor and three full-time clerks. To meet deadlines caused by an unbalanced work flow, an additional part-time clerk is used for 20 to 30 hours per week. The average wage for this clerk is $8.00 per hour.

Advantage of Waters's System

The 319 Interactive Terminal can be programmed for entry of payroll data for each employee on Monday mornings by department managers, with the data reflecting actual hours worked. This system would eliminate the need for manual batching, controlling, and data input. The Payroll Department estimates conservatively that this work consumes 40 hours per week.

Hours per week	40
Average wage (part-time clerk)	× 8.00
Weekly payroll cost	$320.00
Annual savings	$16,640

Elimination of the manual tasks of tabulating, batching, and controlling can save 0.25 hourly unit. Improved work flow resulting from timely data in the system without data-input processing will allow more efficient use of clerical hours. This would reduce payroll by the 0.50 hourly unit currently required to meet weekly check disbursement.

Eliminate manual tasks	0.25
Improve work flow	0.75
40-hour unit reduction	1.00
Hours per week	40
Average wage (full-time clerk)	9.00
Savings per week	$360.00
Annual savings	$18,720

TOTAL SAVINGS: $35,360

FIGURE P–15. Detailed Solution for a Sales Proposal

COST ANALYSIS

This section of our proposal provides detailed cost information for the Waters 319 Interactive Terminal and the 615 Waters Retail Processor. It then multiplies these major elements by the quantities required at each of your four locations.

319 Interactive Terminal

	Price	Maint. (1 yr.)
Terminal	$2,895	$167
Journal Printer	425	38
Receipt Printer	425	38
Forms Printer	525	38
Software	220	—
TOTALS	$4,490	$281

The following summarizes all costs.

Location	Hardware	Maint. (1 yr.)	Software
Store No. 1	$72,190	$4,975	$3,520
Store No. 2	89,190	6,099	4,400
Store No. 3	76,380	5,256	3,740
Store No. 4	80,650	5,537	3,960
Data Center	63,360	6,679	12,480
Subtotals	$381,770	$28,546	$28,100

TOTAL $438,416

DELIVERY SCHEDULE

Waters is normally able to deliver 319 Interactive Terminals and 615 Retail Processors within 30 days of the date of the contract. This can vary depending on the rate and size of incoming orders.

All the software recommended in this proposal is available for immediate delivery. We do not anticipate any difficulty in meeting your tentative delivery schedule.

FIGURE P-16. Cost Analysis and Delivery Schedule for a Sales Proposal

SITE PREPARATION

Waters will work closely with Cookson's to ensure that each site is properly prepared prior to system installation. You will receive a copy of Waters's installation and wiring procedures manual, which lists the physical dimensions, service clearance, and weight of the system components in addition to the power, logic, communications-cable, and environmental requirements. Cookson's is responsible for all building alterations and electrical facility changes, including the purchase and installation of communications cables, connecting blocks, and receptacles.

Wiring

For the purpose of future site considerations, Waters's in-house wiring specifications for the system call for two twisted-pair wires and twenty-two shielded gauges. The length of communications wires must not exceed 2,500 feet.

As a guide for the power supply, we suggest that Cookson's consider the following.

1. The branch circuit (limited to 20 amps) should service no equipment other than 319 Interactive Terminals.
2. Each 20-amp branch circuit should support a maximum of three Interactive Terminals.
3. Each branch circuit must have three equal-size conductors—one hot leg, one neutral, and one insulated isolated ground.
4. Hubbell IG 5362 duplex outlets or the equivalent should be used to supply power to each terminal.
5. Computer-room wiring will have to be upgraded to support the 615 Retail Processor.

P

FIGURE P–17. **Site-Preparation Section for a Sales Proposal**

TRAINING

To ensure a successful installation, Waters offers the following training
course for your operators.

Interactive Terminal/Retail Processor Operations

Course number: 8256
Length: three days
Tuition: $500.00

This course provides the student with the skills, knowledge, and
practice required to operate an Interactive Terminal/Retail Processor
System. Online, clustered, and stand-alone environments are covered.

We recommend that students have a department-store background and
that they have some knowledge of the system configuration with which
they will be working.

P

FIGURE P–18. Training Section for a Sales Proposal

RESPONSIBILITIES

On the basis of its years of experience in installing information-processing systems, Waters believes that a successful installation requires a clear understanding of certain responsibilities.

Generally, it is Waters's responsibility to provide its users with needed assistance during the installation so that live processing can begin as soon thereafter as is practical.

Waters's Responsibilities

- Provide operations documentation for each application that you acquire from Waters
- Provide forms and other supplies as ordered
- Provide specifications and technical guidance for proper site planning and installation
- Provide adviser assistance in the conversion from your present system to the new system

Customer's Responsibilities

- Identify an installation coordinator and system operator
- Provide supervisors and clerical personnel to perform conversion to the system
- Establish reasonable time schedules for implementation
- Ensure that the physical site requirements are met
- Provide personnel to be trained as operators and ensure that other employees are trained as necessary
- Assume the responsibility for implementing and operating the system

FIGURE P–19. Statement of Responsibilities for a Sales Proposal

DESCRIPTION OF VENDOR

The Waters Corporation develops, manufactures, markets, installs, and services total business information-processing systems for selected markets. These markets are primarily in the retail, financial, commercial, industrial, health-care, education, and government sectors.

The Waters total system concept encompasses one of the broadest hardware and software product lines in the industry. Waters computers range from small business systems to powerful general-purpose processors. Waters computers are supported by a complete spectrum of terminals, peripherals, data-communication networks, and an extensive library of software products. Supplemental services and products include data centers, field service, systems engineering, and educational centers.

The Waters Corporation was founded in 1934 and presently has approximately 26,500 employees. The Waters headquarters is located at 17 North Waterloo Boulevard, Tampa, Florida, with district offices throughout the United States and Canada.

WHY WATERS?

Corporate Commitment to the Retail Industry

Waters's commitment to the retail industry is stronger than ever. We are continually striving to provide leadership in the design and implementation of new retail systems and applications that will ensure our users of a logical growth pattern.

Research and Development

Over the years, Waters has spent increasingly large sums on research and development efforts to ensure the availability of products and systems for the future. In 2002, our research and development expenditure for advanced systems design and technological innovations reached the $70-million level.

Leading Point-of-Sale Vendor

Waters is a leading point-of-sale vendor, having installed over 150,000 units. The knowledge and experience that Waters has gained over the years from these installations ensure well-coordinated and effective systems implementations.

FIGURE P–20. Description of Vendor and Organizational Sales Pitch for a Sales Proposal

Writer's Checklist: Writing Persuasive Proposals

☑ Analyze your audience carefully to determine how to best persuade your readers.

☑ Emphasize the readers' benefits and anticipate their questions or objections.

☑ Incorporate evidence and claims effectively.

☑ Write a concise purpose statement to clarify your proposal's goals.

☑ Break down the writing task into manageable segments.

☑ Develop a work schedule.

☑ Understand proposal sections and their uses.

☑ Select an appropriate, visually appealing format. (See **layout and design**.)

☑ Use a confident, upbeat **tone**.

WEB LINK PROPOSALS

For useful Web sites related to proposals, see <*www.bedfordstmartins.com /alred*> and select *Links for Technical Writing.*

---

proved / proven

Both *proved* and *proven* are past participles of *prove*, although *proved* is currently more widely accepted.

* They had *proved* more receptive to change than expected.

* Our attorney had *proven* that we did not violate fair use.

Proven is more commonly used as an **adjective**.

* She was hired because of her *proven* competence as a technical writer.

pseudo- / quasi-

As a **prefix**, *pseudo-*, meaning "false or counterfeit," is joined to the root word without a **hyphen** unless the root word begins with a capital letter (*pseudo*science, *pseudo*-Keynesian). *Pseudo-* is sometimes confused with *quasi-*, meaning "somewhat" or "partial." Unlike *semi-*, *quasi-* does not mean "half." *Quasi-* is usually hyphenated in combinations (*quasi*-scientific theories). See also <u>bi-/semi-</u> and <u>capitalization</u>.

punctuation

Punctuation helps readers understand the structural relationships within a sentence. Marks of punctuation link, separate, enclose, indicate omissions, terminate, and classify. Most punctuation marks can perform more than one function. See also <u>sentence construction</u>.

The use of punctuation is determined by grammatical conventions and the writer's intention. Understanding punctuation is essential for writers because it enables them to communicate with <u>clarity</u> and precision. See also <u>grammar</u>.

Detailed information on each mark of punctuation is given in its own entry. The following are the 13 marks of punctuation.

<u>apostrophe</u>	'	<u>parentheses</u>	( )
<u>brackets</u>	[ ]	<u>period</u>	.
<u>colon</u>	:	<u>question mark</u>	?
<u>comma</u>	,	<u>quotation marks</u>	" "
<u>dash</u>	—	<u>semicolon</u>	;
<u>exclamation mark</u>	!	<u>slash</u>	/
<u>hyphen</u>	-		

See also <u>abbreviations</u>, <u>capitalization</u>, <u>contractions</u>, <u>dates</u>, <u>ellipses</u>, <u>italics</u>, and numbers.

> **WEB LINK PRACTICING PUNCTUATION**
>
> For electronic exercises on punctuation, see *<www.bedfordstmartins.com /alred>* and select *Exercise Central.*

purpose

What do you want your <u>readers</u> to know, believe, or do when they have read your document? When you answer that question, you have determined the primary purpose, or objective, of your document. Too often, beginning writers state their purposes too broadly. A purpose such as "to explain a fax machine" is too general to be helpful to you as you write. In contrast, "to instruct the reader on how to use a fax machine to send and retrieve faxes" is a specific purpose that will help you focus on what you need your document to accomplish.

However, the writer's primary purpose is often more complex than simply to "explain" something. To fully understand this complexity, you need to ask yourself not only *why* you are writing the document but *what* you want to influence your reader to believe or do after reading your document. Suppose a writer for a newsletter has been assigned to write an article about cardiopulmonary resuscitation (CPR). In answer to the question *What?*, the writer could state the purpose as "to emphasize the importance of CPR." To the question *Why?*, the writer might respond, "To encourage employees to sign up for evening CPR classes." Putting the answers to the two questions together, the writer's purpose might be stated as, "To write a document that will emphasize the importance of CPR and encourage employees to sign up for evening CPR classes." Note that the primary purpose of the document on CPR was to persuade the readers of the importance of CPR, and the secondary goal was to motivate them to register for a class. Secondary goals often involve such abstract notions as to motivate, persuade, reassure, or inspire your reader. See also **persuasion**.

If you answer the questions *what* and *why* and put the answers into writing as a stated purpose that includes both primary and secondary goals, you will simplify your writing task and achieve your purpose in writing the document. But remember that even a specific purpose is of no value unless you keep it in mind as you work. Be careful not to lose sight of your purpose as you become engrossed in the other steps of the writing process. See also "Five Steps to Successful Writing."

P

Q

question marks

The *question mark* (?) has several uses. Use a question mark to end a sentence that is a direct question. (Where did you put the specifications?)

Never use a question mark to end a sentence that is an indirect question.

- He asked me whether sales had increased this year.

Use a question mark to end a statement that has an interrogative meaning—a statement that is declarative in form but asks a question. (The laboratory report is finished?)

Use a question mark to end an interrogative clause within a declarative sentence.

- It was not until July (or was it August?) that we submitted the report.

When a directive is phrased as a question, a question mark is usually not used. However, a request (to a customer or a superior, for instance) almost always requires a question mark.

- Will you make sure that the system is operational by August 15. [directive]

- Will you e-mail me if your entire shipment does not arrive by June 10? [request]

Question marks may follow a series of separate items within an interrogative sentence.

- Do you remember the date of the contract? its terms? whether you signed it?

Retain the question mark in a title that is being cited, even though the sentence in which it appears has not ended.

- *Can Technology Be Controlled?* is the title of her book.

When used with quotations, the placement of the question mark is important. When the writer is asking a question, the question mark belongs outside the quotation marks.

- Did she say, "I don't think the project should continue"?

If the quotation itself is a question, the question mark goes inside the quotation marks.

- She asked, "When will we go?"

If both cases apply — the writer is asking a question and the quotation itself is a question — use a single question mark inside the quotation marks.

- Did she ask, "Will you go in my place?"

questionnaires

A *questionnaire* — a series of questions on a particular topic sent out to a number of people — serves the same purpose as an interview but does so on paper, by <u>e-mail</u>, or on the Web. See also <u>research</u> and <u>interviewing for information</u>.

Questionnaires have several advantages over the personal interview as well as several disadvantages.

Q

ADVANTAGES

- A questionnaire allows you to gather information from more people more quickly than you could by conducting personal interviews.
- It enables you to obtain responses from people who are difficult to reach or who are in various geographical locations.
- Those responding to a questionnaire have more time to think through their answers than when faced with the pressure of composing thoughtful and complete answers to an interviewer.
- The questionnaire may yield more objective data because it reduces the possibility that the interviewer's tone of voice or facial expressions might influence an answer.
- The cost of distributing and tabulating a questionnaire is lower than the cost of conducting numerous personal interviews.

DISADVANTAGES

- The results of a questionnaire may be slanted in favor of those people who have strong opinions on a subject because they are more likely to respond than those with only moderate views.

- The questionnaire does not allow specific follow-up to answers; at best, a questionnaire can be designed to let one question lead logically to another.

- Distributing questionnaires and waiting for replies may take considerably longer than conducting a personal interview.

Selecting the Recipients

Selecting the proper recipients for your questionnaire is crucial if you are to gather representative and usable data. If you wanted to survey the opinions of large groups in the general population—for example, all medical technologists working in private laboratories or all independent garage owners—your task would not be easy. Because you cannot include everybody in your survey, you need to choose a representative cross section. For example, you would want to include enough people from around the country, respondents of both genders, and people with an assortment of educational training. Only then could you make a generalized statement based on your findings from the sample. (The best sources of information on sampling techniques are marketing-research and statistics texts.)

> ⊛ WEB LINK MARKETING VIRTUAL LIBRARY
>
> KnowThis.com's Marketing Virtual Library, from Professor Paul Christ of West Chester University of Pennsylvania, is a subject directory of sites related to marketing, market research, and much more. See <www.bedford stmartins.com/alred> and select *Links for Technical Writing*.

Preparing the Questions

A key goal in designing the questionnaire is to keep it as brief as possible. The longer a questionnaire is, the less likely the recipient will be to complete and return it. Next, the questions should be easy to understand. A confusing question will yield confusing results, whereas a carefully worded question will be easy to answer. Ideally, recipients should be able to answer most questions with a "yes" or "no" or by checking or circling a choice among several options. Such answers are easy to tabulate and require minimum effort on the part of the respondent, thus increasing your chances of obtaining a response. See also **forms design**.

- Do you recommend that the flextime program be made permanent?

  ☐ Yes   ☐ No   ☐ No opinion

If you need more information than such questions produce, provide an appropriate range of answers, as in the following example.

- How many hours of overtime would you be willing to work each week?

  ☐ 4 hours   ☐ 8 hours   ☐ Over 10 hours
  ☐ 6 hours   ☐ 10 hours   ☐ No overtime

Questions should be neutral; they should not be worded in such a way as to lead respondents to give a particular answer, which can result in inaccurate or skewed data.

SLANTED   Would you prefer the freedom of a four-day workweek?

NEUTRAL   Would you choose to work a four-day workweek, ten hours a day, with every Friday off?

Writer's Checklist: Designing a Questionnaire

☑ Prepare a **cover letter** or an e-mail explaining who you are, the questionnaire's purpose, the date by which you need a response, and how and where to send the completed questionnaire.

☑ Include a stamped, self-addressed envelope if you are using regular mail.

☑ Construct as many questions as possible for which the recipient does not have to compose an answer.

☑ Include a section on the questionnaire for additional comments, where recipients may clarify their overall attitude toward the subject.

☑ State whether the information provided as well as the recipients' identity will be kept confidential.

☑ Include questions about the respondent's age, gender, education, occupation, and so on, only if such information will be of value in interpreting the answers.

☑ Include your contact information (mailing address, phone number, and e-mail address).

☑ Consider offering some tangible appreciation to those who answer the questionnaire by a specific date, such as a copy of the results or, for a marketing questionnaire, a gift certificate.

The sample cover letter and questionnaire in Figures Q–1 and Q–2 were sent to employees in a large organization who had participated in a six-month program of flexible working hours.

Luxwear Products Corporation
MEMO

To: All Company Employees
From: Nelson Barrett, Director   *NB*
Date: October 17, 2003
Subject: Review of Flexible Working Hours Program

Please complete and return the questionnaire enclosed regarding Luxwear's trial program of flexible working hours. Your answers will help us decide whether we should make the program permanent.

Return the completed questionnaire to Ken Rose, Mail Code 12B, by October 27. Your signature on the questionnaire is not necessary. All responses will be confidential and given serious consideration. Feel free to raise additional issues pertaining to the program.

If you want to discuss any item in the questionnaire, call Pam Peters in the Human Resources Department at extension 8812 or e-mail at pp1@lpc.com.

Enclosure: Questionnaire

Q

FIGURE Q–1. Questionnaire Cover Letter

Flexible Working Hours Program
Questionnaire

1. What kind of position do you occupy?

   ☐ Supervisory
   ☐ Nonsupervisory

2. Indicate to the nearest quarter of an hour your starting time under flextime.

   ☐ 7:00 a.m.   ☐ 8:15 a.m.
   ☐ 7:15 a.m.   ☐ 8:30 a.m.
   ☐ 7:30 a.m.   ☐ 8:45 a.m.
   ☐ 7:45 a.m.   ☐ 9:00 a.m.
   ☐ 8:00 a.m.   ☐ Other (specify) _____

3. Where do you live?

   ☐ Talbot County   ☐ Greene County
   ☐ Montgomery County   ☐ Other (specify) _____

4. How do you usually travel to work?

   ☐ Drive alone   ☐ Walk
   ☐ Bus   ☐ Car pool
   ☐ Train   ☐ Motorcycle
   ☐ Bicycle   ☐ Other (specify) _____

5. Has flextime affected your commuting time?

   ☐ Increase: Approximate number of minutes _____
   ☐ Decrease: Approximate number of minutes _____
   ☐ No change

6. If you drive alone or in a car pool, has flextime increased or decreased the amount of time it takes you to find a parking space?

   ☐ Increased   ☐ Decreased   ☐ No change

7. Has flextime had an effect on your productivity?

   a. Quality of work
      ☐ Increased   ☐ Decreased   ☐ No change

   b. Accuracy of work
      ☐ Increased   ☐ Decreased   ☐ No change

   c. Quiet time for uninterrupted work
      ☐ Increased   ☐ Decreased   ☐ No change

FIGURE Q–2. Questionnaire

8. Have you had difficulty getting in touch with coworkers who are on different work schedules from yours?

☐ Yes   ☐ No

9. Have you had trouble scheduling meetings within flexible starting and quitting times?

☐ Yes   ☐ No

10. Has flextime affected the way you feel about your job?

☐ Yes   ☐ No

If yes, please answer (a) or (b):

a. Feel better about job
☐ Slightly   ☐ Considerably

b. Feel worse about job
☐ Slightly   ☐ Considerably

11. How important is it for you to have flexibility in your working hours?

☐ Very   ☐ Not very   ☐ Somewhat   ☐ Not at all

12. Has flextime allowed you more time to be with your family?

☐ Yes   ☐ No

13. If you are responsible for the care of a young child or children, has flextime made it easier or more difficult for you to obtain babysitting or day-care services?

☐ Easier   ☐ More difficult   ☐ No change

14. Do you recommend that the flextime program be made permanent?

☐ Yes   ☐ No

15. Please describe below or attach any major changes you recommend for the program.

Thank you for your assistance.

FIGURE Q-2. Questionnaire *(continued)*

quid pro quo

Quid pro quo, which is Latin for "one thing for another," suggests mutual cooperation or "tit for tat" in a relationship between two groups or individuals. The term may be appropriate to business and legal contexts if you are sure your readers understand its meaning. (Before approving the plan, we insisted on a fair *quid pro quo.*)

quotation marks

Quotation marks (" ") are used to enclose a direct quotation of spoken or written words. Quotation marks have other special uses, but they should not be used for <u>emphasis</u>. A variety of guidelines govern the use of quotation marks.

　　Enclose in quotation marks anything that is quoted word for word (a direct quotation) from speech or written material.

- She said clearly, "I want the progress report by three o'clock."

Do not enclose indirect quotations—usually introduced by the word *that*—in quotation marks. Indirect quotations are paraphrases of a speaker's words or ideas. (See also <u>paraphrasing</u>.)

- She said that she wanted the progress report by three o'clock.

　　When you use quotation marks to indicate that you are quoting, do not make any changes in the quoted material unless you clearly indicate what you have done. For further information on using and incorporating quoted material, see <u>quotations</u>.

　　Use single quotation marks to enclose a quotation that appears within a quotation.

- John said, "Jane told me that she was going to 'stay with the project if it takes all year.'"

　　Use quotation marks to set off special words or terms only to point out that the term is used in context for a unique or special purpose (that is, in the sense of the term *so-called*).

- What chain of events caused the sinking of the "unsinkable" *Titanic* on its maiden voyage?

Slang, colloquial expressions, and attempts at humor, although infrequent in workplace writing, should seldom be set off by quotation marks.

• Our first six months amounted to a "shakedown cruise." *shakedown cruise.*

Use quotation marks to enclose titles of reports, short stories, articles, essays, single episodes of radio and television programs, short musical works, paintings, and other works of art.

• Did you see the article "No-Fault Insurance and Your Motorcycle" in last Sunday's *Journal?*

Do not use quotation marks for titles of books and periodicals, which should appear in italics. Some titles, by convention, are not set off by quotation marks, underlining, or italics, although they are capitalized.

• Professional Writing [college course title], the Bible, the Constitution, Lincoln's Gettysburg Address, the Lands' End Catalog

Commas and periods always go inside closing quotation marks.

• "Reading *Computor World* gives me the insider's view," he says, adding, "it's like a conversation with the top experts."

Semicolons and colons always go outside closing quotation marks.

• He said, "I will pay the full amount"; this statement surprised us.

All other punctuation follows the logic of the context: If the punctuation is a part of the material quoted, it goes inside the quotation marks; if the punctuation is not part of the material quoted, it goes outside the quotation marks.

Q

ESL TIPS FOR USING QUOTATION MARKS AND PUNCTUATION

When making choices about using quotation marks with other punctuation, keep the following examples in mind.

Correct use of a comma with quotation marks
• "as a last resort,"   (*not* "as a last resort",)

Correct use of a period with quotation marks
• "to the bitter end."   (*not* "to the bitter end".)

Correct use of a semicolon or colon with quotation marks
• "there is no doubt";   (*not* "there is no doubt;")

quotations

Using direct and indirect quotations is an effective way to make or support a point. However, avoid the temptation to overquote during the note-taking phase of your research; concentrate on summarizing what you read. When you do use a quotation (or an idea of another writer), cite your source properly. If you do not, you will be guilty of plagiarism. For specific details on APA, *CMS,* and MLA citation systems, see documenting sources.

Direct Quotations

A *direct quotation* is a word-for-word copy of the text of an original source. Choose direct quotations (which can be of a word, a phrase, a sentence, or, occasionally, a paragraph) carefully and use them sparingly. Enclose direct quotations in quotation marks and separate them from the rest of the sentence by a comma or colon. The initial capital letter of a quotation is retained if the quoted material originally began with a capital letter.

- The economist stated, "Regulation cannot supply the dynamic stimulus that in other industries is supplied by competition."

When a quotation is divided, the material that interrupts the quotation is set off, before and after, by commas, and quotation marks are used around each part of the quotation.

- "Regulation," he said in a recent interview, "cannot supply the dynamic stimulus that in other industries is supplied by competition."

Indirect Quotations

An *indirect quotation* is a paraphrased version of an original text. It is usually introduced by the word *that* and is not set off from the rest of the sentence by punctuation marks. See also paraphrasing.

- In a recent interview he said that regulation does not stimulate the industry as well as competition does.

Deletions or Omissions

Deletions or omissions from quoted material are indicated by three ellipsis points (. . .) within a sentence and a period plus three ellipsis points (. . . .) at the end of a sentence.

- "If monopolies could be made to respond . . . we would be able to enjoy the benefits of . . . large-scale efficiency. . . ."

If you are following the MLA guidelines, enclose the ellipsis points in brackets.

- "If monopolies could be made to respond [. . .] we would be able to enjoy the benefits of [. . .] large-scale efficiency [. . .]."

When a quoted passage begins in the middle of a sentence rather than at the beginning, ellipsis points are not necessary; the fact that the first letter of the quoted material is not capitalized tells the reader that the quotation begins in midsentence. See also capitalization and ellipses.

- Rivero goes on to conclude that "coordination may lessen competition within a region."

Inserting Material into Quotations

When it is necessary to insert a clarifying comment within quoted material, use brackets.

- "The industry is an integrated system that serves an extensive [geographic] area, with divisions existing as islands within the larger system's sphere of influence."

When quoted material contains an obvious error or might be questioned in some other way, the expression *sic* (Latin for "thus"), in italic type and enclosed in brackets, follows the questionable material to indicate that the writer has quoted the material exactly as it appeared in the original.

- The company considers the Baker Foundation to be a "guilt-edged [*sic*] investment."

Incorporating Quotations into Text

Quote word-for-word only when your source concisely sums up a great deal of information or reinforces a point you are making. Quotations must also relate logically, grammatically, and syntactically to the rest of the sentence and surrounding text.

Depending on the length, there are two mechanical methods of handling quotations in your text. For *CMS* system and MLA style, a quotation of three or fewer lines is incorporated into the text and enclosed in quotation marks. For APA style, a quotation of fewer than 40 words is incorporated into the text and enclosed in quotation marks.

Material that runs four lines or longer (*CMS* system and MLA style) or at least 40 words (APA style) is usually inset; that is, set off from the body of the text by being indented from the left margin ten spaces (MLA style) or five to seven spaces (APA style and *CMS* system). The quoted passage is spaced the same as the surrounding text (APA and MLA styles) or single-spaced and set in a smaller font size (*CMS* system), and is not enclosed in quotation marks, as shown in Figure Q–3, which uses MLA style. If you are not following a specific style manual, you may block indent ten spaces from both the left and right margins for reports and other documents.

After reviewing a large number of works in business and technical communication, Alred sees an inevitable connection between theory, practice, and pedagogy:

> Therefore, theory is necessary to prevent us from being overwhelmed by what is local, particular, and temporal. In turn, pedagogy both mediates practice and transforms our theory. Indeed, one reason I find this work rewarding is that I sense it puts me at the intersection of theory, practice, and pedagogy as they are involved with writing in the workplace. (ix–x)

The use of the Web today has reinforced this connection because it calls on the Web-page designer to engage in a teaching function as well as reflect on the practice of Web design. For example, the widespread use of . . .

FIGURE Q–3. Long Quotation (MLA Style)

Notice in Figure Q–3 that the quotation blends with the content of the surrounding text, which uses transitions to introduce and comment on the quotation. At the end of the document, the following entry appears in the MLA-style list of works cited as the source of the quotation in Figure Q–3.

- Alred, Gerald J. The St. Martin's Bibliography of Business and Technical Communication. New York: St. Martin's, 1997.

Do not rely too heavily on the use of quotations in the final version of your document. Generally, avoid quoting anything that is more than one paragraph.

R

raise / rise

Both *raise* and *rise* mean "move to a higher position." However, *raise* is a transitive <u>verb</u> and always takes an <u>object</u> (*raise* crops), whereas *rise* is an intransitive verb and never takes an object (heat *rises*).

re

Re (and its variant form, *in re*) is business and legal <u>jargon</u> meaning "in reference to" or "in the case of." Although *re* is sometimes used in <u>memos</u> and <u>e-mail</u>, *subject* is a preferable term.

readers

The first rule of effective writing is to *help your readers*. If you overlook this commitment, your writing will not achieve its <u>purpose</u>, either for you or for your business or organization.

Determining Your Readers' Needs

To be able to help your readers, you need to determine their needs relative to your purpose and goals by asking some key questions.

- Who specifically is your reader? Do you have multiple readers? Who needs to see or use the document?
- What do your readers already know about your subject? What are your readers' attitudes about the subject? (Skeptical? Supportive? Anxious? Bored?)
- Do you need to adapt your message for international readers? If so, see <u>global communication</u> and <u>international correspondence</u>.

In the workplace, your readers are usually less familiar with the subject than you are. You have to be careful, therefore, when writing on

R

a topic that is unique to your area of specialization. Be sensitive to the needs of those whose training or experience lies in other areas; provide definitions of nonstandard terms and explanations of principles that you, as a specialist, take for granted. Note that even if you write a <u>trade journal article</u> for others in your field, you should explain new or special uses of standard terms and principles.

Writing to Multiple Readers

For documents with groups of readers who have different needs, such as <u>formal reports</u>, you might design parts of a document to reach different groups of readers: an executive summary for top managers, an appendix with detailed data for technical specialists, and the body of a report or <u>proposal</u> for those readers who need to make decisions based on the details.

When you have multiple readers with various needs but cannot segment your document, first determine your primary or most important readers — such as those who will make decisions based on the document — and be sure to meet their needs. Then, meet the needs of secondary readers, such as those who need only some of the document's contents, as long as you do not sacrifice the needs of your primary readers.

Writing to Individual Readers

If your audience is relatively homogeneous, you might combine all your readers into one composite reader and write for *that* reader. You might also make a list of that reader's characteristics (experience, training, attitudes, and work habits, for example) to help you write at the appropriate level. You may find it useful to visualize that person sitting across from you as you write. Doing so can help you to empathize with the reader and write in an appropriate <u>style</u> and <u>tone</u>. Keep in mind that both electronic and paper <u>correspondence</u> might be read by others. Always maintain a style and tone that are appropriate to a wide professional audience. See also "Five Steps to Successful Writing," <u>persuasion</u>, <u>visuals</u>, and <u>"you" viewpoint</u>.

really

Really is an <u>adverb</u> meaning "actually" or "in fact." Although both *really* and *actually* are often used as <u>intensifiers</u> for <u>emphasis</u> or sarcasm in speech, avoid such use in writing.

- Did he ~~really~~ finish the report on time?

reason is [because]

Replace the redundant phrase *the reason is because* with *the reason is that* or simply *because*. See also **conciseness/wordiness**.

reference letters

Writing a *reference letter* (or letter of recommendation) can range from completing an admission form for a prospective student to composing a detailed description of professional accomplishments and personal characteristics for someone seeking employment.

To write an effective letter of recommendation, you must be familiar enough with the applicant's abilities and performance to offer an evaluation, and you must keep in mind the following.

- Communicate truthfully and without embellishment to the inquirer.
- Address specifically the applicant's skills, abilities, knowledge, and personal characteristics.
- Respond directly to the inquiry, being careful to address the specific questions asked.
- Identify yourself: name, title or position, employer, and address.

You could begin by stating the circumstances of your acquaintance and how long you have known the person for whom you are writing the letter. You should mention, with as much substantiation as possible, one or two outstanding characteristics of the applicant. Organize the details in your letter in decreasing **order of importance**. Conclude with a brief summary of the applicant's qualifications and a clear statement of recommendation. Figure R–1 on page 474 is a typical reference letter.

When you are asked to serve as a reference or to supply a letter of reference, be aware that applicants have a legal right to examine the materials in an organization's files that concern them, unless they sign a waiver. See **correspondence** for letter format and general advice.

IVY COLLEGE
DEPARTMENT OF BUSINESS
WEST LAFAYETTE, IN 47906
(691) 423-1719   (691) 423-2239 (FAX)
IVCO@IC.EDU (E-MAIL)

January 14, 2003

Mr. Phillip Lester
Human Resources Director
Thompson Enterprises
201 State Street
Springfield, IL 62705

Dear Mr. Lester:

How long writer has known applicant and the circumstances

As her employer and her former professor, I am happy to have the opportunity to recommend Kerry Hawkins. I've known Kerry for the last four years, first as a student in my class and for the last year as a research assistant.

Outstanding characteristics of applicant

Kerry is an excellent student, with above-average grades in our program. On the basis of a GPA of 3.6 (A = 4.0), Kerry was offered a research assistantship to work on a grant under my supervision. In every instance, Kerry completed her library search assignment within the time agreed upon. The material provided in the reports Kerry submitted met the requirements for my work and more. These reports were always well written. While working 15 hours a week on this project, Kerry has maintained a class load of 12 hours per semester.

Recommendation and summary of qualifications

I strongly recommend Kerry for her ability to work independently, to organize her time efficiently, and to write clearly and articulately. Please let me know if I can be of further service.

Sincerely yours,

Michael Paul
Professor of Business

R

FIGURE R–1. Reference Letter

refusal letters

When you must deliver a negative message (or bad news), you may need to write a refusal letter, **memo,** or **e-mail** message.

The ideal refusal letter says *no* in such a way that you not only avoid antagonizing your <u>reader</u> but also maintain goodwill. To do so, you must convince your reader that your reasons for refusing are logical or understandable *before* you present the bad news. Stating a negative message in your opening may cause your reader to react too quickly and dismiss your logic. The following pattern is an effective way to deal with this problem.

1. In the opening, provide a context (often called a "buffer").
2. Review the facts leading to the refusal or bad news.
3. Give the negative message based on the facts.
4. In the closing, establish or reestablish a positive relationship.

Your opening can establish a positive and professional tone, for example, by expressing appreciation for your reader's time, effort, or interest.

- The Screening Procedures Committee appreciates the time and effort you spent on your proposal for a new security-clearance procedure.

Next, review the circumstances of the situation sympathetically by placing yourself in the reader's position. Clearly establish the reasons you cannot do what the reader wants—even though you have not yet said you cannot do it. A good explanation, as shown in the following example, should detail the reasons for your refusal so thoroughly that the reader will accept the negative message as a logical conclusion.

- We reviewed the potential effects of implementing your proposed security-clearance procedure company-wide. We asked the Security Systems Department to review the data, surveyed industry practices, sought the views of senior management, and submitted the idea to our legal staff. As a result of this process, we have reached the following conclusions:
  - The cost savings you project are correct only if the procedure could be required universally.
  - The components of your procedure are legal, but most are not widely accepted by our industry.
  - Based on our survey, some components could alienate employees who would perceive them as violating an individual's rights.
  - Enforcing companywide use would prove costly and impractical.

Do not belabor the negative message—state your refusal quickly, clearly, and as positively as possible.

- For those reasons, the committee recommends that divisions continue their current security-screening procedures.

Close your message in a way that reestablishes goodwill—do not repeat the bad news (avoid writing "Again, we're sorry we can't use your idea"). You might provide an option, offer a friendly remark, assure the reader of your high opinion of his or her product or service, or merely wish the reader success.

- Because some components of your procedure may apply in certain circumstances, we would like to feature your ideas in the next issue of *The Guardian*. I have asked the editor to contact you next week. On behalf of the committee, thank you for the thoughtful proposal.

The refusal letter in Figure R–2 rejects an invitation to speak.

If your refusal is in response to a <u>complaint letter</u>, see <u>adjustment letters</u>. For refusing a job offer, see <u>acceptance/refusal letters</u>. See also <u>correspondence</u> and <u>inquiries and responses</u>.

Javier A. Lopez, President
TNCO Engineering Consultants
9001 Cummings Drive
St. Louis, MO 63129

Dear Mr. Lopez:

Context

I am honored to have been invited to address your regional meeting in St. Louis on May 17. That you would consider me as a potential contributor to such a gathering of experts is indeed flattering.

Review of facts and refusal

On checking my schedule, I find that I will be attending the annual meeting of our parent corporation's Board of Directors on that date. Therefore, as much as I would enjoy addressing your members, I must decline.

Goodwill close

I have been very favorably impressed over the years with your organization's contributions to the engineering profession, and I would welcome the opportunity to participate in a future meeting.

Sincerely,

FIGURE R–2. Refusal Letter

regarding / with regard to

In regards to and *with regards to* are incorrect <u>idioms</u> for *in regard to* and *with regard to*. Both *as regards* and *regarding* are acceptable variants.

- In ~~regards~~ ^{regard} to your last e-mail, I think a meeting is a good idea.
- ~~With regards to~~ ^{Regarding} your last e-mail, I think a meeting is a good idea.

regardless

Irregardless is nonstandard English because it expresses a double negative. The prefix *ir-* renders the base word negative, but *regardless* is already negative, meaning "unmindful." Always use *regardless* or *irrespective*.

repetition

The deliberate use of repetition to build a sustained effect or to emphasize a feeling or an idea can be a powerful device. See also <u>emphasis</u>.

- Similarly, atoms *come and go* in a molecule, but the molecule *remains;* molecules *come and go* in a cell, but the cell *remains;* cells *come and go* in a body, but the body *remains;* persons *come and go* in an organization, but the organization *remains.*
  —Kenneth Boulding, *Beyond Economics*

Repeating keywords from a previous sentence or paragraph can also be used effectively to achieve <u>transition</u>.

- For many years, *oil* has been a major industrial energy source. However, *oil* supplies are limited, and other sources of energy must be developed.

Be consistent in the word or phrase you use to refer to something. In technical writing, it is generally better to repeat a word (so there will be no question in the reader's mind that you mean the same thing) than to use synonyms to avoid repetition.

R

SYNONYMS	Several recent *analyses* support our conclusion. These *studies* cast doubt on the feasibility of long-range forecasting. The *reports*, however, are strictly theoretical.
CONSISTENT TERMS	Several recent theoretical *studies* support our conclusion. These *studies* cast doubt on the feasibility of long-range forecasting. They are, however, strictly theoretical.

Purposeless repetition, however, makes a sentence awkward and hides its key ideas. See also conciseness/wordiness.

- She said that the customer ~~said that the order was to be canceled.~~ *canceled the order.*

---

reports

This book includes many entries for typical reports, which may be formal or informal and may vary in length and complexity. (See also proposals.) Following is a list of relevant entries:

A *report* is an organized presentation of factual information, often aimed at multiple readers, that may present the results of an investigation, a trip, or a research project. For any report—whether formal or informal—assessing the readers' needs is important.

Formal reports present the results of long-term projects that may involve large sums of money. Such projects may be done either for your own organization or as a contractual requirement for another organization. Formal reports generally follow a precise format and include some or all of the report elements discussed in formal reports. See also abstracts and executive summaries.

Informal and short reports normally run from a few paragraphs to a few pages and include only the essential elements of a report: introduction, body, conclusions, and recommendations. Because of their brevity, informal reports are customarily written as correspondence: letters (if sent outside your organization) and memos or e-mail (if internal).

The introduction announces the subject of the report, states the purpose, and gives any essential background information. It may also summarize the conclusions, findings, or recommendations made in the report. The body presents a clearly organized account of the report's subject—the results of a test carried out, the status of a project, and so on. The amount of detail to include depends on the complexity of the subject and on your readers' familiarity with it.

The conclusion summarizes your findings and tells readers what you think their significance may be. In some reports, a final, separate section gives recommendations; in others, the conclusions and the recommendations sections are combined into one section. This final section makes suggestions for a course of action based on the data you have presented. See also persuasion.

research

Research is the process of investigation—the discovery of facts. To be focused, research must be preceded by preparation, especially consideration of your readers, purpose, and scope. See "Five Steps to Successful Writing."

In an academic setting, your primary resources include conversations with your peers, your instructors, and especially your research librarian. On the job, your main resources are your own knowledge and experience and that of your colleagues. In this setting, begin by brainstorming with colleagues about what sources will be most useful to your topic and how you can track them down.

Primary Research

Primary research is the gathering of raw data compiled from interviews, direct observation, surveys and questionnaires, experiments, recordings, and the like. In fact, direct observation and interaction are the only ways to obtain certain kinds of information, such as behavior, certain natural phenomena, and the operation of systems and equipment.

If you are planning research that involves observation, choose your sites and times carefully, and be sure to obtain permission in advance. During your observations, remain as unobtrusive as possible and keep accurate, complete records that indicate date, time of day, duration of the observation, and so on. Save interpretations of your observations for future analysis. Be aware that observation can be valuable research, but it can also be time-consuming, complicated, and expensive, and you may inadvertently influence the subjects you are observing.

You may want to conduct a usability test of written instructions that you have created. First, using the instructions, try to perform the

R

task yourself, and create a rough outline of your experience. After you write a draft, have someone unfamiliar with the task test your instructions by using them to operate the equipment or perform the procedure. If the tester has a problem, rewrite the passage until it is clear and easy to follow.

Secondary Research

Secondary research is the gathering of information that has been analyzed, assessed, evaluated, compiled, or otherwise organized into accessible form. Sources include books, articles, reports, Web documents, e-mail discussions, business letters, minutes of meetings, operating manuals, brochures, and so forth. The entries <u>Internet research</u> and <u>library research</u> provide strategies for finding and evaluating these sources.

Research Strategies

As you seek information, keep in mind that in most cases the more recent the information, the better. Articles in periodicals and newspapers are current sources because they are published frequently. Academic (.edu), organizational (.org), and government (.gov) Web sites can be good sources of current information about recent research or works in progress and can include interviews, articles, papers, and conference proceedings. Be sure to consider authorship and other aspects of a text or document as outlined in the Writer's Checklists: Evaluating Library Resources (page 323) and Evaluating Internet Resources (page 284). See also <u>documenting sources</u>, <u>paraphrasing</u>, and <u>plagiarism</u>.

When a resource seems useful, read it carefully and take notes that include any additional questions about your topic. Some of your questions may eventually be answered in other sources; those that remain unanswered can guide you to further research. For example, you may discover that you need to talk with an expert. Not only can someone skilled in a field answer many of your questions, but he or she can also suggest further sources of information. See <u>interviewing for information</u>, <u>listening</u>, and <u>note-taking</u>.

R

---

resignation letters or memos

When you are planning to leave a job, for whatever reason, you usually write a resignation letter to your supervisor or to an appropriate person in the Human Resources Department.

- Start on a positive note, regardless of the circumstances under which you are leaving.

- Consider pointing out how you have benefitted from working for the company or say something complimentary about the company.
- Comment on something positive about the people with whom you have been associated.
- Explain why you are leaving in an objective, factual tone.
- Avoid angry recriminations because your resignation will remain on file with the company and could haunt you in the future when you need references.

Your letter or **memo** should give enough notice to allow your employer time to find a replacement. It might be no more than two weeks, or it might be enough time to enable you to train your replacement. Some organizations may ask for a notice equivalent to the number of weeks of vacation you receive. Check the policy of your employer before you begin your letter.

The sample resignation memo in Figure R–3 is from an employee who is leaving to take a job offering greater opportunities. The memo of resignation in Figure R–4 on page 482 is written by an employee who is

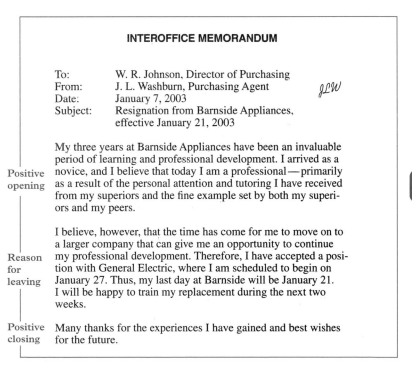

INTEROFFICE MEMORANDUM

To: W. R. Johnson, Director of Purchasing
From: J. L. Washburn, Purchasing Agent   *JLW*
Date: January 7, 2003
Subject: Resignation from Barnside Appliances, effective January 21, 2003

Positive opening
My three years at Barnside Appliances have been an invaluable period of learning and professional development. I arrived as a novice, and I believe that today I am a professional — primarily as a result of the personal attention and tutoring I have received from my superiors and the fine example set by both my superiors and my peers.

Reason for leaving
I believe, however, that the time has come for me to move on to a larger company that can give me an opportunity to continue my professional development. Therefore, I have accepted a position with General Electric, where I am scheduled to begin on January 27. Thus, my last day at Barnside will be January 21. I will be happy to train my replacement during the next two weeks.

Positive closing
Many thanks for the experiences I have gained and best wishes for the future.

FIGURE R–3. Resignation Memo (to Accept a Better Position)

INTEROFFICE MEMORANDUM

To: T. W. Haney, Vice President, Administration
From: L. R. Rupp, Executive Assistant *LRR*
Date: February 12, 2003
Subject: Resignation from Winterhaven, effective
           March 3, 2003

Positive opening

My five-year stay with the Winterhaven Company has been a very pleasant experience, and I believe that it has been mutually beneficial.

Reason for leaving

Because the recent restructuring of my job leaves no career path open to me, I have accepted a position with another company that I feel will offer me greater advancement opportunities. I am, therefore, submitting my resignation, to be effective on March 3, 2003.

Positive closing

I have enjoyed working with my coworkers at Winterhaven and wish the company success in the future.

FIGURE R–4. Resignation Memo (Under Negative Conditions)

leaving under unhappy circumstances; notice that it opens and closes positively and that the reason for the resignation is stated without apparent anger or bitterness. For strategies concerning negative messages, see <u>correspondence</u> and <u>refusal letters</u>.

respective / respectively

R

Respective is an <u>**adjective**</u> that means "pertaining to two or more things regarded individually." (The committee members returned to their *respective* offices.) *Respectively* is the <u>**adverb**</u> form of *respective*, meaning "singly, in the order designated."

- The first, second, and third prizes in the sales contest were awarded to Maria Juarez, Dan Wesp, and Simone Luce, *respectively*.

Respective and *respectively* are unnecessary if the meaning of individuality is made clear.

                Each   *member*   *a report.*
- ~~The~~ committee ~~members~~ prepared ~~their respective reports.~~
     ^   ^   ^

restrictive and nonrestrictive elements

Modifying __phrases__ and __clauses__ may be either restrictive or nonrestrictive. A *nonrestrictive phrase* or clause provides additional information about what it modifies, but it does not restrict the meaning of what it modifies. The nonrestrictive phrase or clause can be removed without changing the essential meaning of the sentence. It is, in effect, a parenthetical element that is set off by __commas__ to show its loose relationship with the rest of the sentence.

NONRESTRICTIVE   This instrument, *which is called a backscatter gauge,* fires beta particles at an object and counts the particles that bounce back.

A *restrictive phrase* or clause limits, or restricts, the meaning of what it modifies. If it were removed, the essential meaning of the sentence would be changed. Because a restrictive phrase or clause is essential to the meaning of the sentence, it is never set off by commas.

RESTRICTIVE   All employees *wishing to donate blood* may take Thursday afternoon off.

It is important for writers to distinguish between nonrestrictive and restrictive elements. The same sentence can take on two entirely different meanings, depending on whether a modifying element is set off by commas (because it is nonrestrictive) or not (because it is restrictive). A slip by the writer can not only mislead __readers__ but also embarrass the writer.

MISLEADING   He gave a poor performance evaluation to the staff members who protested to the Human Resources Department.
[This suggests that he gave the poor evaluation because the staff members had protested.]

ACCURATE   He gave a poor performance evaluation to the staff members, who protested to the Human Resources Department.
[This suggests that the staff members protested because of the poor evaluations.]

Use *which* to introduce nonrestrictive clauses and *that* to introduce restrictive clauses.

NONRESTRICTIVE   After John left the restaurant, *which* is one of the finest in New York, he came directly to my office.

RESTRICTIVE   Companies *that* diversify usually succeed.

résumés

A résumé is the key tool of the job search; it itemizes the qualifications that you summarize in your application letter. A résumé* should be limited to one page — or two pages if you have substantial experience. On the basis of the information in the résumé and application letter, prospective employers decide whether to ask you to come in for an interview. If you are invited to an interview, the interviewer can base specific questions on the contents of your résumé. See also interviewing for a job.

Because résumés affect a potential employer's first impression, take the time to make sure that yours is well organized, well designed, easy to read, and free of errors. Proofreading is essential. Verify the accuracy of the information, and have someone else review it. Experiment to determine a layout and design that is attractive and highlights your strengths. Use a quality printer and high-grade paper.

Analyzing Your Background

In preparing to write your résumé, determine what kind of job you are seeking. Then ask yourself what information about you and your back-

---

*A detailed résumé for someone in an academic and scientific area is often called a *curriculum vitae* (also *vita* or *c.v.*). It may include education, publications, projects, grants, and awards as well as a full work history. Outside the United States, the term *curriculum vitae* is often a synonym for the term *résumé*.

ground would be most important to a prospective employer. List the
following:

- schools you attended, degrees you hold, your major field of study,
  academic honors you were awarded, your grade point average, particular academic projects that reflect your best work
- jobs you have held, your principal and secondary duties in each of
  them, when and how long you held each job, promotions, skills
  you developed in your jobs that potential employers value and seek
  in ideal job candidates, projects or accomplishments that reflect
  your important contributions
- other experiences and skills you have developed that would be of
  value in the kind of job you are seeking; extracurricular activities
  that have contributed to your learning experience; leadership, interpersonal, and communication skills you have developed; any
  collaborative work you have performed; computer skills you have
  acquired

Use this information to brainstorm any further key details. Then, based
on all the details, decide which to include in your résumé and how you
can most effectively present your qualifications.

Organizing Your Résumé

A number of different organizational patterns can be used effectively. A
common one arranges information chronologically in the following categories:

- Heading (name and contact information)
- Job Objective (optional)
- Education
- Employment Experience
- Computer Skills (optional)
- Honors and Activities
- References

Whether you place education or employment experience first depends
on the job you are seeking and on which credentials would strengthen
your résumé the most. If you are a recent graduate without much work
experience, you would probably list education first. If you have years of
job experience, including jobs directly related to the kind of position
you are seeking, you would probably list employment experience first.
In your education and employment sections, use a chronological sequence and list the most recent experience first, the next most recent
experience second, and so on.

Heading.   At the top of your résumé, include your name, address, telephone and fax numbers, and e-mail address. Make sure that your name stands out on the page. If you have just one address, center it at the top of the page, as shown in Figure R–6 on page 493. If you have both a school and a permanent home address, place your school address on the left side of the page and your permanent home address on the right side of the page. Place both underneath your name, as shown in Figure R–5 on page 492. Indicate the dates you can be reached at each address (but do not date the résumé itself).

Job Objective.   Many potential employers prefer to see a clear employment objective in résumés. If you decide to include an objective, use a heading such as "Objective," "Employment Objective," "Career Objective," or "Job Objective." State not only your immediate job objective but also the direction you hope your career will take. A job objective introduces the material in a résumé and helps the reader quickly understand your goal. Try to write your objective in no more than three lines, as illustrated in the following examples.

- A full-time computer science position aimed at solving engineering problems. Special interest in the potential to gain valuable management experience.

- A position involving meeting the concerns of women, such as family planning, career counseling, or crisis management.

- Full-time management of a high-quality, local restaurant.

- A summer research or programming position providing opportunities to further develop problem-solving skills.

Education.   List the school(s) you have attended, the degrees you received and the dates you received them, your major field(s) of study, and any academic honors you have earned. A list of computer skills in this section or elsewhere in the résumé can be especially effective with prospective employers, as shown in Figure R–6 on page 493. Include your grade point average only if it is 3.0 or higher. List courses only if they are unusually impressive or if your résumé is otherwise sparse (see Figure R–6). Mention your high school only if you want to call attention to special high-school achievements, awards, projects, programs, internships, or study abroad.

Employment Experience.   You can organize your employment experience chronologically, starting with your most recent job and working backward under a single major heading called "Experience," "Employment," "Professional Experience," or the like. You could also organize your experience functionally by clustering similar types of jobs into one or several sections with specific headings such as "Management Experience" or "Major Accomplishments," as shown in Figure R–9 on page 496.

One type of arrangement might be more persuasive than the other, depending on the situation. For example, if you are applying for an accounting job but have no specific background in accounting, you would probably do best to list past and present jobs in chronological order, from most to least recent. If you are applying for a supervisory position and have had three supervisory jobs in addition to two nonsupervisory positions, you might choose to create a single section called "Supervisory Experience" and list only your three supervisory jobs. Or you could create two sections — "Supervisory Experience" and "Other Experience" — and include the three supervisory jobs in the first section and your nonsupervisory jobs in the second section.

The functional résumé might be especially useful for applicants who have been employed at only one job and want to demonstrate the diversity of their experience in that position (as shown in Figure R–8 on page 495) or for applicants with time gaps in their résumé caused by unemployment, illness, or family obligations.

In general, follow these conventions when working on the "Experience" section of your résumé.

- Include jobs or internships when they relate directly to the position you are seeking. Although some applicants choose to omit internships and temporary or part-time jobs, including such experiences can make a résumé more persuasive if they have helped you develop specific related skills.

- Include extracurricular experiences, such as taking on a leadership position in a fraternity or sorority or directing a community-service project, if they demonstrate that you have developed skills valued by potential employers.

- List military service as a job; give the dates served, the duty specialty, and the rank at discharge. Discuss military duties if they relate to the job you are seeking.

- For each job or experience, list both the job and company titles. Throughout each section, consistently begin with either the job or the company title, depending on which will likely be more impressive to potential employers.

- Under each job or experience, provide a concise description of your primary and secondary duties. If a job is not directly relevant, provide only a job title and a brief description of duties that helped you develop skills valued in the position you are seeking. For example, if you were a lifeguard and now seek a management position, focus on supervisory experience or even experience in averting disaster to highlight your management, decision-making, and crisis-control skills.

- Use action verbs (for example, "managed" rather than "was the manager") and state ideas succinctly, as shown in Figure R–7 on page 494. Even though the résumé is about you, do not use "I" (for example, instead of "I was promoted to Section Leader," use "Promoted to Section Leader").

Computer Skills.   Employers are interested in hiring applicants with knowledge of a variety of computer skills or who can learn new ones fairly quickly. If you are applying for a job that will require computer knowledge, include a section on computer skills that includes specific languages, software, and hardware.

Honors and Activities.   If you have room on your résumé, it can be persuasive to list any honors and unique activities near the end. Include items such as fluency in foreign languages, writing and editing abilities, specialized technical knowledge, student or community activities, professional or club memberships, and published works. Be selective: Do not duplicate information given in other categories and include only information that supports your employment objective. Provide a heading for this section that fits its contents, such as "Activities," "Honors," "Professional Affiliations," or "Publications and Memberships."

References.   The primary purpose of the résumé is to obtain a job interview. At the end of a successful interview, a potential employer is likely to ask for a list of your references. Unless your résumé is sparse, avoid listing references. Instead, just write a phrase such as "References available upon request" at the bottom of your résumé, either centered or flush left, or write "Available upon request" after the heading "References." Have a printed list of references available for a prospective employer; your list should include the main heading "References for [your name]." See also **reference letters**.

Special Advice for Résumé Preparation

Be truthful. The consequences of giving false information in your résumé are serious. In fact, the truthfulness of your résumé reflects not only your own ethical stance but also on the integrity with which you would represent the organization. See also **ethics in writing**.

Salary.   Avoid listing the salary you desire in the résumé. On the one hand, you may price yourself out of a job you want if the salary you list is higher than a potential employer is willing to pay. On the other hand, if you list a low salary, you may not get the best possible offer. See **salary negotiations**.

Returning Job Seekers.   If you are returning to the workplace after an absence, most career experts say that it is important to acknowledge the gap in your career. That is particularly true if, for example, you are re-entering the workforce because you have devoted a full-time period to care for children or dependent adults. Do not undervalue such work. Although unpaid, it often provides experience that develops important time-management, problem-solving, organizational, and interpersonal skills. The following examples illustrate how you might reflect such experiences in a résumé.

- **Primary Child-Care Provider, 2001 to 2003**

  Provided full-time care to three preschool children at home. Instructed in beginning scholastic skills, time management, basics of nutrition, arts, and swimming. Organized activities, managed household, and served as neighborhood-watch captain.

- **Home Caregiver, 2001 to 2003**

  Provided 60 hours per week in-home care to Alzheimer's patient. Coordinated medical care, developed exercise programs, completed and processed complex medical forms, administered medications, organized budget, and managed home environment.

If you have participated in volunteer work during such a period, list that experience. Volunteer work often results in the same experience as does full-time, paid work, a fact that your résumé should reflect, as in the following example.

- **School Association Coordinator, 2001 to 2003**

  Managed special activities of the Briarwood High School Parent-Teacher Association. Planned and coordinated meetings, scheduled events, and supervised fund-drive operations. Raised $70,000 toward refurbishing the school auditorium.

Electronic Résumés

In addition to the traditional paper résumé, you can post a Web-based résumé. You may also need to submit a résumé on disk or through e-mail to a potential employer to be included in an organization's database. As Internet and database technologies converge, remain current with the forms and protocols that employers prefer by reviewing popular job-search sites, such as Monster.com and others listed in the Web Link box "Popular Job Hunting Sites" in the entry job search.

Web Résumés.   If you plan to post your résumé on your own Web site, keep the following points in mind.

- Follow the general advice for Web design, such as viewing your résumé on several browsers to see how it looks.

- Just below your name, you may wish to provide a series of internal page jumps to such important categories as "experience" and "education."
- Use a counter to keep track of the number of times your résumé Web page has been visited.
- If privacy is an issue, include an e-mail link ("mailto") at the top of the résumé rather than your home address and phone number.

The disadvantage of posting a résumé at your own Web site is that you must attract the attention of employers on your own. As discussed in the entry <u>job search</u>, commercial services can attract recruiters with their large databases.

Scannable and Plain Text Résumés.   A paper or electronic version of a résumé that will be scanned or downloaded into a company's searchable database differs from the traditional résumé. For example, such résumés often use nouns rather than verbs to describe experience and skills (*designer* and *management* rather than *designed* and *managed*). You may also include a section in such a résumé titled "Keywords." This section, which can include up to 50 terms, can be placed to follow the main heading of your résumé or near the end of your résumé (for example, before the "References" heading). Figure R–10 (on page 498) is a sample of an electronic résumé that demonstrates the use of keywords. Keywords (also called *descriptors*) allow potential employers to search the database for qualified candidates. So be sure to use keywords that are the same as those used in the employer's description of the jobs that best match your interests and qualifications.

You may need to submit a résumé electronically as an e-mail attachment that will simply be printed out by the employer. If so, format your résumé as a plain text file (.txt) or use ASCII-compatible characters so that it can be read accurately across most applications and systems. Keep such a résumé as simple, clear, and concise as possible. If you need to submit an attached résumé with special design features, however, consider scanning it as a PDF file. You can attach this file to your e-mail, which will then serve as your <u>application letter</u>.

Sample Résumés

This section includes sample résumés that are formatted for paper, e-mail attachments, or PDF documents. An ASCII-compatible version is also included. These samples are here to stimulate your thinking; your own résumé, tailored to your own job search, will look quite different. Before you design and write your résumé, look at as many samples as possible, and then organize and format your own to best suit

---

DIGITAL TIP PREPARING AN ASCII RÉSUMÉ

When preparing an ASCII (American Standard Code for Information Interchange) document, keep the following points in mind.

- Do not use word-processing elements, such as graphic lines or boxes, bullets, underlining, italics, or boldface.

- Avoid uncommon typefaces; use simple font styles and sizes between 10 and 14 points.

- Create horizontal lines by using a series of hyphens or asterisks (up to 60 characters).

- Replace bullets with hyphens or asterisks, and use white space between sections rather than indenting.

- Set the margins so that you do not exceed 65 characters per line, including spaces.

- Save your document as an ASCII or MS-DOS text document using the *.txt* extension (for example, "resume.txt").

For more on this topic, see *<www.bedfordstmartins.com/alred>* and select *Digital Tips.*

---

your professional goals and to make the most persuasive case to your target employers.

- Figure R–5 presents a conventional student résumé in which the student is seeking an entry-level business position.

- Figure R–6 presents a student résumé with a format that is appropriately a bit nonconventional because this student needs to demonstrate skills in graphic design for his potential audience.

- Figure R–7 shows a résumé that focuses on the applicant's management experience.

- Figure R–8 focuses on how the applicant advanced and was promoted within a single company.

- Figure R–9 illustrates how an applicant can organize a résumé according to skills or functions, instead of using the more conventional chronological order.

- Figure R–10 presents an ASCII résumé. Notice how this résumé emphasizes keywords so that potential employers searching online for applicants will be able to find it easily.

R

CAROL ANN WALKER

SCHOOL
148 University Drive
Bloomington, Indiana 47405
(812) 652-4781
caw2@iu.edu

HOME (after June 2004)
1436 W. Schantz Avenue
Laurel, Pennsylvania 17322
(717) 399-2712
caw@yahoo.com

OBJECTIVE

Position in financial research, leading to management in corporate finance.

EDUCATION

Bachelor of Science in Business Administration, expected June 2004
Indiana University

Emphasis: Finance   Minor: Professional Writing
Grade Point Average: 3.88 out of possible 4.0
Senior Honor Society

FINANCIAL EXPERIENCE

FIRST BANK, INC., of Bloomington, Indiana, 2003
Research Assistant, Summer and Fall Quarters
    Assisted manager of corporate planning and developed long-range
    planning models.

MARTIN FINANCIAL RESEARCH SERVICES, Bloomington, Indiana, 2002
Financial Audit Intern
    Developed a design concept for in-house financial audits and provided
    research assistance to staff.
Associate Editor, *Martin Client Newsletter,* 2001–2002
    Wrote articles on financial planning with computer models; surveyed
    business periodicals for potential articles; edited submissions.

COMPUTER SKILLS

Software: Microsoft Word, Excel, PowerPoint, Pagemaker, Quark XPress
Hardware: Macintosh, IBM-PC, scanners
Languages: FORTRAN, PASCAL

REFERENCES

Available upon request.

FIGURE R–5. Student Résumé (for an Entry-Level Position)

JOSHUA S. GOODMAN
222 Morewood Avenue
Pittsburgh, PA 15212
Jgoodman@aol.com

OBJECTIVE
A position as a graphic designer with responsibilities in information design, packaging, and media presentations.

EDUCATION
Carnegie Mellon University, Pittsburgh, Pennsylvania
BFA in Graphic Design—May 2004

Graphic Design	Corporate Identity
Industrial Design	Graphic Imaging Processes
Color Theory	Computer Graphics
Typography	Serigraphy
Photography	Video Production

GRAPHIC DESIGN EXPERIENCE

Assistant Designer

Dyer/Khan, Los Angeles, California, Summer 2002, Summer 2003
Assistant Designer in a versatile design studio.
Responsibilities: design, layout, comps, mechanicals, and project management.
Clients: Paramount Pictures, Mattel Electronics, and Motown Records.

Photo Editor

Paramount Pictures Corporation, Los Angeles, California, Summer 2001
Photo Editor for merchandising department.
Responsibilities: establish art files for movie and television properties, edit images used in merchandising, maintain archive and database.

Production Assistant

Grafis, Los Angeles, California, Summer 2000
Production assistant and miscellaneous studio work at fast-paced design firm.
Responsibilities: comps and mechanicals.
Clients: ABC Television, A&M Records, and Ortho Products Division.

COMPUTER SKILLS
XML, HTML, JavaScript, Forms, Macromedia Dreamweaver 3, Macromedia Flash 4, Photoshop 5.5, Image Ready (Animated GIFs), CorelDRAW, DeepPaint, iGrafx Designer, MapEdit (Image Mapping), Scanning, Microsoft Access/Excel, Quark XPress.

ACTIVITIES
Member, Pittsburgh Graphic Design Society; Member, The Design Group.

REFERENCES AND PORTFOLIO
Available upon request.

R

FIGURE R–6. Student Résumé (for a Graphic Design Job)

ROBERT MANDILLO
7761 Shalamar Drive
Dayton, Ohio 45424

Home: (513) 255-4137   Fax: (513) 255-3117
Business: (513) 543-3337   mand@juno.com

MANAGEMENT EXPERIENCE

Manager, Engineering Drafting Department — May 1995–Present
Wright-Patterson Air Force Base, Dayton, Ohio

Supervise 17 Drafting Mechanics in support of the engineering design staff. Develop, evaluate, and improve materials and equipment for the design and construction of exhibits. Write specifications, negotiate with vendors, and initiate procurement activities for exhibit design support.

Supervisor, Graphics Illustrators —June 1982–April 1995
Henderson Advertising Agency, Cincinnati, Ohio

Supervised five Illustrators and four Drafting Mechanics after promotion from Graphics Technician; analyzed and approved work-order requirements; selected appropriate media and techniques for orders; rendered illustrations in pencil and ink; converted department to CAD system.

EDUCATION

Bachelor of Science in Mechanical Engineering Technology, 1982
Edison State College, Wooster, Ohio

Associate's Degree in Mechanical Drafting, 1980
Wooster Community College, Wooster, Ohio

PROFESSIONAL AFFILIATIONS

National Association of Mechanical Engineers and Drafting Mechanics

REFERENCES

References, letters of recommendation, and a portfolio of original designs and drawings available upon request.

R

FIGURE R–7. Résumé (by an Applicant with Management Experience)

CAROL ANN WALKER
1436 W. Schantz Avenue
Laurel, Pennsylvania 17322
(717) 399-2712
caw@yahoo.com

FINANCIAL EXPERIENCE

KERFHEIMER CORPORATION, Dayton, Ohio

Senior Financial Analyst, June 2000–Present
Report to Senior Vice President for Corporate Financial Planning.
Develop manufacturing cost estimates totaling $30 million annually for
mining and construction equipment with Department of Defense.

Financial Analyst, November 1997–June 2000
Developed $50-million funding estimates for major Department of
Defense contracts for troop carriers and digging and earth-moving
machines. Researched funding options, recommending those with most
favorable rates and terms.

FIRST BANK, INC., Bloomington, Indiana

Planning Analyst, September 1992–November 1997
Developed successful computer models for short- and long-range planning.

EDUCATION

Ph.D. in Finance: expected, June 2004
The Wharton School of the University of Pennsylvania

M.S. in Business Administration, 1996
University of Wisconsin–Milwaukee
"Executive Curriculum" for employees identified as promising by their
employers.

B.S. in Business Administration (*magna cum laude*), 1992
Indiana University
Emphasis: Finance   Minor: Professional Writing

PUBLISHING AND MEMBERSHIP

Published "Developing Computer Models for Financial Planning," *Midwest
Finance Journal* (Vol. 34, No. 2, 2002), pp. 126–136.

Association for Corporate Financial Planning, Senior Member.

REFERENCES

References and a portfolio of financial plans are available upon request.

R

FIGURE R–8. Advanced Résumé (Showing Promotion Within Single Company)

CAROL ANN WALKER
1436 W. Schantz Avenue
Laurel, Pennsylvania 17322
(717) 399-2712
caw@yahoo.com

MAJOR ACCOMPLISHMENTS

FINANCIAL PLANNING
- Researched funding options to achieve a 23% return on investment.
- Developed long-range funding requirements for over $1 billion in government and military contracts.
- Developed a computer model for long- and short-range planning that saved 65% in proposal-preparation time.
- Received the Financial Planner of the Year Award from the Association of Financial Planners, a national organization composed of both practitioners and academics.

CAPITAL ACQUISITION
- Developed strategies to acquire over $1 billion at 3% below market rate.
- Secured over $100 million through private and government research grants.
- Developed computer models for capital acquisition that enabled the company to decrease its long-term debt during several major building expansions.

RESEARCH AND ANALYSIS
- Researched and developed computer models applied to practical problems of corporate finance.
- Functioned primarily as a researcher at two different firms for over 11 years.
- Published research in financial journals while pursuing an advanced degree at the Wharton School.

EDUCATION

Ph.D. in Finance: expected, June 2004
The Wharton School of the University of Pennsylvania

M.S. in Business Administration, 1996
University of Wisconsin–Milwaukee
"Executive Curriculum" for employees identified as promising by their employers.

B.S. in Business Administration (*magna cum laude*), 1992
Indiana University
Emphasis: Finance   Minor: Professional Writing

FIGURE R–9. Advanced Résumé (Organized by Function)

Carol Ann Walker   Page 2

EMPLOYMENT EXPERIENCE

KERFHEIMER CORPORATION, Dayton, Ohio
November 1997–Present
   Senior Financial Analyst
   Financial Analyst

FIRST BANK, INC., Bloomington, Indiana
September 1992–November 1997
   Planning Analyst

PUBLICATIONS AND MEMBERSHIPS

Published "Developing Computer Models for Financial Planning," *Midwest Finance Journal* (Vol. 34, No. 2, 2002), pp. 126–136.

Association for Corporate Financial Planning, Senior Member.

REFERENCES

References and a portfolio of financial plans are available upon request.

R

FIGURE R–9. Advanced Résumé (Organized by Function) *(continued)*

DAVID B. EDWARDS
6819 Locustview Drive
Topeka, Kansas 66614
(913) 233-1552
dedwards@cpu.fairview.edu

JOB OBJECTIVE
Work as a programmer with writing, editing, and training responsibilities, leading to a
career in information design management.

KEYWORDS
Programmer, Operating Systems, Unipro, Newsletter, Graphics, Cybernetics, Listserv,
Professional Writer, Editor, Trainer, Teacher, Instructor, Technical Writer, Tutor,
Designer, Manager, Information Design.

EDUCATION
** Fairview Community College, Topeka, Kansas
** Associate's Degree, Computer Science, June 2002
** Dean's Honor List Award (six quarters)

RELEVANT COURSE WORK
** Operating Systems Design
** Database Management
** Introduction to Cybernetics
** Technical Writing

EMPLOYMENT EXPERIENCE
** Computer Consultant: September 2002 to Present
Fairview Community College Computer Center: Advised and trained novice computer
users; wrote and maintained Unipro operating system documentation.
** Tutor: January 2001 to June 2002
Fairview Community College: Assisted students in mathematics and computer
programming.

SKILLS AND ACTIVITIES
** Unipro Operating System: Thorough knowledge of word-processing, text-editing,
and file-formatting programs.
** Writing and Editing Skills: Experience in documenting computer programs for
beginning programmers and users.
** Fairview Community Microcomputer Users Group: Cofounder and editor of
monthly newsletter ("Compuclub"); listserv manager.

FURTHER INFORMATION
** References, college transcripts, a portfolio of computer programs, and writing
samples available upon request.

R

FIGURE R–10. Electronic Résumé (in ASCII Format)

revision

The more natural a piece of writing seems to the reader, the more effort the writer has probably put into its revision. If possible, put your draft away for a day or two before you begin to revise it. If that is not possible, take a break from your writing and try to do something else before you revise. Without such a cooling period, you are too close to the draft to evaluate it objectively.

When you return to revise your draft, read and evaluate it primarily from the point of view of your <u>readers</u>. In fact, revising requires a different frame of mind than <u>writing a draft</u>. To achieve that frame of mind, experienced writers have developed the following tactics.

- Print out your draft and mark up the paper copy; it is often difficult to revise on-screen.

- Read your draft aloud—often, hearing the text will enable you to spot problem areas that need improvement.

- Revise in passes by reading through your draft several times, each time searching for and correcting a different set of problems.

- Print out your draft on colored paper to reorient your perception.

When you can no longer spot improvements, you may wish to give the draft to a colleague for review—especially for writing projects that are important for you or your organization.

Writer's Checklist: Revising Your Draft

☑ *Completeness.* Does the document achieve its primary purpose? Will it fulfill the readers' needs? Your writing should give readers exactly what they need but not overwhelm them.

☑ *Appropriate introduction and conclusion.* Check to see that your <u>**introduction**</u> frames the rest of the document and your <u>**conclusion**</u> ties the main ideas together. Both should account for revisions in the body of the document.

☑ *Accuracy.* No matter how careful and painstaking you may have been to this point, you need to look for any inaccuracies that may have crept into your draft. See also <u>**quotations**</u> and <u>**documenting sources**</u>.

☑ *Unity and coherence.* Check to see that sentences and ideas are closely tied together (<u>**coherence**</u>) and contribute directly to the main idea expressed in the topic sentence of each <u>**paragraph**</u> (<u>**unity**</u>). Provide <u>**transitions**</u> where they are missing and strengthen those that are weak.

R

Writer's Checklist: Revising Your Draft (continued)

☑ *Consistency.* Make sure that <u>layout and design</u>, <u>visuals</u>, and use of language are consistent. Do not call the same item by one term on one page and a different term on another page.

☑ *Conciseness.* Tighten your writing so that it says exactly what you mean. Prune unnecessary words, phrases, sentences, and even paragraphs. See <u>conciseness/wordiness</u>.

☑ *Awkwardness.* Look for <u>awkwardness</u> in <u>sentence construction</u> — especially passive-voice constructions. The active <u>voice</u> and strong <u>verbs</u> make your writing more direct.

☑ *Word choice.* Delete or replace <u>vague words</u> and unnecessary <u>intensifiers</u>. Check for <u>affectation</u> and unclear <u>pronoun references</u>. See also <u>word choice</u>.

☑ *Ethical language.* Check for <u>ethics in writing</u>, including <u>biased language</u>.

☑ *Jargon.* If you have any doubt that *all* your readers will understand any terms or <u>jargon</u> you have used, eliminate it.

☑ *Clichés.* Replace <u>clichés</u> with fresh <u>figures of speech</u> or direct statements.

☑ *Grammar.* Check your draft for possible errors in <u>grammar</u>.

☑ *Typographical errors.* Check your final draft for typographical errors both with your spell checker and through <u>proofreading</u>.

rhetorical questions

A *rhetorical question* is a question to which a specific answer is neither needed nor expected. The question is often intended to make an <u>audience</u> think about the subject from a different perspective; the writer or speaker then answers the question in the article or <u>presentation</u>. (Is space exploration worth the cost?) The answer to this rhetorical question may not be a simple yes or no; it might be a detailed explanation of the pros and cons of the value of space exploration.

The rhetorical question can be an effective opening, and it is often used as a <u>title</u>. By its nature, it is somewhat journalistic and indirect, and therefore should be used judiciously in technical writing. For example, while a rhetorical question might be an appropriate opening for a <u>newsletter article</u>, it would not be as effective for a <u>report</u> or <u>e-mail</u> addressed to a busy manager. When you do use a rhetorical question, be sure it is not trivial, obvious, or forced. The rhetorical question requires that you know your <u>readers</u>.

run-on sentences

A *run-on sentence,* sometimes called a *fused sentence,* is two or more sentences without punctuation to separate them. The term is also sometimes applied to a pair of independent <u>clauses</u> separated by only a <u>comma,</u> although this variation is usually called a <u>comma splice</u>. (See also <u>sentence construction</u> and <u>sentence faults</u>.) Run-on sentences can be corrected, as shown in the following examples, by (1) making two sentences, (2) joining the two clauses with a <u>semicolon</u> (if they are closely related), (3) joining the two clauses with a comma and a coordinating <u>conjunction,</u> or (4) subordinating one clause to the other.

- The client suggested several solutions *. Some* some are impractical.

- The client suggested several solutions *;* some are impractical.

- The client suggested several solutions *, but* some are impractical.

- The client suggested several solutions *, although* some are impractical.

R

S

Salary negotiations usually take place either at the end of an interview or after a formal job offer has been made. If possible, delay discussing salary until after you receive a formal written job offer because you will have more negotiating power at that point.

Before <u>interviewing for a job</u>, prepare for possible salary negotiations by researching the following:

- The company's range of salaries for the position you are seeking. Call the company and ask to talk with the human-resources manager. Explain that you will be interviewing for a particular position and ask about the salary range for that job.

- The current range of salaries for the work you hope to do at your level (entry? intermediate? advanced?) in your region of the country. Check trade journals and organizations in your field, or ask a reference librarian for help in finding this information. Job listings that include salary can also be helpful.

- Salaries made by last year's graduates from your college or university at your level and in your line of work. Your campus career-development office should have these figures.

- Salaries made by people you know at your level and in your line of work. Attend local organizational meetings in your field or contact officers of local organizations who might have this information or steer you to useful contacts.

If a potential employer requests your salary requirements with a <u>résumé</u>, consider your options carefully. If you provide a salary that is too high, the company might never interview you; if you provide a salary that is too low, you may have no opportunity later in the hiring process to negotiate for a higher salary. However, if you fail to follow the potential employer's directions and omit the requested information, an employer may disqualify you on principle. If you choose to provide salary requirements, always do so in a range (for example, $30,000 to $35,000).

If an interviewer asks your salary requirements toward the end of the job interview, you can try these strategies to delay salary negotiations.

- Say something like "I am sure that this company always pays a fair salary for a person with my level of experience and qualifications" or "I am ready to consider your best offer."
- Indicate that you would like to learn more details about the position before discussing salary; point out that your primary goal is to work in a stimulating environment with growth potential, not to earn a specific salary.
- Express a strong interest in the position and the organization.
- Emphasize your unique qualifications (or combination of skills) for the job and what you can do for the company that other candidates cannot.

If the interviewer or company demands to know your salary requirements during a job interview, provide a wide salary range that you know would be reasonable for someone at your level in your line of work in that region of the country. For example, you could say, "I would hope for a salary somewhere between $28,000 and $38,000, but of course this is negotiable."

Once salary negotiations begin, resist the temptation to accept, immediately, the first salary offer you receive. If you have done thorough research, you'll know if the first salary offer is at the low, middle, or high end of the salary range for your level of experience in your line of work. If you have little or no experience and receive an offer for a salary at the low end of the range, you will realize that the offer probably is fair and reasonable. Yet, if you receive the same low-end offer but bring considerable experience to the job, you can negotiate for a higher salary that is more reasonable for someone with your background and credentials in your line of work in your region of the country.

Never say that you are unable to accept a salary below a particular figure. To keep negotiations going, simply indicate that you would have trouble accepting the first offer because it was smaller than you had expected.

Remember that you are negotiating a package and not just a starting salary. For example, benefits can offer you substantial value. If the starting salary seems low, consider negotiating for some of these possible job perks:

- The chance for an early promotion, thus higher salary within a few years
- A particular job title or special job responsibilities that would provide you with impressive chances for career growth
- Tuition credits for continued education
- Payment of relocation costs

- Paid personal leave or paid vacations
- Personal or sick days
- Overtime potential
- Flexible hours
- Health, dental, optical/eye care, disability, and life insurance
- Retirement plans, such as 401(k) and pension plans
- Profit sharing; investment or stock options
- Bonuses
- Commuting or parking cost reimbursement
- Child or elder care
- Discounts on company products and services

You might find it most comfortable to respond to an initial offer in writing, and then meet later with the potential employer for further negotiation. If possible, indicate all of your preferences and requirements at one time instead of continually asking for new and different benefits during negotiations. Throughout this process, focus on what is most important to you (which might differ from what is most important to your friends) and on what you would find acceptable and comfortable. See also **application letters** and **job search**.

 WEB LINK **SALARY INFORMATION RESOURCES**

Many Web sites, such as Salary.com, Wageweb.com, and that of the Bureau of Labor Statistics, offer useful resources for salary negotiations. See *<www.bedfordstmartins.com/alred>* and select *Links for Technical Writing*.

---

schematic diagrams (*see* drawings)

scope

Scope is the depth and breadth of detail needed to cover a subject. Determine the scope during the preparation stage of the writing process, even though you may refine it later. Your **readers'** needs and your primary **purpose** determine the kind of information and the amount of detail you will need to include. Defining your scope will expedite your **research**.

Your scope will also be affected by the type of document you are writing as well as the medium you select for your message. For example, government agencies often prescribe the general content and length for <u>proposals,</u> and some organizations set limits for the length of <u>memos</u> and <u>e-mail</u>. See <u>selecting the medium</u> and "Five Steps to Successful Writing."

search engines (*see* Internet research)

selecting the medium

You must select the most appropriate medium for communicating your message early in your <u>preparation</u>. You can choose from such relatively recent technologies as e-mail, fax, voice mail, and videoconferencing to more traditional means such as letters and memos, telephone calls, and face-to-face meetings.

The most important considerations in selecting the medium are <u>audience</u> expectations and the <u>purpose</u> of the communication. For example, if you need to collaborate with someone to solve a problem or if you need to establish rapport, written exchanges (even by e-mail) may be far less efficient than a phone call or a face-to-face meeting. However, if you need precise wording or a record of a complex message, communicate in writing. Understanding the typical means of communicating on the job will help you select the most appropriate medium.

Letters on Organizational Stationery

Letters are often the most appropriate choice for initial contacts with new business associates or customers and for other formal communications. Letters written on your organization's letterhead communicate formality, respect, and authority. See also <u>correspondence</u>.

Memos

<u>Memos</u> on printed company stationery or attached to an e-mail are appropriate for internal communication among members of the same organization, even when offices are geographically separated. They have many of the same characteristics as letters, such as formality and authority, but they are used for a wider variety of functions — from reminders to short reports. The use of memos must follow the organization's protocol.

S

E-mail

E-mail can replace letters and memos or deliver them as attachments. It can be a less-formal medium to send information, elicit discussions, collect opinions, and transmit many other kinds of messages quickly. Because e-mail recipients can print copies of messages and attachments they receive and easily forward them to others, always write your e-mail with care, and reread the message carefully before you send it.

Faxes

A fax is most useful when speed is essential and when the information—a drawing or contract, for example—must be viewed in its original form. Faxes are also useful when the recipient does not have e-mail or when the material is not available in electronic form.

Telephone Calls

Information exchanged through telephone calls can range from a brief call to answer a question to a lengthy call to negotiate or clarify the conditions of a contract. Because phone calls enable participants to interpret tone of voice, they can make it easier to resolve misunderstandings, although they do not provide the visual cues possible during face-to-face meetings.

In addition, when there are three or more participants, a conference call is a less-expensive alternative to face-to-face meetings requiring travel. To ensure maximum efficiency, the person coordinating the call works from an agenda shared by all the participants and directs the discussion as though he or she were leading a meeting. Telephone calls in which a decision has been reached should be followed up with written confirmation.

Voice Mail

Voice mail allows callers to leave a brief message ("Call me about the deadline for the new project" or "I got the package, so you don't need to call the distributor"). If the message is complicated or contains numerous details, use another medium, such as an e-mail message or a letter. If you want to discuss a subject, let the recipient know the subject so that he or she can prepare a response before returning your call. When you leave a message, give your name, phone number, the date, and the time of the call.

Face-to-Face Meetings

Face-to-face meetings are most appropriate for initial or early contacts with associates and clients with whom you intend to develop an important, long-term relationship. Meetings may also be best for brainstorming, negotiating, solving a technical problem, or handling a controversial issue. For advice on how to record discussions and decisions, see **minutes of meetings**.

Videoconferencing

Videoconferencing is particularly useful for meetings when travel is impractical. Unlike telephone conference calls, videoconferences have the advantage of allowing participants to see as well as to hear one another. Videoconferences work best with participants who are at ease in front of the camera.

semicolons

The *semicolon* (;) links independent **clauses** or other sentence elements of equal weight and grammatical rank when they are not joined by a **comma** and a **conjunction**. The semicolon indicates a greater pause between clauses than a comma but not as great a pause as a **period**.

Independent clauses joined by a semicolon should balance or contrast with each other, and the relationship between the two statements should be so clear that further explanation is not necessary.

- The new Web site was a success; every division reported increased online sales.

Be careful not to use a semicolon between a dependent clause and its main clause.

- No one applied for the ~~position;~~ *position,* even though it was heavily advertised.

With Strong Connectives

In complicated sentences, a semicolon may be used before transitional words or **phrases** (*that is, for example, namely*) that introduce examples or further explanation. See also **transition**.

- The study group was aware of her position on the issue; that is, federal funds should not be used for the housing project.

A semicolon should also be used before conjunctive **adverbs** (*therefore, moreover, consequently, furthermore, indeed, in fact, however*) that connect independent clauses.

- The test results are not complete; *therefore*, I cannot make a recommendation.
  [The semicolon shows that *therefore* belongs to the second clause.]

For Clarity in Long Sentences

Use a semicolon between two independent clauses connected by a coordinating conjunction (*and, but, for, or, nor, so, yet*) if the clauses are long and contain other **punctuation**.

- In most cases, these individuals are executives, bankers, lawyers; *but* they do not, as the press seems to believe, simply push the button of their economic power to affect local politics.

A semicolon may also be used if any items in a series contain commas.

- Among those present were John Howard, president of the Omega Paper Company; Carol Delgado, president of Environex Corporation; and Larry Stanley, president of Stanley Papers.

Do not use semicolons to enclose a parenthetical element that contains commas. Use **parentheses** or **dashes** for that purpose.

- All affected job classifications (receptionists, secretaries, transcriptionists, and clerks) will be upgraded this month.

Do not use a semicolon as a mark of anticipation or enumeration. Use a **colon** for that purpose.

- Three decontamination methods are under ~~consideration;~~ *consideration:* a zeolite-resin system, an evaporation system, and a filtration system.

The semicolon always appears outside closing **quotation marks**.

- The attorney said, "You must be accurate"; her client replied, "I will."

S

sentence construction

A sentence is the most fundamental and versatile tool available to writers. Sentences generally flow from a subject to a <u>verb</u> to any objects, <u>complements</u>, or <u>modifiers</u>, but they can be ordered in a variety of ways to achieve <u>emphasis</u>. When shifting word order for emphasis, however, be aware that word order can make a great difference in the meaning of a sentence.

- He was *only* the service technician.

- He was the *only* service technician.

Subjects

The most basic components of sentences are subjects and predicates. The subject of a sentence is a <u>noun</u> or <u>pronoun</u> (and its modifiers) about which the predicate of the sentence makes a statement. Although a subject may appear anywhere in a sentence, it most often appears at

---

ⓔ TIPS FOR UNDERSTANDING THE SUBJECT OF A SENTENCE

In English, every sentence, except commands, must have an explicit subject.

- *Ozzie* worked fast. ~~Established~~ the parameters for the project.
  He established ^

In commands, the subject *you* is understood and is used only for emphasis.

- (*You*) Show up at the airport at 6:30 tomorrow morning.

- *You* do your homework, young man. [parent to child]

If you move the subject from its normal position (subject-verb-object), English often requires you to replace the subject with an expletive (*there, it*). In this construction, the verb agrees with the subject that follows it.

- *There are* two files on the desk. [The subject is *files*.]

- *It is* presumptuous for me to speak for Jim.
  [The subject is *to speak for Jim*.]

Time, distance, weather, temperature, and environmental expressions use *it* as their subject.

- *It* is ten o'clock.

- *It* is ten miles down the road.

- *It* seldom snows in Florida.

- *It* is very hot in Jorge's office.

S

the beginning. (*The wiring* is defective.) Grammatically, a subject must agree with its verb in <u>number</u>.

- These *departments have* much in common.
- This *department has* several functions.

The subject is the actor in sentences using the active <u>voice</u>.

- The *Web master* reported a record number of hits in November.

A compound subject has two or more <u>substantives</u> as the subject of one verb.

- *The president* and *the treasurer* agreed to begin the audit.

Predicates

The *predicate* is the part of a sentence that contains the main verb and any other words used to complete the thought of the sentence (the verb's modifiers and <u>complements</u>). The principal part of the predicate is the verb, just as a noun (or noun substitute) is the principal part of the subject.

The *simple predicate* is the verb (or verb phrase) alone. The *complete predicate* is the verb and its modifiers and complements.

- Bill *piloted the airplane.*

A *compound predicate* consists of two or more verbs with the same subject. It is an important device for <u>conciseness</u> in writing.

- The company *tried* but *did not succeed* in that field.

A *predicate nominative* is a noun construction that follows a linking verb and renames the subject.

- She is my *attorney.* [noun]
- His excuse was *that he had been sick.* [noun clause]

Sentence Types

Sentences may be classified according to *structure* (simple, compound, complex, compound-complex); *intention* (declarative, interrogative, imperative, exclamatory); and *stylistic use* (loose, periodic, minor).

Structure.   A *simple sentence* consists of one independent clause. At its most basic, the simple sentence contains only a subject and a predicate.

- Profits [subject] rose [predicate].
- The storm [subject] finally ended [predicate].

A *compound sentence* consists of two or more independent clauses connected by a comma and a coordinating <u>conjunction</u>, by a <u>semicolon</u>, or by a semicolon and a conjunctive <u>adverb</u>.

- Drilling is the only way to collect samples of the layers of sediment below the ocean floor, *but* it is not the only way to gather information about these strata. [comma and coordinating conjunction]

- There is little similarity between the chemical composition of seawater and that of river water; the various elements are present in entirely different proportions. [semicolon]

- It was 500 miles to the site; *therefore,* we made arrangements to fly. [semicolon and conjunctive adverb]

A *complex sentence* contains one independent clause and at least one dependent clause that expresses a subordinate idea.

- The generator will shut off automatically [independent clause] if the temperature rises above a specified point [dependent clause].

A *compound complex* sentence consists of two or more independent clauses plus at least one dependent clause.

- Productivity is central to controlling inflation [independent clause]; when productivity rises [dependent clause], employers can raise wages without raising prices [independent clause].

Intention.   A *declarative sentence* conveys information or makes a factual statement. (The motor powers the conveyor belt.) An *interrogative sentence* asks a direct question. (Does the conveyor belt run constantly?) An *imperative sentence* issues a command. (Restart in MS-DOS mode.) An *exclamatory sentence* is an emphatic expression of feeling, fact, or opinion. It is a declarative sentence that is stated with great feeling. (The files were deleted!)

Stylistic Use.   A *loose sentence* makes its major point at the beginning and then adds subordinate phrases and clauses that develop or modify that major point. A loose sentence might seem to end at one or more points before it actually does end, as the periods in brackets illustrate in the following sentence.

- It went up[.], a great ball of fire about a mile in diameter[.], an elemental force freed from its bonds[.] after being chained for billions of years.

A *periodic sentence* delays its main ideas until the end by presenting subordinate ideas or modifiers first.

- During the last century, the attitude of the American citizen toward automation underwent a profound change.

A *minor sentence* is an incomplete sentence. It makes sense in its context because the missing element is clearly implied by the preceding sentence.

- In view of these facts, is the service contract really useful? *Or economical?*

Constructing Effective Sentences

The subject-verb-object pattern is effective because it is most familiar to readers. In "The company dismissed Joe," we know the subject and the object by their positions relative to the verb.

An *inverted sentence* places the elements in unexpected order, thus emphasizing the point by attracting the readers' attention.

- A better job I never had. [direct object-subject-verb]
- More optimistic I have never been.
  [subjective complement-subject-linking verb]
- A poor image we presented. [complement-subject-verb]

Use uncomplicated sentences to state complex ideas. If readers have to cope with a complicated sentence in addition to a complex idea, they are likely to become confused. Just as simpler sentences make complex ideas more digestible, a complex sentence construction makes a series of simple ideas more smooth and less choppy.

Avoid loading sentences with a number of thoughts carelessly tacked together. Such sentences are monotonous and hard to read because all ideas seem to be of equal importance. Rather, distinguish the relative importance of sentence elements with **subordination**.

LOADED	We started the program three years ago, there were only three members on the staff, and each member was responsible for a separate state, but it was not an efficient operation.
IMPROVED	When we started the program three years ago, there were only three members on the staff, each having responsibility for a separate state; however, that arrangement was not efficient.

Express coordinate or equivalent ideas in similar form. The very construction of the sentence helps the reader grasp the similarity of its components, as illustrated in **parallel structure**. See also **garbled sentences**.

> **ⒺⓈⓁ TIPS FOR UNDERSTANDING THE REQUIREMENTS OF A SENTENCE**
>
> - A sentence must start with a capital letter.
> - A sentence must end with a period, a question mark, or an exclamation point.
> - A sentence must have a subject.
> - A sentence must have a verb.
> - A sentence must conform to subject-verb-object word order (or inverted word order for questions or emphasis).
> - A sentence must express an idea that can stand on its own (called the main or independent clause).

sentence faults

A number of problems can create sentence faults, including faulty <u>subordination</u>, <u>clauses</u> with no subjects, rambling sentences, and omitted <u>verbs</u>.

Faulty subordination occurs when a grammatically subordinate element actually contains the main idea of the sentence or when a subordinate element is so long or detailed that it obscures the main idea. Both of the following sentences are logical, depending on what the writer intends as the main idea and as the subordinate element.

- Although the new filing system saves money, many of the staff are unhappy with it.
  [If the main point is that *many of the staff are unhappy,* this sentence is correct.]

- The new filing system saves money, although many of the staff are unhappy with it.
  [If the writer's main point is that *the new filing system saves money,* this sentence is correct.]

In the following example, the subordinate element overwhelms the main point.

FAULTY   Because the noise level in the assembly area on a typical shift is as loud as a smoke detector's alarm ten feet away, employees often develop hearing problems.

IMPROVED   Because the noise level in the assembly area is so high, employees often develop hearing problems. In fact, the noise level on a typical shift is as loud as a smoke detector's alarm ten feet away.

S

Writers sometimes inappropriately assume a subject that is not stated in a clause—thus the clause has no subject.

CONFUSING   Your application program can request to end the session after the next command.
[Your application program can request *who* or *what* to end the session?]

CLEAR   Your application program can request *the host program* to end the session after the next command.

Rambling sentences contain more information than the reader can comfortably absorb. The obvious remedy for a rambling sentence is to divide it into two or more sentences. When you do that, put the main message of the rambling sentence into the first of the revised sentences.

RAMBLING   The payment to which a subcontractor is entitled should be made promptly in order that in the event of a subsequent contractual dispute we, as general contractors, may not be held in default of our contract by virtue of nonpayment.

DIRECT   Pay subcontractors promptly. Then if a contractual dispute occurs, we cannot be held in default of our contract because of nonpayment.

Do not omit a required verb.

                          written
- I never have and probably never will write the annual report.
                    ʌ

The assertion that a sentence's predicate makes about its subject must be logical. "Mr. Wilson's *job* is a sales representative" is not logical, but "*Mr. Wilson* is a sales representative" is. "Jim's *height* is six feet tall" is not logical, but "*Jim* is six feet tall" is. See also **run-on sentences** and **sentence fragments**.

S

sentence fragments

A *sentence fragment* is an incomplete grammatical unit that is punctuated as a sentence.

SENTENCE   He quit his job.

FRAGMENT   And quit his job.

A sentence fragment lacks either a subject or a **verb** or is a subordinate **clause** or **phrase**. Sentence fragments are often introduced by relative pronouns (*who, which, that*) or subordinating conjunctions (*although, because, if, when, while*).

- The new manager instituted several new procedures .~~Although~~ , *although*
she didn't train her staff first. ^

A sentence must contain a finite verb; verbals do not function as verbs. The following examples are sentence fragments because their verbals (*providing, to work*) cannot function as finite verbs.

FRAGMENT   *Providing* all employees with disability insurance.

SENTENCE   The company *must provide* all employees with disability insurance.

FRAGMENT   *To work* a 40-hour week.

SENTENCE   All employees *are expected* to work a 40-hour week.

Explanatory phrases beginning with *such as, for example,* and similar terms often lead writers to create sentence fragments.

- The staff wants additional benefits .~~For~~ example, the use of company cars. ^ , *such as*

A hopelessly snarled fragment simply has to be rewritten. To rewrite such a fragment, pull the main points out of the fragment, list them in the proper sequence, and then rewrite the sentence as illustrated in garbled sentences. See also run-on sentences, sentence construction, and sentence faults.

sentence variety

Sentences can vary in length, structure, and complexity. As you revise, make sure your sentences have not become tiresomely alike.

S

Sentence Length

Because a series of sentences of the same length is monotonous, varying sentence length makes writing less tedious to the reader. For example, avoid stringing together a number of short independent clauses. Either connect them with subordinating connectives, thereby making some dependent clauses, or make some clauses into separate sentences. See also subordination.

STRING	The river is 60 miles long, and it averages 50 yards in width, and its depth averages 8 feet.
IMPROVED	The river, which is 60 miles long and averages 50 yards in width, has an average depth of 8 feet.
IMPROVED	The river is 60 miles long. It averages 50 yards in width and 8 feet in depth.

You can often effectively combine short sentences by converting verbs into adjectives.

- The digital shift indicator failed. It was pulled from the market.

failed

Although too many short sentences make your writing sound choppy and immature, a short sentence can be effective following a long one.

- During the past two decades, many changes have occurred in American life, the extent, durability, and significance of which no one has yet measured. *No one can.*

In general, short sentences are good for emphatic, memorable statements. Long sentences are good for detailed explanations and support. Nothing is inherently wrong with a long sentence, or even with a complicated one, as long as its meaning is clear and direct. Sentence length becomes an element of style when varied for emphasis or contrast; a conspicuously short or long sentence can be used to good effect.

Word Order

When a series of sentences all begin in exactly the same way (usually with an article and a noun) the result is likely to be monotonous. You can make your sentences more interesting by occasionally starting with a modifying word, phrase, or clause. However, overuse of this technique itself can be monotonous, so use it in moderation.

- *To salvage the project,* she presented alternatives when existing policies failed to produce results. [modifying phrase]

Inverted sentence order can be an effective way to achieve variety, but do not overdo it.

CONFUSING	Then occurred the event that gained us the contract.
EFFECTIVE	Never have sales been so good.

For variety, you can alter normal sentence order by inserting a phrase or clause.

- Titanium fills the gap, *both in weight and in strength,* between aluminum and steel.

The technique of inserting a phrase or clause is good for achieving emphasis, providing detail, breaking monotony, and regulating pace.

Loose and Periodic Sentences

A loose sentence makes its major point at the beginning and then adds subordinate phrases and clauses that develop or modify the point. A loose sentence could end at one or more points before it actually does end, as the periods in brackets illustrate in the following example.

- It went up[.], a great ball of fire about a mile in diameter[.], an elemental force freed from its bonds[.] after being chained for billions of years.

A periodic sentence delays its main idea until the end by presenting modifiers or subordinate ideas first, thus holding the readers' interest until the end.

- During the last century, the attitude of Americans toward technology underwent a profound change.

Experiment with shifts from loose sentences to periodic sentences in your own writing, especially during revision. Avoid the singsong monotony of a long series of loose sentences, particularly a series containing coordinate clauses joined by <u>conjunctions</u>. Subordinating some thoughts to others makes your sentences more interesting. See also <u>sentence construction</u>.

sequential method of development

The sequential, or step-by-step, <u>method of development</u> is especially effective for explaining a process or describing a mechanism in operation. It is also the logical method for writing <u>instructions</u>, as shown in Figure S–1.

    The main advantage of the sequential method of development is that it is easy to follow because the steps correspond to the elements of the process or operation being described. The disadvantages are that it can become monotonous and does not lend itself well to achieving <u>emphasis</u>.

    Practically all methods of development have elements of sequence. The <u>chronological method of development,</u> for example, is also sequential: To describe a trip chronologically, from beginning to end, is also to describe it sequentially.

S

MANUALLY PROCESSING FILM

Developing. In total darkness, load the film on the spindle and enclose it in the developing tank. Be careful not to allow the film to touch the tank walls or other film. Add the developing solution, turn the lights on, and set the timer for seven minutes. Agitate for five seconds initially and then every half minute.

Stopping. When the timer sounds, drain the developing solution from the tank and add the stop bath. Agitate continuously for 30 seconds.

Fixing. Drain the stop bath and add the fixing solution. Allow the film to remain in the fixing solution for two to four minutes. Agitate for five seconds initially and then every half minute.

Washing. Remove the tank top and wash the film for at least 30 seconds under running water.

Drying. Suspend the film from a hanger to dry. It is generally advisable to place a drip pan below the rack. Sponge the film gently to remove excess water. Allow the film to dry completely.

FIGURE S-1. Sequential Method of Development

service

When used as a <u>verb</u>, *service* means "keep up or maintain" as well as "repair." (Our company will *service* your equipment.) If you mean "providing a more general benefit," use *serve*.

- Our company ~~services~~ the northwest area of the state.

  serves

set / sit

Sit is an intransitive <u>verb</u>; it does not, therefore, require an <u>object</u>. (I *sit* by a window in the office.) Its past <u>tense</u> is *sat*. (We *sat* around the conference table.) *Set* is usually a transitive verb, meaning "put or place," "establish," or "harden." Its past tense is *set*.

- Please *set* the trophy on the shelf.
- The jeweler *set* the stone beautifully.
- Can we *set* a date for the tests?
- The high temperature *sets* the epoxy quickly.

Set is occasionally an intransitive verb.

- The sun *sets* a little earlier each day.

sexism in language (*see* biased language)

shall / will

Although traditionally *shall* was used to express the future tense with *I* and *we, will* is generally accepted with all persons. *Shall* is commonly used today only in questions requesting an opinion or a preference ("Shall we go?") rather than a prediction ("Will we go?"). It is also used in statements expressing determination ("I shall return") or in a formal regulation ("Applicants shall provide a proof of certification").

sic

Latin for "thus," *sic* is used in <u>quotations</u> to indicate that the writer has quoted the material exactly as it appears in the original source. It is most often used when the original material contains an obvious error or might be questioned in some other way. *Sic* is placed in <u>brackets</u>.

- In the textbook *Basic Astronomy,* the author notes that the "earth does not revolve around the son [*sic*] at a constant rate."

similes (*see* figures of speech)

slashes

The *slash* (/)—called a variety of names, including *slant line, diagonal, virgule, bar, solidus,* and *shilling*—both separates and shows omission.

The slash is often used to separate parts of addresses in continuous writing.

- The return address on the envelope was Ms. Rose Howard/3555 Market St./Brookville/Ohio/45424.

The slash can indicate alternative items or numbers.

- David's telephone number is 549-2278/2335.

The slash often indicates omitted words and letters.

- miles/hour (miles per hour), w/o (without)

In fractions, the slash separates the numerator from the denominator (3/4 for three-quarters).

The slash also separates items in the URL (Uniform Resource Locator) address for sites on the World Wide Web <http://www.bedford stmartins.com/alred>.

In informal writing, the slash separates day from month and month from year in dates (5/29/03). Do not use this form for <u>international correspondence</u> because the order of the items varies (29 May 2003).

so / so that / such

So is often vague or weak and should be avoided if another word would be more precise.

- *Because she*
  ~~She~~ reads faster, ~~so~~ she finished before I did.
  ^

Another problem occurs with the phrase *so that,* which should never be replaced with *so* or *such that.*

- The report should be written ~~such that~~ *so that* it can be copied.
  ^

Such, an <u>adjective</u> meaning "of this or that kind," should never be used as a <u>pronoun</u>.

- Our company provides on-site day care, but I do not anticipate
  it.
  using ~~such.~~
  ^

some / somewhat

When *some* functions as an indefinite <u>pronoun</u> for a plural count <u>noun</u> or as an indefinite <u>adjective</u> modifying a plural count noun, use a plural <u>verb</u>.

- *Some* of us *are* prepared to work overtime.

- *Some* people *are* more productive than others.

Some is singular, however, when used with mass nouns.

- *Some* sand *has* trickled through the crack.

- Most of the water evaporated, but *some remains.*

When *some* is used as an <u>adjective</u> or a <u>pronoun</u> meaning "an undetermined quantity" or "certain unspecified persons," it should be replaced by the <u>adverb</u> *somewhat*, which means "to some extent."

- His writing has improved ~~some.~~ *somewhat.*

some time / sometime / sometimes

Some time refers to a duration of time. (We waited for *some time* before calling the customer.) *Sometime* refers to an unknown or unspecified time. (We will visit with you *sometime*.) *Sometimes* refers to occasional occurrences at unspecified times. (He *sometimes* visits the branch offices.)

spatial method of development

The spatial <u>method of development</u> describes an object or a process according to the physical arrangement of its features. Depending on the subject, you describe its features from bottom to top, side to side, east to west, outside to inside, and so on. Descriptions of this kind rely mainly on dimension (height, width, length), direction (up, down, north, south), shape (rectangular, square, semicircular), and proportion (one-half, two-thirds). Features are described in relation to one another or to their surroundings, as illustrated in Figure S–2 on page 522.

The spatial method of development is commonly used in descriptions of laboratory equipment, <u>proposals</u> for landscape work, construction-site <u>progress and activity reports</u>, and, in combination with a step-by-step sequence, in many types of <u>instructions</u>. The description for a house inspection in Figure S-2 relies on a bottom-to-top, clockwise (south to west to north to east) sequence, beginning with the front door.

S

Interior First Floor of a Two-Story House

The front door faces south and opens into a hallway 7 feet deep and 10 feet wide. At the end of the hallway is the stairwell, which begins on the right (east) side of the hallway, rises five steps to a landing, and reverses direction at the left (west) side of the hallway. To the left (west) of the hallway is the dining room, which measures 15 feet along its southern exposure and 10 feet along its western exposure. Directly to the north of the dining room is the kitchen, which measures 10 feet along its western exposure and 15 feet along its northern exposure. To the east of the kitchen, along the northern side of the house, is a bathroom that measures 10 feet (west to east) by 5 feet. Parallel to the bathroom is a passageway with the same dimensions as the bathroom and leading from the kitchen to the living room. The living room, which measures 15 feet (west to east) by 20 feet (north to south), occupies the entire eastern end of the floor.

FIGURE S-2. Spatial Method of Development

specie / species

Specie means "coined money" or "in coin." (The salvage included gold coin and other *specie*.) *Species* means a category of animals, plants, or things having some of the same characteristics or qualities. (The wolf is a member of the canine *species*.) *Species* is the correct spelling for both the singular and the plural. (Many animal *species* are represented in the Arctic.)

specific-to-general method of development
(*see* general and specific method of development)

specifications

A *specification* is a detailed and exact statement that prescribes materials, dimensions, and quality of something to be built, installed, or manufactured.

The two broad categories of specifications—industrial and government—both require precision. A specification must be written clearly and precisely and state *explicitly* what is needed. Because of the stringent requirements of specifications, careful <u>research</u> and <u>preparation</u> are especially important before you begin to write, as is careful <u>revision</u> after you have completed the draft. See also <u>clarity</u> and <u>ambiguity</u>.

Industrial Specifications

Industrial specifications are used, for example, in software development, in which there are no engineering drawings or other means of documentation. An industrial specification is a permanent record that documents the item being developed so that it can be maintained by someone other than the person who designed it and provides detailed technical information about the item to all who need it (engineers, technical writers, technical instructors, etc.).

The industrial specification describes a planned project, a newly completed project, or an old project. All three types must contain detailed technical descriptions of all aspects of the item: what was done, how it was done, what is required to use the item, how it is used, what is its function, who would use it, and so on.

Government Specifications

Government agencies are required by law to contract for equipment strictly according to definitions provided in formal specifications. A government specification is a precise definition of exactly what the contractor is to provide. In addition to a technical description of the item to be purchased, the specification normally includes an estimated cost; an estimated delivery date; and standards for the design, manufacture, quality, testing, training of government employees, governing codes, inspection, and delivery of the item.

Government specifications contain details on the scope of the project; documents the contractor is required to furnish with the device; required product characteristics and functional performance of the device; required tests, test equipment, and test procedures; required preparations for delivery; notes; and <u>appendixes</u>. Government specifications are often used to prescribe the content and deadline for government <u>proposals</u> submitted by vendors bidding on a project.

S

🌟 **WEB LINK GOVERNMENT SPECIFICATIONS**

The Business and Contracting Opportunities portion of the U.S. Government Printing Office (GPO) Web site provides detailed military and nonmilitary specifications. See <*www.bedfordstmartins.com/alred*> and select *Links for Technical Writing.*

spelling

The use of a spell checker is crucial; however, it will not catch all spelling mistakes. It cannot detect a spelling error if the error results in a valid word; for example, if you mean *to* but inadvertently type *too,* the spell checker will not detect the error. Likewise, spell checkers will not detect errors in the names of people, places, and organizations. Because the misspelling of someone's personal name or the name of an organization will damage your credibility, careful <u>proofreading</u> is essential.

strata / stratum

Strata is the plural form of *stratum,* meaning a "layer of material."

- The land's *strata* are exposed by erosion.
- Each *stratum* is clearly visible in the cliff.

style

A dictionary definition of *style* is "the way in which something is said or done, as distinguished from its substance." Writers' styles are determined by the way writers think and transfer their thoughts to paper — the way they use words, sentences, images, <u>figures of speech,</u> and so on.

A writer's style is the way his or her language functions in particular situations. For example, an <u>e-mail</u> to a friend would be relaxed, even chatty, in <u>tone,</u> whereas a job <u>application letter</u> would be more restrained and formal. Obviously, the style appropriate to one situation would not be appropriate to the other. In both situations, the <u>readers</u> and the <u>purpose</u> determine the manner or style the writer adopts. Beyond an individual's personal style, various kinds of writing have distinct stylistic traits, such as <u>technical writing style.</u>

Standard English can be divided into two broad categories of style — formal and informal — according to how it functions in certain situations. Understanding the distinction between formal and informal writing styles helps writers use the appropriate style. We must recognize, however, that no clear-cut line divides the two categories and that some writing may call for a combination of the two.

Formal Writing Style

A formal writing style can perhaps best be defined by pointing to certain material that is clearly formal, such as scholarly and scientific articles in professional journals, lectures read at meetings of professional societies, and legal documents. Material written in a formal style is usually the work of a specialist writing to other specialists or writing that embodies laws or regulations. As a result, the vocabulary is specialized and precise. The writer's tone is impersonal and objective because the subject matter looms larger in the writing than does the author's personality. (See point of view.) A formal writing style does not use contractions, slang, or dialect. (See English, varieties of.) Because the material generally examines complex ideas, the sentence construction may be elaborate.

Formal writing need not be dull and lifeless. By using such techniques as the active voice whenever possible, sentence variety, and subordination, a writer can make formal writing lively and interesting, especially if the subject matter is inherently interesting to the reader.

- Although a knowledge of the morphological chemical constitution of cells is necessary to the proper understanding of living things, in the final analysis it is the activities of their cells that distinguish organisms from all other objects in the world. Many of these activities differ greatly among the various types of living things, but some of the basic sorts are shared by all, at least in their essentials. It is these fundamental actions with which we are concerned here. They fall into two major groups — those that are characteristic of the cell in the *steady state,* that is, in the normally functioning cell not engaged in reproducing itself, and those that occur during the process of *cellular reproduction.*
  —Lawrence S. Dillon, *The Principles of Life Sciences*

Whether you should use a formal style in a particular instance depends on your readers and purpose. When writers attempt to force a formal style when it should not be used, their writing is likely to fall victim to affectation, awkwardness, and gobbledygook.

S

Informal Writing Style

An informal writing style is a relaxed and colloquial way of writing standard English. It is the style found in most private letters and in some business correspondence, memos, e-mail, nonfiction books of general interest, and mass-circulation magazines. There is less distance between the writer and the reader because the tone is more personal than in a formal writing style. Contractions and elliptical constructions are common. Consider the following passage, written in an informal style, from a nonfiction book.

- Business, like art and science, has been revealed and conceived through the intellect and imagination of people, and it develops or declines because of the intellect and imagination of people.

  In fact, there is no business; there are only people. Business exists only *among* people and *for* people.

  Seems simple enough, and it applies to every aspect of business, but not enough businesspeople seem to get it.

  Reading the economic forecasts and the indicators and the ratios and the rates of this or that, someone from another planet might actually believe that there really are invisible hands at work in the marketplace.

  It's easy to forget what the measurements are measuring. Every number—from productivity rates to salaries—is just a device contrived by people to measure the results of the enterprise of other people. For managers, the most important job is not measurement but motivation. And you can't motivate numbers.

  —James A. Autry, *Love and Profit: The Art of Caring Leadership*

As the example illustrates, the vocabulary of an informal writing style is made up of generally familiar rather than unfamiliar words and expressions, although slang and dialect are usually avoided. An informal style approximates the cadence and structure of spoken English while conforming to the grammatical conventions of written English.

Writers who consciously attempt to create a style usually defeat the purpose. Attempting to impress readers with a flashy writing style can lead to affectation; attempting to impress them with scientific objectivity can produce a style that is dull and lifeless. Technical writing need be neither affected nor dull. It can and should be simple, clear, direct, even interesting—the key is to master basic writing skills and always to keep your readers in mind. What will be both informative and interesting to your **audience**? When that question is uppermost in your mind as you apply the steps of the writing process, you will achieve an interesting and informative writing style. See "Five Steps to Successful Writing."

S

Writer's Checklist: Developing an Effective Style

☑ Use the active voice—not exclusively but as much as possible without becoming awkward or illogical.

☑ Use **parallel structure** whenever a sentence presents two or more thoughts of equal importance.

☑ Use a variety of sentence structures to avoid a monotonous style.

☑ Avoid stating positive thoughts in negative terms (write "40 percent responded" instead of "60 percent failed to respond"). See also **positive writing** and **ethics in writing**.

☑ Concentrate on achieving the proper balance between **emphasis** and subordination.

subordination

Writers use subordination to show, by the structure of a sentence, the appropriate relationship between ideas of unequal importance by subordinating the less-important ideas to the more-important ideas.

- Pacific Enterprises now employs 500 people. It was founded just three years ago. [The two ideas are equally important.]

- Pacific Enterprises, *which now employs 500 people*, was founded just three years ago. [The number of employees is subordinated.]

- Pacific Enterprises, *which was founded just three years ago*, now employs 500 people. [The founding date is subordinated.]

Effective subordination can be used to achieve **sentence variety**, **conciseness**, and **emphasis**. For example, consider the following sentences.

DEPENDENT CLAUSE	The regional manager's report, *which covered five pages*, was carefully illustrated.
SINGLE MODIFIER	The regional manager's report, *covering five pages*, was carefully illustrated.
SINGLE MODIFIER	The regional manager's *five-page* report was carefully illustrated.

Use a coordinating **conjunction** (*and, but, for, nor, or, so, yet*) to concede that an opposite or balancing fact is true; however, a subordinating conjunction (*although, since, while*) can often make the point more smoothly.

- *Although* their lab is well-funded, ours is better equipped.

The relationship between a conditional statement and a statement of consequences is clearer if the condition is expressed as a subordinate clause.

- Because the data was inaccurate, the client was angry.

S

Subordinating conjunctions (*because, if, while, when, though*) achieve subordination effectively.

- A buildup of deposits is impossible *because* the pipes are flushed with water every day.

Relative <u>pronouns</u> (*who, whom, which, that*) can be used effectively to combine related ideas that would be less smooth as independent clauses or sentences.

- The generator, *which* is the most common source of electric current, uses mechanical energy to produce electricity.

Avoid overlapping subordinate constructions that depend on the preceding construction. Overlapping can make the relationship between a relative pronoun and its antecedent less clear.

OVERLAPPING	Shock, *which* often accompanies severe injuries and infections, is a failure of the circulation, *which* is marked by a fall in blood pressure *that* initially affects the skin (*which* explains pallor) and later the vital organs such as the kidneys and brain.
CLEAR	Shock often accompanies severe injuries and infections. It is a failure of the circulation, first to the skin (thus producing pallor) and later to the vital organs like the kidneys and the brain.

substantives

A *substantive* is a <u>noun</u> or noun equivalent (<u>pronoun</u>, <u>verbal</u>, noun <u>phrase</u>, or noun <u>clause</u>) that functions as the subject of a sentence.

NOUN	The *report* is due today.
PRONOUN	*We* must finish the project on schedule.
GERUND	*Drilling* is expensive.
INFINITIVE	*To succeed* will require hard work.
NOUN PHRASE	*Several local college graduates* applied for the job.
NOUN CLAUSE	*What I think* is unimportant.

S

suffixes

A *suffix* is a letter or letters added to the end of a word to change its meaning in some way. Suffixes can change the part of speech of a word.

NO SUFFIX	The market survey was *thorough*. [adjective]
SUFFIX	The *thoroughness* should be obvious to all. [noun]
SUFFIX	The report *thoroughly* described the problem. [adverb]

The suffix *-like* is sometimes added to <u>nouns</u> to make them into <u>adjectives</u>. The resulting compound word is hyphenated only if it is unusual or might not immediately be clear (childlike, lifelike, dictionary-like, robot-like).

surveys (*see* questionnaires)

symbols

From highway signs to <u>mathematical equations</u>, people communicate with written symbols. When a symbol seems appropriate in your writing, either be certain that your <u>readers</u> understand its meaning or place the symbol in <u>parentheses</u> following the spelled-out term the first time it appears. Never use a symbol when readers would more readily understand the full term. See also <u>abbreviations</u>, <u>numbers</u>, and <u>global graphics</u>.

synonyms

S

A *synonym* is a word that means nearly the same thing as another word does (seller, vendor, supplier). The dictionary definitions of synonyms are very similar, but the connotations may differ. (A *seller* may be the same thing as a *supplier*, but the term *supplier* does not suggest a retail transaction as strongly as *seller* does.)

Do not try to impress your <u>readers</u> by finding fancy or obscure synonyms in a thesaurus; the result is likely to be <u>affectation</u>. See also <u>connotation/denotation</u> and <u>antonyms</u>.

syntax

Syntax refers to the way that words, phrases, and clauses are combined to form sentences. In English, the most common structure is the subject-verb-object pattern. For more information about the word order of sentences, see <u>sentence construction</u>, <u>sentence faults</u>, <u>sentence fragments</u>, and <u>sentence variety</u>.

S

T

tables of contents

A *table of contents* is typically included in a document longer than ten pages. It previews what the work contains and how it is organized, and it allows **readers** looking for specific information to locate sections quickly and easily. See also **outlining**.

When creating a table of contents, use the major **headings** and subheadings of your document exactly as they appear in the text, as shown in the entry **formal reports**. (See Figure F-8 on page 214.) Note that the table of contents is typically placed in the front matter so that it follows the title page and **abstract**, and precedes the list of tables or figures, the **foreword**, and the preface.

tables

A table can present data, such as statistics, more concisely than text and more accurately than **graphs**. A table facilitates comparisons among data by organizing it into rows and columns. However, overall trends are more easily conveyed in charts, graphs, and other **visuals**.

Table Elements

Tables typically include the elements shown in Figure T–1 on page 532.

Table Number.   Table numbers are usually Arabic and should be assigned sequentially to the tables throughout the document.

Table Title.   The title, which is normally placed just above the table, should describe concisely what the table represents.

Box Head.   The box head contains the column headings, which should be brief but descriptive. Units of measurement, where necessary, should be either specified as part of the heading or enclosed in

Table number → Table 1.   Estimated Emissions from Electric Power ← Table title
Generation (tons per gigawatthour)

Box head

Fuel	Sulphur Dioxide	Nitrogen Oxides	Particulate Matter	Carbon Dioxide	Volatile Organic Compounds
Eastern coal	1.74	2.90	0.10	1,000	0.06
Western coal	0.81	2.20	0.06	1,039	0.09
Gas	0.003	0.57	0.02	640	0.05
Biomass	0.06	1.25	0.11	0*	0.61
Oil	0.51	0.63	0.02	840	0.03
Wind	0	0	0	0	0
Geothermal	0	0	0	0	0
Hydro	0	0	0	0	0
Solar	0	0	0	0	0
Nuclear	0	0	0	0	0

Column headings

Stub

Body

Rule

Footnote → *Net emissions.
Source line → SOURCE: Department of Energy

FIGURE T–1. Elements of a Table

parentheses beneath the heading. Standard <u>abbreviations</u> and <u>symbols</u> are acceptable. Avoid vertical lettering whenever possible.

Stub.   The stub, the left vertical column of a table, lists the items about which information is given in the body of the table.

Body.   The body comprises the data below the column headings and to the right of the stub. Within the body, arrange columns so that the items to be compared appear in adjacent rows and columns. Where no information exists for a specific item, substitute a row of dots or a dash to acknowledge the gap.

Rules.   Rules are the lines that separate the table into its various parts. Horizontal rules are placed below the title, below the body of the table, and between the column headings and the body of the table. Tables should be open at the sides. The columns within the table may be separated by vertical rules only if such lines aid clarity.

Footnotes.   Footnotes are used for explanations of individual items in the table. Symbols (such as * and †) or lowercase letters (sometimes in parentheses) rather than numbers are ordinarily used to key table footnotes because numbers might be mistaken for numerical data or could be confused with the numbering system for text footnotes. See also <u>documenting sources</u>.

Source Line.   The source line identifies where the data originated. When a source line is appropriate, it appears below the table. Many organizations place the source line below the footnotes. See also <u>copyright</u> and <u>plagiarism</u>.

Continued Tables.   When a table must be divided so that it can be continued on another page, repeat the column headings and give the table number at the head of each new page with a "continued" label (for example, "Table 3, *continued*").

Informal Tables

To list relatively few items that would be easier for the reader to grasp in tabular form, you can use an informal table, as long as you introduce it properly. Although informal tables do not need titles or table numbers to identify them, they do require column headings that accurately describe the information listed, as shown in Figure T–2.

The sound-intensity levels (decibels) for the three frequency bands (in hertz) were determined to be the following:

Frequency Band (Hz)	Decibels
600–1200	68
1200–2400	62
2400–4800	53

FIGURE T–2. Informal Table

technical manuals (*see* manuals)

technical writing style

The goal of technical writing is to enable <u>readers</u> to use a technology or understand a process or concept. Because the subject matter is more important than the writer's voice, technical writing style uses an objective, not subjective, <u>tone</u>. The writing <u>style</u> is direct and utilitarian, emphasizing exactness and <u>clarity</u> rather than elegance or allusiveness. A technical writer uses figurative language only when a <u>figure of speech</u> would facilitate understanding.

Technical writing is often—but not always—aimed at readers who are not experts in the subject, such as consumers or employees learning to operate unfamiliar equipment. Because such audiences are inexperienced and the procedures described may involve hazardous material or equipment, clarity becomes an ethical as well as a stylistic concern. (See also ethics in writing.) Figure T–3 is an excerpt from a technical manual that instructs eye-care specialists about operating testing equipment. As the figure illustrates, visuals as well as layout and design enhance clarity, providing the reader with both text and visual information.

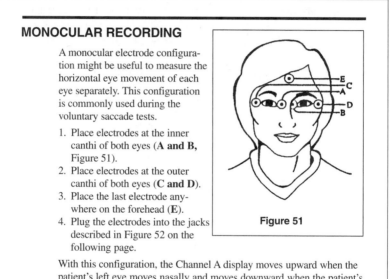

MONOCULAR RECORDING

A monocular electrode configuration might be useful to measure the horizontal eye movement of each eye separately. This configuration is commonly used during the voluntary saccade tests.

1. Place electrodes at the inner canthi of both eyes (**A and B,** Figure 51).
2. Place electrodes at the outer canthi of both eyes (**C and D**).
3. Place the last electrode anywhere on the forehead (**E**).
4. Plug the electrodes into the jacks described in Figure 52 on the following page.

Figure 51

With this configuration, the Channel A display moves upward when the patient's left eye moves nasally and moves downward when the patient's left eye moves temporally. The Channel B display moves upward when the patient's right eye moves temporally and moves downward when the patient's right eye moves nasally.

NOTE: *During all testing, the Nystar Plus identifies and eliminates eye blinks without vertical electrode recordings.*

T

FIGURE T–3. Technical Writing Style

Technical writing style may use a technical vocabulary appropriate for the reader, as Figure T–3 shows (*monocular electrode, saccade*), but it avoids affectation. Good technical writing also avoids overusing the passive voice. See also instructions, organization, and process explanation.

telegraphic style

Telegraphic style condenses writing by omitting <u>articles</u>, <u>pronouns</u>, <u>conjunctions</u>, and <u>transitions</u>. Although <u>conciseness</u> is important, especially in <u>instructions</u>, writers sometimes try to achieve conciseness by omitting necessary words. Telegraphic style forces <u>readers</u> to supply the missing words mentally, thus creating the potential for misunderstandings. Compare the following two passages and notice how much easier the revised version reads (the added words are italicized).

TELEGRAPHIC	Take following action when treating serious burns. Remove loose clothing on or near burn. Cover injury with clean dressing and wash area around burn. Secure dressing with tape. Separate fingers/toes with gauze/cloth to prevent sticking. Do not apply medication unless doctor prescribes.
CLEAR	Take *the* following action when treating *a* serious burn. Remove *any* loose clothing on or near *the* burn. Cover *the* injury with *a* clean dressing and wash *the* area around *the* burn. *Then* secure *the* dressing with tape. Separate *the* fingers *or* toes with gauze *or* cloth to prevent *them from* sticking *together*. Do not apply medication unless *a* doctor prescribes *it*.

Telegraphic style can also produce <u>ambiguity</u>, as the following example, written for a crane operator, demonstrates.

AMBIGUOUS	Grasp knob and adjust lever before raising boom. [Does this sentence mean that the reader should *adjust the lever* or also *grasp an "adjust lever"*?]
CLEAR	Grasp *the* knob and adjust *the* lever before raising *the* boom.

Although you may save yourself work by writing telegraphically, your readers will have to work that much harder to decipher your meaning.

T

tenant / tenet

A *tenant* is one who holds or temporarily occupies a property owned by another person. (The *tenant* was upset by the rent increase.) A *tenet* is an opinion or belief held by a person, an organization, or a system. (Competition is a central *tenet* of capitalism.)

tense

Tense is the grammatical term for <u>verb</u> forms that indicate time distinctions. There are six tenses in English: past, past perfect, present, present perfect, future, and future perfect. Each tense also has a corresponding progressive form.

TENSE	BASIC FORM	PROGRESSIVE FORM
Past	I began	I was beginning
Past perfect	I had begun	I had been beginning
Present	I begin	I am beginning
Present perfect	I have begun	I have been beginning
Future	I will begin	I will be beginning
Future perfect	I will have begun	I will have been beginning

Perfect tenses allow you to express a prior action or condition that continues in a present, past, or future time.

PRESENT PERFECT	*I have begun* to compile the survey results and will continue for the rest of the month.
PAST PERFECT	*I had begun* to read the manual when the lights went out.
FUTURE PERFECT	*I will have begun* this project by the time funds are allocated.

Progressive tenses allow you to describe some ongoing action or condition in the present, past, or future.

PRESENT PROGRESSIVE	*I am beginning* to be concerned that we will not meet the deadline.
PAST PROGRESSIVE	*I was beginning* to think we would not finish by the deadline.
FUTURE PROGRESSIVE	*I will be requesting* a leave of absence when this project is finished.

T

Past Tense

The simple *past tense* indicates that an action took place in its entirety in the past. The past tense is usually formed by adding -*d* or -*ed* to the root form of the verb. (We *closed* the office early yesterday.)

Past Perfect Tense

The *past perfect tense* (also called *pluperfect*) indicates that one past event preceded another. It is formed by combining the helping verb *had* with the past participle form of the main verb. (He *had finished* by the time I arrived.)

Present Tense

The simple *present tense* represents action occurring in the present, without any indication of time duration. (I *ride* the train.)

A general truth is always expressed in the present tense. (He believes the saying "time *heals* all wounds.") The present tense can be used to present actions or conditions that have no time restrictions. (Water *boils* at 212 degrees Fahrenheit.)

The present tense can be used to indicate habitual action. (I *pass* the coffee shop every day.) The present tense can be used as a "historical present" to make things that occurred in the past more vivid.

- "FDA Approves New Cancer Drug"
  [In this news headline, *Approves* is in the present tense.]

Present Perfect Tense

The *present perfect tense* describes something from the recent past that has a bearing on the present—a period of time before the present but after the simple past. The present perfect tense is formed by combining a form of the helping verb *have* with the past participle form of the main verb. (We *have finished* the draft and can now revise it.)

Future Tense

The simple *future tense* indicates a time that will occur after the present. It uses the helping verb *will* (or *shall*) plus the main verb. (I *will finish* the job tomorrow.)

Do not use the future tense needlessly; doing so merely adds complexity.

T

- This system ~~will be~~ *is* explained on page 3.

moves
- When you press this button, the hoist ~~will move~~ the plate into position.
  ∧

Future Perfect Tense

The *future perfect tense* indicates action that will have been completed at a future time. It is formed by linking the helping verbs *will have* to the past participle form of the main verb. (She *will have driven* 1,400 miles by the time she returns.)

 TIPS FOR USING THE PROGRESSIVE FORM

The progressive form of the verb is composed of two features: a form of the helping verb *be* and the *ing* form of the base verb.

> **PRESENT PROGRESSIVE**   I *am rewriting* the memo.
>
> **PAST PROGRESSIVE**   I *was rewriting* the memo for several days.
>
> **FUTURE PROGRESSIVE**   I *will be rewriting* that memo forever.

The present progressive is used in three ways:

1. To refer to an action that is in progress at the moment of speaking or writing:
   - The parliamentarian *is taking* the meeting minutes.
2. To highlight that a state or action is not permanent:
   - The office temp *is helping* us for a few weeks.
3. To express future plans:
   - The summer intern *is leaving* to return to school this Friday.

The past progressive is used to refer to a continuing action or condition in the past, usually with specified limits.

- I *was failing* calculus until I got a tutor.

The future progressive is used to refer to a continuous action or condition in the future.

- We *will be monitoring* his condition all night.

Verbs that express mental activity (*believe, know, see,* and so on) are generally not used in the progressive.

believe
- I ~~am believing~~ the defendant's testimony.
  ∧

Shift in Tense

Be consistent in your use of tense. The only legitimate shift in tense records a real change in time. Illogical shifts in tense will only confuse your readers.

- Before he *installed* the circuit, the technician ~~cleans~~ the contacts.
  cleaned

---

test reports

The test report differs from the more formal <u>laboratory report</u> in both size and <u>scope</u>. Considerably smaller, less formal, and more routine than the laboratory report, the test report can be presented as a <u>memo</u> or a letter, depending on whether its recipient is inside or outside the organization. Either way, the <u>report</u> should have a subject line at the beginning to identify the test being discussed. See also <u>correspondence</u>.

The opening of a test report should state the test's purpose, unless it is obvious. The body of the report presents the data and, if relevant, describes procedures used to conduct the test. State the results of the test and, if necessary, interpret them. Conclude the report with any recommendations made as a result of the test. Figure T–4 on page 540 shows a typical test report.

---

that / which / who

The word *that* is often overused.

- You will note *that,* ~~as you assume greater responsibility,~~ *that* your
  benefits will increase ~~accordingly.~~
  as you assume greater responsibility.

However, include *that* in a sentence if it avoids <u>ambiguity</u> or it improves the <u>pace</u>. (See also <u>conciseness/wordiness</u>.)

- Some designers fail to appreciate the human beings who operate
  that
  equipment constitute an important safety system.

Use *which,* not *that,* with nonrestrictive clauses (clauses that do not change the meaning of the basic sentence). See also <u>restrictive and nonrestrictive elements</u>.

Biospherics, Inc.

4928 Wyaconda Road
Rockville, MD 20852

Phone: 301-598-9011
Fax: 301-598-9570

September 9, 2003

Mr. Leon Hite, Administrator
The Angle Company, Inc.
1869 Slauson Boulevard
Waynesville, VA 23927

Subject: Monitoring Airborne Asbestos at the Route 66 Site

Dear Mr. Hite:

On August 29, Biospherics, Inc., performed asbestos-in-air monitoring at your Route 66 construction site, near Front Royal, Virginia. Six people and three construction areas were monitored.

All monitoring and analyses were performed in accordance with "Occupational Exposure to Asbestos," U.S. Department of Health and Human Services, Public Health Service, National Institute for Occupational Safety and Health, 1999. Each worker or area was fitted with a battery-powered personal sampler pump operating at a flow rate of approximately two liters per minute. The airborne asbestos was collected on a 37-mm Millipore-type AA filter mounted in an open-face filter holder. Samples were collected over an 8-hour period.

In all cases, the workers and areas monitored were exposed to levels of asbestos fibers well below the standard set by OSHA. The highest exposure found was that of a driller exposed to 0.21 fibers per cubic centimeter. The driller's sample was analyzed by scanning electron microscopy followed by energy dispersive X-ray techniques that identify the chemical nature of each fiber, to identify the fibers as asbestos or other fiber types. Results from these analyses show that the fibers present were tremolite asbestos. No nonasbestos fibers were found.

If you need more details, please let me know.

Yours truly,

Gary Geirelach

Gary Geirelach
Chemist
gg2@Lios.org

FIGURE T–4.   Test Report

| NONRESTRICTIVE | After John left the facility, *which* is one of the best in the region, he came directly to my office. |
| RESTRICTIVE | Companies *that* diversify usually succeed. |

That and *which* should refer to animals and things; *who* should refer to people.

- Companies *that* fund basic research must not expect immediate results.
- The jet stream, *which* is approximately eight miles above the earth, blows at an average of 64 miles per hour.
- Diane Stoltzfus, *who* retires tomorrow, has worked for the company for 20 years.

there / their / they're

There is an <u>expletive</u> (a word that fills the position of another word, phrase, or clause) or an <u>adverb</u>.

| EXPLETIVE | *There* were more than 1,500 people at the conference. |
| ADVERB | More than 1,500 people were *there*. |

Their is the <u>possessive case</u> form of *they*. (Managers should check *their* e-mail regularly.) *They're* is a <u>contraction</u> of *they are*. (Clients tell us *they're* pleased with our services.)

thesaurus

A *thesaurus* is a book or electronic file of words and their <u>synonyms</u> and <u>antonyms</u>, arranged or retrievable by categories. Thoughtfully used, a thesaurus can help you with <u>word choice</u> during the <u>revision</u> phase of the writing process. However, the variety of words it offers may tempt you to choose inappropriate or obscure synonyms just because they are available. Use a thesaurus only to clarify your meaning, not to impress your <u>readers</u>. (See also <u>affectation</u>.) Never use a word unless you are sure of its meanings; <u>connotations</u> of the word that might be unknown to you could mislead your readers.

titles

Titles are important because many __readers__ decide to read a document, such as a __trade journal article__, based on the title. The title of a __report__ or other document should both state its topic and indicate its __scope__ and __purpose__, as in the following example.

- Effects of 60-Hertz Electric Fields on Embryo Chick Development, Growth, and Behavior

A title should be concise but not so short that it is not specific. For example, the title "Electric Fields and Living Organisms" announces the topic of a report, but it does not answer important questions, such as "What is the relationship between electric fields and living organisms?" and "What stage of the organism's life cycle does the report discuss?"

Writer's Checklist: Creating Titles

☑ Avoid titles with redundancies that begin with "Notes on," "Studies on," or "A Report on." However, works like annual reports or __feasibility reports__ should be identified as such in the title because this information specifies the purpose and scope of the report.

☑ Do not indicate dates in the title of a periodic or __progress report__; put them in a subtitle ("Quarterly Report on the Hospital Admission Rates: January–March 2003").

☑ Use the active __voice__ as much as possible.

☑ Do not put titles in sentence form, except for titles that ask a question.

☑ Avoid including __abbreviations__, chemical formulas, and the like, in your title unless the work is addressed exclusively to specialists in the field.

☑ For multivolume publications, repeat the title on each volume and include the subtitle and number of each volume.

For guidelines on how to capitalize titles and when to use __italics__ and __quotation marks__, see those entries and __capitalization__.

T

to / too / two

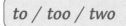

To, *too*, and *two* are confused only because they sound alike. *To* is used as a __preposition__ or to mark an infinitive. See __verbs__.

- Send the report *to* the district manager. [preposition]
- I do not wish *to* go. [mark of the infinitive]

Too is an <u>adverb</u> meaning "excessively" or "also."

- The price was *too* high. [excessively]
- I, *too*, thought it was high. [also]

Two is a number (*two* buildings, *two* concepts).

tone

Tone is the writer's attitude toward the subject and his or her <u>readers</u>. In workplace writing, tone may range widely—depending on the <u>purpose</u>, situation, context, <u>audience</u>, and even the medium of a communication. For example, in an <u>e-mail</u> message to be read only by an associate who is also a friend, your tone might be casual.

- Your proposal to Smith and Kline is super. We'll just need to hammer out the schedule. If we get the contract, I owe you lunch!

In a <u>memo</u> to your manager or superior, however, your tone might be more formal and respectful.

- I think your proposal to Smith and Kline is excellent. I have marked a couple of places where I'm concerned that we are committing ourselves to a schedule that we might not be able to keep. If I can help in any other way, please let me know.

In a message that serves as a <u>report</u> to numerous readers, the tone would be professional, without a casual <u>style</u> that could be misinterpreted.

- The Smith and Kline proposal appears complete and thorough, based on our department's evaluation. Several small revisions, however, would ensure that the company is not committing itself to an unrealistic schedule. These are marked on the copy of the report being circulated.

The <u>word choice</u>, the <u>introduction</u>, and even the <u>title</u> contribute to the overall tone of your document. For instance, a title such as "Ecological Consequences of Diminishing Water Resources in California" clearly sets a different tone from "What Happens When We've Drained California Dry?" The first title would be appropriate for a report; the second title would be appropriate for a popular magazine or <u>newsletter article</u>. See also <u>correspondence</u>.

T

topics

On the job, the topic of a writing project is usually determined by need. In a college writing course, you may have to select your own topic. If you do, keep the following points in mind.

- Select a topic that interests you so that you will maintain your enthusiasm.
- Select a topic you can **research** adequately with the facilities available.
- Limit your topic so its **scope** is small enough to handle in the time you are given.

For example, you might be enthusiastic about a topic like "Environmentalism," but it would be too broad. However, the topic "Water Purification Options for Cities" would be broad enough to write about and could offer a manageable research project. See also **formal reports** and **proposals**.

trade journal articles

A *trade journal article* is an article written on a specific subject for a professional periodical. Such periodicals, commonly known as trade journals (or professional or scholarly journals), are often the official publications of professional societies. *Technical Communication,* for example, is an official voice of the Society for Technical Communication. Other professional publications include *Electrical Engineering Review, Chemical Engineering, Nucleonics Week,* and hundreds of similar titles. Professional staff people, such as engineers, scientists, educators, and legal professionals, regularly contribute articles to trade journals.

Writing a trade journal article in your field can make your work more widely known, provide publicity for your employer, give you a sense of satisfaction, and even improve your chances for professional advancement.

Planning the Article

When you are thinking about writing an article for a trade journal, consider the following questions:

* Is your work or your knowledge of the subject original? If not, what is there about your approach that justifies publication?
* Will the significance of the article justify the time and effort needed to write it?
* What parts of your work, project, or study are most appropriate to include in the article?

To help you answer those questions, learn as much as possible about the periodical or periodicals to which you plan to submit an article and consult your colleagues for advice. Once you have decided on several journals, consider the following factors about each one:

* the professional interests and size of its readership
* the professional reputation of the journal
* the appropriateness of your article to the journal's goals, as stated on its masthead page or in a mission statement on its Web site
* the frequency with which its articles are cited in other journals

Next, read back issues of the journal or journals to find out information such as the amount and kind of details that the articles include, their length, and the typical writing style.

If your subject involves a particular project, begin work on your article when the project is in progress. That allows you to write the draft in manageable increments and record the details of the project while they are fresh in your mind. It also makes the writing integral to the project and may even reveal any weaknesses in the design or details, such as the need for more data.

As you plan your article, decide whether to invite one or more coauthors to join you. Doing so can add strength and substance to an article. However, as with all collaborative writing, you should establish a schedule, assign tasks, and designate a primary author to ensure that the finished article reads smoothly.

Gathering the Data

As you gather information, take notes from all the sources of primary and secondary research available to you. Begin your research with a careful review of the literature to establish what has been published about your topic. A review of the relevant information in your field can be insurance against writing an article that has already been published. (Some articles, in fact, begin with a <u>literature review</u>.) As you compile that information, record your references in full; include all the information you need to document the source. See also <u>documenting sources</u>.

Organizing the Draft

Some trade journals use a prescribed <u>organization</u> for the major sections. The following organization is common in scientific journals: introduction, materials and methods, results, discussion, and sources cited. Look closely at several issues of the journals that you target to determine how those or other sections are developed. If the major organization is not prescribed, choose and arrange the various sections of the draft in a way that shows your results to best advantage.

The best guarantee of a logically organized article is a good outline. (See <u>outlining</u>.) If you have coauthors, work from a common, well-developed outline to coordinate the various writers' work and make sure the parts fit together logically.

Preparing Sections of the Article

As you are preparing an article, pay particular attention to a number of key sections and elements: the <u>abstract</u>, the <u>introduction</u>, the <u>conclusion</u>, <u>visuals</u> and <u>tables</u>, and <u>headings</u>.

Abstract.   Although it will appear at the beginning of the article, you should write the abstract only after you have finished writing the body of the manuscript. Be sure to follow any instructions provided by the journal on writing abstracts and review abstracts previously published in that journal. Prepare your abstract carefully—it will be the basis on which many other researchers decide whether to read your article in full. Abstracts are often published independently in abstract journals and are a source of terms (called *keywords*) used to index, by subject, the original article for computerized information-retrieval systems.

Introduction.   The introduction should discuss these aspects of the article:

- the <u>purpose</u> of the article
- a definition of the problem examined
- the <u>scope</u> of the article
- the rationale for your approach to the problem or project and the reasons you rejected alternative approaches

- previous work in the field, including other approaches described in previously published articles

Above all, your introduction should emphasize what is new and different about your approach, especially if you are not dealing with a new concept. It should also demonstrate the overall significance of your project or approach by explaining how it fills a need, solves a current problem, or offers a useful application.

Conclusion.   The conclusion section pulls together your results and interprets them in relation to the purpose of your study and the methods used to conduct it. Your conclusion must grow out of the evidence for the findings in the body of the article.

Illustrations and Tables.   Use illustrations and tables wherever they are appropriate, but design each for a specific purpose: to describe a function, to show an external appearance, to show internal construction, to display statistical data, or to indicate trends. Used effectively and appropriately, visuals can clarify information and reinforce the point you are making in the article.

Headings.   Headings are important for an article because they break the text into manageable portions. They also allow journal readers to understand the development of your topic and pinpoint sections of particular interest to them. Check the use of headings when you review back issues of the journal.

References.   The references section (often titled "Works Cited") of the article lists the sources you used in the article. The specific format for listing sources varies from field to field. Usually the journal to which you submit an article will specify the form the editors require for citing sources. (For a detailed discussion of using and citing sources, see <u>documenting sources</u> and <u>quotations</u>.)

Preparing the Manuscript

Some trade journals recommend a particular style guide like the *Chicago Manual of Style* or offer a style sheet with detailed guidelines on style and format. Such style sheets often include specific instructions about how to format the manuscript, how many copies to submit, and how to handle <u>abbreviations</u>, <u>symbols</u>, <u>mathematical equations</u>, and the like. The following guidelines are typical.

- Double-space the manuscript, leaving one-inch margins all around, and number each page.

- Provide specific, accurate, and self-explanatory captions for all figures and tables. Add *callouts* (labels) to those illustrations that need them.

- Provide clear and accurately worded labels for drawings and other illustrations.

- Place any mathematical equations on separate lines in the text and number them consecutively.

- Include only high-quality <u>photographs</u> or scanned images. If you submit an original photograph, check with the editor about any special requirements. Identify each photograph by writing lightly in pencil on the back. If necessary, indicate which edge of the photograph is the top.

Obtaining Publication Clearance

After you have finalized your article, submit a copy to your employer for review before sending it to the journal. A review will ensure that you have not inadvertently revealed any proprietary information. Likewise, secure permission ahead of time to print information for which someone else holds the <u>copyright</u>. See also <u>plagiarism</u>.

transition

Transition is the means of achieving a smooth flow of ideas from sentence to sentence, paragraph to paragraph, and subject to subject. Transition is a two-way indicator of what has been said and what will be said: It provides <u>readers</u> with guideposts for linking ideas and clarifying the relationship between them. You can achieve transition with a word, a phrase, a sentence, or even a <u>paragraph</u>.

Transition can be obvious.

- *Having considered* the technical problems outlined in this proposal, *we move next* to the question of adequate staffing.

Transition can be subtle.

- *Even if* the technical problems could be solved, there *still remains* the problem of adequate staffing.

Either way, you now have your readers' attention fastened on the problem of adequate staffing, exactly what you set out to do.

Methods of Transition

Transition can be achieved in many ways: (1) using transitional words and phrases, (2) repeating keywords or ideas, (3) using pronouns with clear antecedents, (4) numbering with enumeration, (5) summarizing a previous paragraph, (6) asking a question, and (7) using a transitional paragraph.

Certain words and phrases are inherently transitional. Consider the following terms and their functions:

FUNCTION	TERMS
Result	*therefore, as a result, consequently, thus, hence*
Example	*for example, for instance, specifically, as an illustration*
Comparison	*similarly, likewise*
Contrast	*but, yet, still, however, nevertheless, on the other hand*
Addition	*moreover, furthermore, also, too, besides, in addition*
Time	*now, later, meanwhile, since then, after that, before that time*
Sequence	*first, second, third, then, next, finally*

Within a paragraph, such transitional expressions clarify and smooth the movement from idea to idea. Conversely, the lack of transitional devices can make for disjointed reading.

Transition Between Sentences

You can achieve effective transition between sentences by repeating keywords or ideas from preceding sentences and by using pronouns that refer to antecedents in previous sentences. Consider the following short paragraph, which uses both of those means.

- Representative of many American university towns is Middletown. *This Midwestern town,* formerly a *sleepy farming community,* is today the home of a large and vibrant *academic community.* Attracting students from all over the Midwest, *this university town* has grown very rapidly in the last ten years.

Enumeration is another device for achieving transition.

- The recommendation rests on *two conditions. First,* the department staff must be expanded to handle the increased workload. *Second,* sufficient time must be provided for training the new staff.

Transition Between Paragraphs

The means discussed so far for achieving transition between sentences can also be effective for achieving transition between <u>paragraphs</u>. For paragraphs, however, longer transitional elements are often required. One technique is to use an opening sentence that summarizes the preceding paragraph and then moves on to a new paragraph.

- One property of material considered for manufacturing processes is hardness. Hardness is the internal resistance of the material to the forcing apart or closing together of its molecules. Another property is ductility, the characteristic of material that permits it to be drawn into a wire. The smaller the diameter of the wire into which the material can be drawn, the greater the ductility. Material also may possess malleability, the property that makes it capable of being rolled or hammered into thin sheets of various shapes. When selecting raw materials, consider these properties before choosing the one best suited to production.

  The requirements of hardness, ductility, and malleability account for the high cost of such materials. . . .

Another technique is to ask a question at the end of one paragraph and answer it at the beginning of the next.

- New technology has always been feared because it has at times displaced some jobs. However, it invariably creates many more jobs than it eliminates. Historically, the vast number of people employed in the automobile industry as compared with the number of people who had been employed in the harness-and-carriage-making business is a classic example. Almost always, the jobs eliminated by technological advances have been menial, unskilled jobs, and workers who have been displaced have been forced to increase their skills, which resulted in better and higher-paying jobs for them. *In view of these facts, is new technology really bad?*

  Certainly technology has given us unparalleled access to information and created many new roles for employees. . . .

A purely transitional paragraph may be inserted to aid readability.

- The problem of poor management was a key factor that has caused the weak performance of the company.

  Two other setbacks to the company's fortunes that year also marked the company's decline: the loss of many skilled workers through the early retirement program and the intensification of the devastating rate of employee turnover.

  The early retirement program caused the failure. . . .

If you provide logical <u>organization</u> and you have prepared an outline, your transitional needs will easily be satisfied and your writing will have <u>unity</u> and <u>coherence</u>. During revision, look for places where transition is missing and add it. Look for places where it is weak and strengthen it.

trip reports

Many companies require or encourage employees to prepare <u>reports</u> of the trips they take to branch offices, client locations, and training seminars. A trip report provides a permanent record of a trip and its accomplishments. Further, it can enable many employees to benefit from the information that one employee has gained.

A trip report—typically an internal document—is normally written as a <u>memo</u> or an <u>e-mail</u> and addressed to an immediate superior. The subject line gives the destination and dates of the trip. The body of the report explains why you made the trip, whom you visited, and what you accomplished. The report should devote a brief section to each major event and may include a <u>heading</u> for each section (you need not give equal space to each event—instead, elaborate on the more important events). Follow the body of the report with the appropriate conclusions and recommendations. Finally, an expense report is often attached to a trip report, as is shown in Figure T–5 on page 552.

trite language

Trite language is made up of words, phrases, or ideas that have been used so often they are stale.

TRITE *It may interest you to know* that *all the folks* at the branch office are *doin' fine*. I should finish my report *in a nanosecond,* and the rest of the project should be a *slam dunk.*

IMPROVED Everyone here at the branch office is well. I should finish my report today, and the rest of the project should go smoothly.

See also <u>clichés</u> and <u>word choice</u>.

trouble reports

The *trouble report* is used to report such events as accidents, equipment failures, or health emergencies. The report assesses the causes of the problem and suggests changes necessary to prevent its recurrence. Because it is an internal document, the trouble report normally follows a simple <u>memo</u> format. Usually, trouble reports are not large enough in

---

Subject:	Trip to Smith Electric Co., Huntington, West Virginia, January 2003
To:	Roberto Camacho <rcamacho@smithelec.com>
From:	James D. Kerson <jdkerson@smithelec.com>
Date:	Tues, 14 Jan 2003 12:16:30 EST
Attachments:	📎 Expense Report.xls (25 KB)

I visited the Smith Electric Company in Huntington, West Virginia, to determine the cause of a recurring failure in a Model 247 printer and to fix it.

Problem

The printer stopped printing periodically for no apparent reason. Repeated efforts to bring it back online eventually succeeded, but the problem recurred at irregular intervals. Neither customer personnel operating the printer nor the local maintenance specialist was able to solve the problem.

Action

On January 2, I met with Ms. Ruth Bernardi, the Office Manager, who explained the problem. My troubleshooting did not reveal the cause of the problem then or on January 3.

Only when I tested the logic cable did I find that it contained a broken wire. I replaced the logic cable and then ran all the normal printer test patterns to make sure no other problems existed. All patterns were positive, so I turned the printer over to the customer.

Conclusion

There are over 12,000 of these printers in the field and to my knowledge this is the first occurrence of a bad cable. Therefore, I do not believe the logic cable problem found at Smith Electric Company warrants further investigation.

=================================

James D. Kerson, Maintenance Specialist
Smith Electric Company
1366 Federal St., Allentown PA 18101
(610) 747-9955 Fax: (610) 747-9956
jdkerson@smithelec.com
www.smithelec.com
=================================

FIGURE T–5. Trip Report

---

either size or **scope** to require the format of a **formal report**, as in Figure T–6.

In the subject line of the memo, state the precise problem you are reporting. Then, in the body of the report, provide a detailed, precise description of the problem. What happened? Where and when did the problem occur? Was anybody hurt? Was there any property damage? Was there a work stoppage? Because insurance claims, workers' compensation awards, and even lawsuits may hinge on the information contained in a trouble report, be sure to include precise times, dates, locations, treatment of injuries, names of any witnesses, and any other crucial information. Be thorough and accurate in your analysis of the

Consolidated Energy, Inc.

To: Marvin Lundquist, Vice President
           Administrative Services
From: Kalo Katarlan, Safety Officer   *KK*
           Field Service Operations
Date: August 19, 2003
Subject: Field Service Employee Accident on August 6, 2003

The following is an initial report of an accident on Wednesday, August 6, 2003, involving John Markley that resulted in two days of lost time.

Accident Summary
John Markley stopped by a rewiring job on German Road. Chico Ruiz was working there, stringing new wire, and John was checking with Chico about the materials he wanted for framing a pole. Some tree trimming had been done in the area, and John offered to help remove some of the debris by loading it into the pickup truck he was driving. While John was loading branches into the bed of the truck, a piece broke off in his right hand and struck his right eye.

Accident Details
1. John's right eye was struck by a piece of tree branch. John had just undergone laser surgery on his right eye on Monday, August 4, to reattach his cornea.
2. John immediately covered his right eye with his hand, and Chico Ruiz gave him a paper towel with ice to cover his eye and help ease the pain.

7. On Monday, August 11, John returned to his eye surgeon. Although bruised, his eye was not damaged, and the surgically implanted lens was still in place.

Recommendations
To prevent a recurrence of such an accident, the Safety Department will require the following actions in the future:

- When working around and moving debris, such as tree limbs or branches, all service-crew employees must wear safety eyewear with side shields.
- All service-crew employees must always consider the possibility of shock for an injured employee. If crew members cannot leave the job site to care for the injured employee, someone on the crew must call for assistance from the Service Center. The Service Center phone number is printed in each service-crew member's Handbook.

T

FIGURE T–6. Trouble Report

problem and support any judgments or conclusions with facts. Be objective: Always use a neutral <u>tone</u> and avoid assigning blame. If you speculate about the cause of the problem, make it clear to your reader that you are speculating. See also <u>ethics in writing</u> and <u>reports</u>.

In your <u>conclusion</u>, state what has been or will be done to correct the conditions that led to the problem. That may include training in safety practices, improved equipment, protective clothing, and so on.

The report shown in Figure T–6 on page 553 describes an accident involving personal injury. Notice the careful use of language and factual detail.

---

try to

The phrase *try and* is colloquial for *try to.* For workplace writing, use *try to.*

- Please try ~~and~~ *to* finish the report on time.

T

U

unity

Unity is the cohesive element that holds a **paragraph** or a document together; unity means that everything is essentially related to a central idea.

To achieve unity, you must select one **topic** and then treat it with singleness of **purpose**, without digressing into unrelated or loosely related paths. The logical sequence provided through **outlining** is essential to achieving unity. An outline enables you to lay out the most direct route from **introduction** to **conclusion**, and it enables you to build each paragraph around a topic sentence that expresses a single idea.

Effective **transition** helps build unity, as well as **coherence**, because transitional terms clarify the relationship of each part to what precedes it.

up

Adding the word *up* to **verbs** often creates a redundant phrase. See also **conciseness/wordiness**.

- You must open up the exhaust valve.

usability testing

Usability refers to whether **readers** can use a document to easily fulfill their goals or accomplish tasks. For example, the usability of a technical **manual** might refer to the ability of readers to perform a procedure accurately and smoothly with the aid of the **instructions**. The usability of a **brochure** might refer to the ability of readers to understand the contents quickly and decide whether to buy a product or use a service.

Ideally, usability is built into documents or products from the beginning of their life cycle during the early planning, development, and design phases. To ensure that a document is usable, it is important to focus on users throughout the evolution of a document. Usability

U

engineering, or user-centered design (UCD), refers to the process of designing usable products and systems, with the user at the center of the process. Everything involved with a product or a service, including its goals, context, and environment, are approached from the user's viewpoint.

Usability testing helps produce a document that reduces the learning curve, allows more functionality with less effort, and increases productivity. At the same time, by focusing development on users, companies also reap benefits in reduced costs and increased customer satisfaction in the following ways:

- The documents are functional and more likely to be used.
- Product and document changes can be made before they become expensive.
- Efficient document development and training are facilitated.
- The need for updates and maintenance releases is minimized.
- More documents (and products) are sold.
- The organization's reputation is enhanced.

Usability testing is a common way to involve users, for example, by testing periodic drafts on users and then revising the document in response to the test results. The process typically begins with the establishment of specific, quantitative, and measurable goals for documents. Next, the documents are designed to fulfill those goals. Finally, tests must be conducted with representative users to detect problems and determine whether the established goals have been achieved.

Usability testing involves teams of skilled usability specialists, interface designers, and technical writers. Usability testing has three main goals:

- to create a document that is easy to use and allows users to accomplish the tasks outlined in that document.
- to detect potential problems for users as well as guide designers in resolving such issues. That process minimizes the risk of releasing ineffective documents and poorly designed products.
- to enable companies to avoid repeating mistakes when developing future documents.

Test participants typically are members of the actual target audience for the document. If the participants encounter problems with the document, it is likely that in the future, actual users will experience similar problems. For example, if page 15 of a tax form is unclear to test participants, it is likely to be confusing to most taxpayers.

Usability tests can rely on one or more of the following methods.

- *User testing.* Testers observe and record the actions of test participants who perform real tasks using the document.
- *Protocols.* Test participants make comments aloud as they read documents and perform document tasks, to reveal their thought processes, attitudes, and reasons for decision-making.
- *Comprehension tests.* Users complete tests to determine whether they understand and can recall document features, such as <u>visuals</u> or <u>tables</u>.
- *Surveys and interviews.* Testers interview users both before and after they read documents to determine their comprehension and attitudes and the <u>clarity</u> of the documents.

Analyzing the results of those test methods can help document designers detect problems and determine whether the document <u>purposes</u> have been met. If test participants have trouble navigating a document or quickly locating specific items, for example, the organization as well as <u>layout and design</u> are revised. See also <u>forms design</u>, <u>questionnaires</u>, and <u>interviewing for information</u>.

usage

Usage describes the choices we make among the various words and constructions available in our language. The lines between standard English and nonstandard English and between formal and informal English are determined by those choices. Your guideline in any situation requiring such choices should be appropriateness: Is the word or expression you use appropriate to your <u>readers</u> and your subject? When it is, you are practicing good usage.

This book contains many entries that focus on various usages (they appear in *italics*). A good <u>dictionary</u> is also an invaluable aid in your selection of the right word.

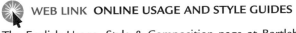 **WEB LINK ONLINE USAGE AND STYLE GUIDES**

The English Usage, Style & Composition page at Bartleby.com provides classic reference books online. For this and additional useful links related to style and usage, see <*www.bedfordstmartins.com/alred*> and select *Links for Technical Writing.*

U

utilize

Do not use *utilize* as a long variant of *use,* which is the general word for "employ for some purpose." *Use* will almost always be clearer and less pretentious. See <u>affectation</u>.

U

V

vague words

A *vague word* is one that is imprecise in the context in which it is used. Some words encompass such a broad range of meanings that there is no focus for their definition. Words such as *real, nice, important, good, bad, contact, thing,* and *fine* are often called "omnibus words" because they can mean everything to everybody. In speech, we sometimes use words that are imprecise, but our vocal inflections and the context of our conversation make their meanings clear. Because you cannot rely on vocal inflections when you are writing, avoid using vague words. Be concrete and specific. See also abstract/concrete words.

VAGUE   It was a *good* meeting.

SPECIFIC   The meeting resolved three questions: pay scales, fringe benefits, and workloads.

verbals

Verbals are derived from verbs and function as nouns, adjectives, and adverbs. There are three types of verbals: gerunds, infinitives, and participles.

Gerunds

A *gerund* is a verbal ending in *-ing* that is used as a noun. A gerund can be used as a subject, a direct object, the object of a preposition, a subjective complement, or an appositive.

- *Estimating* is an important managerial skill. [subject]
- I find *estimating* difficult. [direct object]
- We were unprepared for their *coming*. [object of preposition]
- Seeing is *believing*. [subjective complement]
- My primary departmental function, *programming*, occupies about two-thirds of my time on the job. [appositive]

V

559

Only the possessive form of a noun or <u>pronoun</u> should precede a gerund.

- *John's* working has not affected his grades.
- *His* working has not affected his grades.

Infinitives

An *infinitive* is the bare, or uninflected, form of a verb (for example, *go, run, fall, talk, dress, shout*) without the restrictions imposed by <u>person</u> and <u>number</u>. Along with the gerund and the participle, it is one of the nonfinite verb forms. The infinitive is generally preceded by the word *to*, which, although not an inherent part of the infinitive, is considered to be the sign of an infinitive. An infinitive is a verbal and can function as a noun, an adjective, or an adverb.

- *To expand* is not the only objective. [noun]
- These are the instructions *to follow*. [adjective]
- The company struggled *to survive*. [adverb]

The infinitive can reflect two <u>tenses</u>: the present and (with a helping verb) the present perfect.

- *to go* [present tense]
- *to have gone* [present perfect tense]

The most common mistake made with infinitives is using the present perfect tense when the simple present tense is sufficient.

- I should not have tried to ~~have gone~~ *go* so early.

Infinitives formed with the root form of transitive verbs can express both active and (with a helping verb) passive <u>voice</u>.

- *to hit* [present tense, active voice]
- *to have hit* [present perfect tense, active voice]
- *to be hit* [present tense, passive voice]
- *to have been hit* [present perfect tense, passive voice]

A split infinitive is one in which an adverb is placed between the sign of the infinitive, *to,* and the infinitive itself. Because they make up a grammatical unit, the infinitive and its sign are better left intact than separated by an intervening adverb.

V

- To ~~initially~~ build the table in the file, you could input transaction records containing the data necessary to construct the record and table.

However, it may occasionally be better to split an infinitive than to allow a sentence to become awkward, ambiguous, or incoherent.

AMBIGUOUS	She agreed immediately *to deliver* the toxic materials. [This sentence could be interpreted to mean that she agreed immediately.]
CLEAR	She agreed *to* immediately *deliver* the toxic materials. [This sentence is no longer ambiguous.]

Participles

A *participle* is a verb form that functions as an adjective. Present participles end in *-ing*.

- *Declining* sales forced us to close one branch office.

Past participles end in *-ed, -t, -en, -n,* or *-d.*

- What are the *estimated* costs?

- Repair the *bent* lever.

- Here is the *broken* calculator.

- What are the metal's *known* properties?

- The story, *told* many times before, was still interesting.

The perfect participle is formed with the present participle of the helping verb *have* plus the past participle of the main verb.

- *Having gotten* [perfect participle] a large bonus, the *smiling* [present participle], *contented* [past participle] sales representative worked harder than ever.

A participle cannot be used as the verb of a sentence. Inexperienced writers sometimes make that mistake, and the result is a <u>sentence fragment</u>.

- The committee chairperson was responsible *, his* ~~. His~~ vote being the decisive one.

- The committee chairperson was responsible. His vote *was* ~~being~~ the decisive one.

For information on participial and infinitive phrases, see <u>phrases</u>.

V

verbs

A *verb* is a word or group of words that describes an action ("the copier *jammed* at the beginning of the job"), states the way something or someone is affected by an action ("he *was disappointed* that the proposal was rejected"), or affirms a state of existence ("she *is* a district manager now").

Types of Verbs

Verbs are either transitive or intransitive. A *transitive verb* requires a direct <u>object</u> to complete its meaning.

- They *laid* the foundation on October 24.
  [*Foundation* is the direct object of the transitive verb *laid*.]

- Rosalie Anderson *wrote* the treasurer a letter.
  [*Letter* is the direct object of the transitive verb *wrote*.]

An *intransitive verb* does not require an object to complete its meaning. It makes a full assertion about the subject without assistance (although it may have modifiers).

- The engine *ran*.

- The engine *ran* smoothly and quietly.

A *linking verb* is an intransitive verb that links a complement to the subject. When the complement is a <u>noun</u> or a <u>pronoun</u>, it refers to the same person or thing as the noun or pronoun that is the subject.

- The winch *is* rusted.
  [*Is* is a linking verb; *rusted* is an adjective modifying *winch*.]

Some intransitive verbs, such as *be, become, seem,* and *appear,* are almost always linking verbs. A number of others, such as *look, sound, taste, smell,* and *feel,* can function as either linking verbs or simple intransitive verbs.

If you are unsure about whether one of those verbs is a linking verb, try substituting *seem;* if the sentence still makes sense, the verb is probably a linking verb.

- Their antennae *feel* delicate.
  [*Seem* can be substituted for *feel*—thus *feel* is a linking verb.]

- Their antennae *feel* delicately for their prey.
  [*Seem* cannot be substituted for *feel;* in this case, *feel* is a simple intransitive verb.]

Forms of Verbs

Verbs are described as being either finite or nonfinite.

Finite Verbs.   A *finite verb* is the main verb of a clause or sentence. It makes an assertion about its subject and often serves as the only verb in its clause or sentence. (The receptionist *answered* the phone.) A helping verb (sometimes called an *auxiliary verb*) is used in a verb <u>phrase</u> to help indicate <u>mood</u>, <u>tense</u>, and <u>voice</u>. (The phone *had* rung.)

Phrases that function as helping verbs are often made up of combinations with the sign of the infinitive, *to* (for example, *am going to, is about to, has to,* and *ought to*). For typical verb phrases that give speakers of <u>English as a second language</u> trouble, see <u>idioms</u>.

The helping verb always precedes the main verb, although other words may intervene. (Machines *will* never completely *replace* people.)

Nonfinite Verbs.   *Nonfinite verbs* are <u>verbals</u>—verb forms that function as nouns, <u>adjectives,</u> or <u>adverbs.</u>

A *gerund* is a noun that is derived from the *-ing* form of a verb.

- *Seeing* is *believing.*

An *infinitive,* which uses the root form of a verb (usually preceded by *to*), can function as a noun, an adverb, or an adjective.

- He hates *to complain.* [noun, direct object of *hates*]

- The valve closes *to stop* the flow. [adverb, modifies *closes*]

- This is the proposal *to consider.* [adjective, modifies *proposal*]

A *participle* is a verb form that can function as an adjective.

- The *rejected* proposal was ours.
  [*Rejected* is a verb that is used as an adjective modifying *proposal.*]

V

Properties of Verbs

Verbs must (1) agree in <u>person</u> with personal pronouns functioning as subjects, (2) agree in tense and <u>number</u> with their subjects, and (3) be in the appropriate voice.

Person is the term for the form of a personal pronoun that indicates whether the pronoun refers to the speaker, the person spoken to, or the person (or thing) spoken about. Verbs change their forms to agree in person with their subjects.

- I *see* [first person] a yellow tint, but she *sees* [third person] a yellow-green hue.

Tense refers to verb forms that indicate time distinctions. There are six tenses: present, past, future, present perfect, past perfect, and future perfect.

 TIPS FOR AVOIDING SHIFTS IN VOICE, MOOD, OR TENSE

To achieve clarity in your writing, it is important to maintain consistency and avoid shifts. A shift occurs when there is an abrupt change in voice, mood, or tense. Pay special attention when you edit your writing to check for the following types of shifts.

Voice

- The captain permits his crew to go ashore, but ~~they are not~~ *he does not permit* *them* ~~permitted~~ to go downtown.

  [The entire sentence is now in the active voice.]

Mood

- Reboot your computer, and ~~you should~~ empty the cache.

  [The entire sentence is now in the imperative mood.]

Tense

- I was working quickly, and suddenly a box ~~falls~~ *fell* off the conveyor belt and ~~breaks~~ *broke* my foot.

  [The entire sentence is now in the past tense.]

V

Number refers to the two forms of a verb that indicate whether the subject of a verb is singular ("the copier *was* repaired") or plural ("the copiers *were* repaired").

Most verbs show the singular of the third person, present tense, indicative mood by adding *-s* or *-es* (he *stands*, she *works*, it *goes*). To indicate the plural form, the verb *to be* normally changes from singular ("I *am* ready") to plural ("we *are* ready").

Voice refers to the two forms of a verb that indicate whether the subject of the verb acts or receives the action. The verb is in the *active voice* if the subject of the verb acts ("the bacteria *grow*"); the verb is in the *passive voice* if it receives the action ("the bacteria *are grown*" in a petri dish").

Conjugation of Verbs

The conjugation of a verb arranges all forms of the verb so that the differences caused by the changing of the tense, number, person, and voice are readily apparent. Figure V–1 shows the conjugation of the verb *drive*.

TENSE	NUMBER	PERSON	ACTIVE VOICE	PASSIVE VOICE
Present	Singular	1st	I drive	I am driven
		2nd	You drive	You are driven
		3rd	He drives	He is driven
	Plural	1st	We drive	We are driven
		2nd	You drive	You are driven
		3rd	They drive	They are driven
Progressive present	Singular	1st	I am driving	I am being driven
		2nd	You are driving	You are being driven
		3rd	He is driving	He is being driven
	Plural	1st	We are driving	We are being driven
		2nd	You are driving	You are being driven
		3rd	They are driving	They are being driven
Past	Singular	1st	I drove	I was driven
		2nd	You drove	You were driven
		3rd	He drove	He was driven
	Plural	1st	We drove	We were driven
		2nd	You drove	You were driven
		3rd	They drove	They were driven

(continued)

V

FIGURE V–1. Verb Conjugation Chart

TENSE	NUMBER	PERSON	ACTIVE VOICE	PASSIVE VOICE
Progressive past	Singular	1st	I was driving	I was being driven
		2nd	You were driving	You were being driven
		3rd	He was driving	He was being driven
	Plural	1st	We were driving	We were being driven
		2nd	You were driving	You were being driven
		3rd	They were driving	They were being driven
Future	Singular	1st	I will drive	I will be driven
		2nd	You will drive	You will be driven
		3rd	He will drive	He will be driven
	Plural	1st	We will drive	We will be driven
		2nd	You will drive	You will be driven
		3rd	They will drive	They will be driven
Progressive future	Singular	1st	I will be driving	I will have been driven
		2nd	You will be driving	You will have been driven
		3rd	He will be driving	He will have been driven
	Plural	1st	We will be driving	We will have been driven
		2nd	You will be driving	You will have been driven
		3rd	They will be driving	They will have been driven
Present perfect	Singular	1st	I have driven	I have been driven
		2nd	You have driven	You have been driven
		3rd	He has driven	He has been driven
	Plural	1st	We have driven	We have been driven
		2nd	You have driven	You have been driven
		3rd	They have driven	They have been driven
Past perfect	Singular	1st	I had driven	I had been driven
		2nd	You had driven	You had been driven
		3rd	He had driven	He had been driven
	Plural	1st	We had driven	We had been driven
		2nd	You had driven	You had been driven
		3rd	They had driven	They had been driven
Future perfect	Singular	1st	I will have driven	I will have been driven
		2nd	You will have driven	You will have been driven
		3rd	He will have driven	He will have been driven
	Plural	1st	We will have driven	We will have been driven
		2nd	You will have driven	You will have been driven
		3rd	They will have driven	They will have been driven

FIGURE V–1. Verb Conjugation Chart *(continued)*

very

The temptation to overuse <u>intensifiers</u> like *very* is great. In many sentences, the word can simply be deleted.

* The board was ~~very~~ angry about the newspaper report.

When you do use intensifiers, clarify their meaning. (Retail sales were *very* strong; they were up 43 percent this month.)

via

Via is Latin for "by way of." The term should be used only in routing instructions.

* The package was shipped *via* FedEx.

* Her project was funded ~~via~~ the recent legislation.
  <small>*as a result of*</small>

visuals

Visuals can express ideas or convey information in ways that words alone cannot. They communicate by showing how things look (drawings, photographs, maps), by representing numbers and quantities (graphs, tables), by depicting relationships (flowcharts, schematic diagrams), and by making abstract concepts and relationships concrete (organizational charts). They also highlight the most important information and emphasize key concepts succinctly and clearly, especially in documents such as <u>brochures</u> and <u>newsletters</u>.

Selecting Visuals

Consider your <u>purpose</u> and your <u>audience</u> carefully. For example, you would need different illustrations for an automobile owner's <u>manual</u> or an auto dealer's Web site than you would for a mechanic's diagnostic guide.

Many of the qualities of good writing—simplicity, clarity, conciseness, directness—are equally important in the creation and use of visuals. Presented with clarity and consistency, visuals can help the audience focus on key portions of your document, presentation, or Web site. Be aware, though, that even the best visual only enhances or

V

supports the text. Your writing must provide context for the visual and point out its significance.

The following entries are related to specific visuals and their use in printed and online documents as well as in presentations.

Integrating Visuals with Text

To integrate visuals smoothly with your text, consider your graphics requirements before you begin writing a draft. Jot down visual options when you are considering your scope and organization. Make visuals an integral part of your outline. (See outlining.) At appropriate points in your outline, either make a rough sketch of the visual, if you can, or write "illustration of . . . ," noting the source of the visual and enclosing each suggestion in a text box. Planning your graphics requirements from the beginning stages of your outline ensures their integration throughout all versions of the draft to the finished product.

Writer's Checklist: Using and Integrating Visuals

☑ Clarify why each visual is included in the text. The amount of description varies, depending on the illustration's complexity and the **readers'** backgrounds. Nonexperts usually require lengthier explanations than do experts.

☑ Use consistent terminology. Do not refer to something as a "proportion" in the text and as a "percentage" in the visual.

☑ Define all **abbreviations** the first time they appear in the text and in all figures and tables. If any **symbols** are not self-explanatory, as in graphs, include a key that defines them.

☑ Place a visual as close as possible to the text where it is discussed, especially if an illustration is central to the discussion. No visual should precede its first text mention.

☑ Consider placing a lengthy and detailed visual in an **appendix** and refer to it in the text.

☑ Give each visual a concise title that clearly describes its content.

☑ Assign figure and table numbers, particularly if your document contains more than one illustration or table. The figure or table number

V

Writer's Checklist: Using and Integrating Visuals (continued)

precedes the title (Figure 1. Projected groundwater migration rates for 2004–2014).

☑ Refer to visuals in the text of your document by their figure or table numbers. (Note that in **reports** and many other documents, graphical illustrations — drawings, maps, and photographs — are generically labeled "figures," while tables are labeled "tables.")

☑ In documents with more than five illustrations or tables, include a section titled "List of Figures" or "List of Tables" that identifies each by number, title, and page number. This list should follow the **table of contents**.

☑ Keep visuals simple; include only information necessary to the discussion in the text and eliminate unnecessary labels, arrows, boxes, and lines.

☑ Specify the units of measurement used or include a scale of relative distances, when appropriate. Make sure relative sizes are clear or indicate distance with a scale.

☑ Position the lettering of any explanatory text or labels horizontally for readability.

☑ Allow adequate white space around and within the visual.

☑ Obtain written permission for any visuals that are copyrighted, and acknowledge borrowed material in a source or credit line below the caption for a figure and in a footnote at the bottom of a table. See also **copyright**.

☑ Acknowledge your use of any public (uncopyrighted) information, such as demographic or economic data, from federal government publications by including a credit line. See also **documenting sources** and **plagiarism**.

☑ When writing a **trade journal article**, consult the editorial guidelines of the particular journal or recommended style manual for guidance on the preparation and format of visuals.

vogue words (*see* **buzzwords**)

voice

In grammar, *voice* indicates the relation of the subject to the action of the **verb**. When the verb is in the *active voice,* the subject acts; when it is in the *passive voice,* the subject is acted upon.

ACTIVE   David Cohen *wrote* the newsletter article.
[The subject, *David Cohen,* performs the action; the verb *wrote* describes the action.]

PASSIVE   The newsletter article *was written* by David Cohen.
[The subject, *the newsletter article,* is acted upon; the verb *was written* describes the action.]

The two sentences say the same thing, but each has a different <u>empha-sis</u>: The first emphasizes *David Cohen;* the second emphasizes *the newsletter article.* In technical writing, it is often important to emphasize who or what performs an action. Further, the passive-voice version is indirect because it places the performer of the action behind the verb instead of in front of it. Because the active voice is generally more di-rect, more concise, and easier for <u>readers</u> to understand, use the active voice unless the passive voice is more appropriate, as described on page 571. Whether you use the active voice or the passive voice, be careful not to shift voices in a sentence.

- David Cohen corrected the inaccuracy as soon as ~~it was identified~~
  identified it
  ~~by~~ the editor .
        ^

Using the Active Voice

Improving Clarity.   The active voice improves <u>clarity</u> and avoids confusion, especially in <u>instructions</u>.

PASSIVE   Sections B and C *should be checked* for errors.
[Are they already checked?]

ACTIVE   *Check* sections B and C for errors.
[The performer of the action, *you,* is understood: (You) *Check* the sections.]

Active voice can also help avoid <u>dangling modifiers</u>.

PASSIVE   Hurrying to complete the work, the cables *were connected* improperly. [*Who* was hurrying?]

ACTIVE   Hurrying to complete the work, the technician *connected* the cables improperly.
[Here, *hurrying to complete the work* properly modifies the performer of the action: *the technician.*]

V

Highlighting Subjects.   One difficulty with passive sentences is that they can bury the performer of the action in <u>expletives</u> and prepo-sitional <u>phrases</u>.

PASSIVE	It *was reported by* the agency that the new model is defective.
ACTIVE	The agency *reported* that the new model is defective.

Sometimes writers using the passive voice fail to name the performer—information that might be missed.

PASSIVE	The problem *was discovered* yesterday.
ACTIVE	The Maintenance Department *discovered* the problem yesterday.

Do not use the passive voice to hide information from the reader, evade responsibility, or obscure an issue. (See also <u>ethics in writing</u>.)

- Several mistakes were made. [*Who* made the mistakes?]
- It has been decided. [*Who* has decided?]

Avoiding Wordiness.   The active voice helps avoid wordiness because it eliminates the need for an additional helping verb as well as an extra <u>preposition</u> to identify the performer of the action.

PASSIVE	Changes in policy *are resented by* employees.
ACTIVE	Employees *resent* changes in policy.

The active-voice version takes one verb (*resent*); the passive-voice version takes two verbs (*are resented*) and an extra preposition (*by*). See also <u>conciseness/wordiness</u>.

Using the Passive Voice

There are instances when the passive voice is effective or even necessary. Indeed, for reasons of tact and diplomacy, you might need to use the passive voice to avoid identifying the performer of the action.

ACTIVE	Your staff did not meet the quota last month.
PASSIVE	The quota was not met last month.

When the performer of the action is either unknown or unimportant, use the passive voice.

- The copper mine *was discovered* in 1929.

When the performer of the action is less important than the receiver of that action, the passive voice is sometimes more appropriate.

- Ann Bryant *was presented* with an award by the president.

When you are explaining an operation in which the reader is not actively involved or when you are explaining a process or a procedure, the passive voice may be more appropriate. In the following example, anyone—it really does not matter who—could be the performer of the action.

- Area strip mining *is used* in regions of flat to gently rolling terrain, like that found in the Midwest. Depending on applicable reclamation laws, the topsoil *may be removed* from the area *to be mined, stored,* and later *reapplied* as surface material during reclamation of the mined land. After the removal of the topsoil, a trench *is cut* through the overburden to expose the upper surface of the coal to be mined. The overburden from the first cut *is placed* on the unmined land adjacent to the cut. After the first cut *has been completed,* the coal *is removed.*

Do not, however, simply assume that any such explanation should be in the passive voice. Ask yourself, "Would it be of any advantage to the reader to know the performer of the action?" If the answer is yes, use the active voice, as in the following example.

- In the operation of an internal combustion engine, an explosion in the combustion chamber *forces* the pistons down in the cylinders. The movement of the pistons in the cylinders *turns* the crankshaft.

ESL TIPS FOR CHOOSING VOICE

Different languages place different values on active-voice and passive-voice constructions. In some languages, the passive is used frequently; in others, hardly at all. As a nonnative speaker of English, you may have a tendency to follow the pattern of your native language. But remember, even though technical writing may sometimes require the passive voice, active verbs are highly valued in English.

W

wait for / wait on

Wait on should be restricted in writing to the activities of hospitality and service employees. (We need extra staff to *wait on* customers.) Otherwise, use *wait for*. (Be sure to *wait for* Ms. Sturgess's approval.) See also <u>idioms</u>.

Web design

You can apply many of the principles covered throughout this book to designing Web sites. As when creating other documents, carefully consider your <u>purpose</u> and <u>readers'</u> needs as you prepare Web pages and assemble those pages to create Web sites. Most organizational sites have well-defined goals, such as reference, training, education, publicity, advocacy, or marketing. Before you begin building your Web pages and sites, create a clear statement of purpose that identifies your target audience.

EXTERNAL SITE   The purpose of this site is to enable our customers to locate product information, place online orders, and contact our customer-service department.

INTERNAL SITE   The purpose of this site is to provide SNR Security Corporation employees, suppliers, and partner organizations with a single, consistent, and up-to-date resource for materials about SNR Security.

W

573

As the examples suggest, there are two general kinds of sites: external and internal. External sites target an audience from the entire <u>Internet</u>; internal sites are designed for audiences on an *intranet* (a computer network within an educational institution or company that is not accessible to audiences outside that institution or company) or an *extranet* (a computer network made available exclusively to partners and suppliers to provide shipping logs, inventories, etc.).

Web-Site Navigation

Your main goal when designing a Web page is to establish a predictable environment in which users can comfortably navigate and easily find the information they need. This information should be logically accessible to the user in the fewest possible steps (or clicks). To design an efficient navigation plan, draft a navigation chart or map of your site early in the process. Figure W–1 on page 575 shows an initial navigation plan for a Web site for a U.S. Department of Energy National Laboratory.

In Figure W–1, each higher-level page (for example, Basic Science) is linked to related subject areas (Biology, Chemistry, Sensors, Materials, Mathematics, Physics). Under Scientific Facilities, related content areas are shown, with a specific site for the laboratory's Auto Shredder Facility, which in turn is linked to a site with information about recovering plastics from scrapped autos. That page has an e-mail link to the principal investigator for the Recovering Plastics project. The pattern is repeated, with greater or lesser detail, throughout the site.

Create links to sites related to your visitors' organizational or professional needs and interests. Links to customer service and online order forms are especially useful for cultivating potential customers.

When you identify links, do not write out "Click here for more information." Instead, write the sentence as you normally would, and anchor the link on the most relevant word in the sentence, as illustrated in the following example.

- For information about employment opportunities, visit <u>Human Resources</u>.

Avoid writing paragraphs that are dense with links. Instead, list links alphabetically in groups of about four to seven to make them easier for the viewer to see at a glance.

▶ Basic Science
▶ Energy Resources
▶ Environmental Management
▶ Scientific Facilities

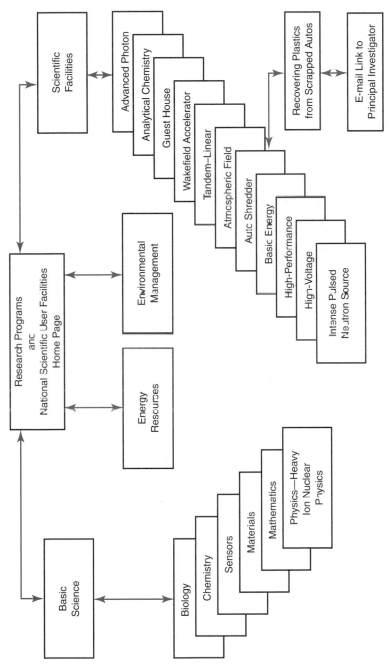

FIGURE W-1. Site Map

In addition to identifying links with words, you can use icons and graphics, as shown in the preceding list; they not only are easy to use but also provide the visual cues Web users expect. Include a link that allows visitors to contact you by e-mail. See also <u>writing for the Web</u>.

Graphics and Typography

Your audience and purpose should determine the graphic style and theme of your Web site. Although graphic elements provide visual relief from dense text, do not overdo graphics, especially those with animation. Complex graphics and motion can clutter or slow access to your site. Avoid bold and multiple colors; instead, use lighter colors, especially for backgrounds. Keep in mind that large or high-resolution graphics, like color photographs, can cause long delays as they download to the user's system. Consider providing a graphics-free option for quicker access.

When you have decided on fonts, spacing between paragraphs, heading sizes, and other elements, consider creating a cascading style sheet to implement these specifications. *Cascading style sheets* are codes, linked from or included in your HTML (hypertext markup language) that work like templates. You can define the style (color, font, size, emphasis, etc.) for a particular set of HTML tags once, and thereafter every time you use those tags in your site they will look the same. Cascading style sheets are especially helpful for large sites with numerous pages, as you can quickly and easily change your headings by simply changing the style sheet.

In general, aim for consistency in your graphics and typography to establish a sense of unity and to provide visual cues that help visitors find information.

Page Layout and Design

Take into account the viewing capabilities of visitors to your site. Many computer monitors cannot display more than half of a typical Web page at any one time; on many screen settings, only the top four or five inches may be visible. For that reason, place important graphic elements and information in the upper half of the page (or top 300 pixels). Then, use short narrative passages or lists so that viewers can scan and access the information quickly. Organize the material in a logical, predictable pattern, as shown in Figure W–1 on page 575. See also <u>methods of development</u>.

Because most users do not like scrolling down long pages, many Web designers recommend that Web pages (homepages in particular) contain no more than roughly one or two screens of information. If a

W

greater amount of information is covered, include a concise table of contents at the top of the page linked to specific sections elsewhere on the page. Viewers can select the section they want without having to scroll through the entire page. You can also provide periodic links that return visitors to the top of the page.

When viewers access Web pages randomly, they often have no context for where they are or who sponsors the site. Therefore, incorporate the name of your company or organization and its logo in a banner at the top of each page. Another way to orient users is to provide a set of links in graphics or words called a *tool bar* or *button bar,* as shown here.

[Home | Search | Order | What's New | About Us]

A tool bar at the top, bottom, or sides of each page serves as a table of contents and shows visitors the structure of your site. Tool bars on each page help you avoid sending users to a dead-end page (one with no link); in fact, every page should, if nothing else, link back to your homepage. For guidance on designing interactive forms on the Web, see forms design.

Because Web sites change constantly, periodically check that your links are still appropriate and working properly. Your Web pages should include the date they were created or last updated (usually in a footer) so that users can determine whether the information is current. You might also place a new icon next to each new or updated item to alert users to these changes. Be sure to remove the icon after a suitable interval, such as 30 days.

Finally, test your design by viewing it on several different Web browsers as well as different platforms (including UNIX, Linux, Mac, and Windows). Different software and different platforms can interact unpredictably, and you may well be surprised at how different your page can look on systems unlike your own.

DIGITAL TIP TESTING YOUR WEB SITE

You might take advantage of Web sites that will test a limited sample of your site's features for free. (They also offer greatly expanded testing for a fee.) The services can test your site's browser and hardware platform compatibility, page-load speed, and links. NetMechanic.com, for example, will test pages at your site using 16 different browser and computer combinations. It will also show page-load speeds in seconds on 14.4K, 28.8K, and 56K modems, as well as on high-speed lines. You can then print the test results and work to eliminate any problems identified. For more on this topic, see <*www.bedfordstmartins.com/alred*> and select *Digital Tips.*

W

Homepages

A *homepage* (or *start page*) is an important focus for all pages in your Web site. Many organizational homepages include an image map with links to other pages at the site. The image map introduces visitors to the overall site design, identifies the purpose of the site, and provides an overview of major content areas. If you do not use an image map, at least display a small graphic banner, tool bar, or menu across the top of the homepage.

Because a complex homepage graphic can take a minute or longer to download, consider using relatively small ones on your homepage, gradually increasing the size of the graphics for pages deeper into your site, if necessary. Users who go beyond a page or two into a site are more committed and therefore more willing to tolerate longer delays, especially if you offer them warnings that particular pages contain graphics that may take time to download.*

Access for People with Disabilities

Many of the advantages of Web sites include colorful graphics, animation, and streaming video and audio. However, these design elements can be barriers to people with impaired vision or hearing or those who are color-blind. Use the following strategies to meet the needs of such audiences.

- Avoid frames, complex tables, animation, JavaScript, and other design elements that are incompatible with text-only browsers and adaptive technologies, such as voice or large-print software.
- Provide HTML versions of pages and documents whenever possible because this format is most compatible with the current generation of screen readers.
- Attach text equivalents for graphic or audio elements.
- Design for the color-blind reader by making meaning independent of color.

You may want to consider offering different options for site visitors, such as full-graphics, light-graphics, and text-only versions.

Finally, test your pages using "Bobby" <www.cast.org/bobby>, a site that will provide an on-screen analysis of a Web page to ensure its accessibility.

W

*The current industry "best practices" recommendation is that a page load in ten or fewer seconds on a 56K modem.

WEB LINK IMPROVING ACCESSIBILITY

Numerous government and private Web sites provide help for both Web designers and those who need greater accessibility. The World Wide Web Consortium (W3C) and DRM Web Watcher provide guidelines, technological information, and additional links. See <*www.bedfordstmartins.com/alred*> and select *Links for Technical Writing.*

Writer's Checklist: Designing Web Pages

Many of the guidelines on typography in **layout and design** are applicable to Web pages; however, not all apply, so keep the following in mind:

- ☑ Limit the number of typeface colors as well as styles.
- ☑ Use typeface colors that contrast (but do not clash) with background colors.
- ☑ Use uppercase letters or boldface type sparingly.
- ☑ Use underlined text only for links.
- ☑ Separate text from graphics with generous blank space (the equivalent of white space).
- ☑ Use **heading** and subheading styles sparingly and consistently.
- ☑ Block indent text sections that you expect viewers to read in detail.
- ☑ Test your design by viewing it on alternate browsers and platforms.
- ☑ Check your links to ensure they are working, particularly after site changes.
- ☑ Test the load time of your pages using a slow modem.

WEB LINK WEB DESIGN TUTORIAL

Mike Markel's Web Design Tutorial provides a concise overview of the process of creating a Web site, introduces important design principles to consider during site design, and presents sample Web pages for analysis. See <*www.bedfordstmartins.com/alred*> and select *Mike Markel's Web Design Tutorial.*

W

when / where / that

When and if (or *if and when*) is a colloquial expression that should not be used in writing.

- When ~~and if~~ your new position is approved, I will see that you receive adequate staff support.

- *If*
  ~~When and if~~ your new position is approved, I will see that you
  ^
  receive adequate staff support.

In phrases using the *where . . . at* construction, *at* is unnecessary and should be omitted.

- Where is his office ~~at?~~

Do not substitute *where* for *that* to anticipate an idea or fact to follow.

- *that*
  I read in the newsletter ~~where~~ semiconductor sales have increased
  ^
  in the fourth quarter.

whether

Whether communicates the notion of a choice. When *whether or not* is used to indicate a choice between alternatives, omit *or not;* it is redundant.

- The project director asked whether ~~or not~~ the proposal was finished.

  The phrase *as to whether* is clumsy and redundant. Either omit it altogether or use only *whether.*

  We have decided to   *contract.*
- ~~As to whether we will~~ commit our firm to a long-term ~~contract,~~
  ^   ^
  ~~we have decided to do so.~~

while

While, meaning "during an interval of time," is sometimes substituted for connectives like *and, but, although,* and *whereas.* Used as a connective in that way, *while* often causes ambiguity.

W

and
- Ian Evans is sales manager, ~~while~~ Joan Thomas is in charge of research.
  ^

Do not use *while* to mean *although* or *whereas.*

Although
- ~~While~~ Ryan Patterson wants the job of engineering manager, he
  ^
  has not yet applied for it.

Restrict *while* to its meaning of "during the time that."

- I'll have to catch up on my reading *while* I am on vacation.

who / whom

Writers are often unsure whether to use *who* or *whom.* *Who* is the subjective case form, and *whom* is the objective case form. When in doubt about which form to use, substitute a personal pronoun to see which one fits. If *he, she,* or *they* fits, use *who.*

- *Who* is the service manager for our area?
  [You would say, "*He* is the service manager for our area."]

If *him, her,* or *them* fits, use *whom.*

- It depends on *whom?*
  [You would say, "It depends on *them.*"]

who's / whose / of which

Who's is the contraction of *who is.* (*Who's* scheduled today?) *Whose* is the possessive case of *who.* (*Whose* office will be relocated?)
    Normally, *whose* is used with persons, and *of which* is used with inanimate objects.

- The man *whose* car had been towed away was angry.
- The mantel clock, the parts *of which* work perfectly, is 100 years old.

If *of which* causes a sentence to sound awkward, *whose* may be used with inanimate objects. (There are added fields *whose* totals should never be zero.)

W

-wise

Although the suffix -wise often seems to provide a tempting shortcut in writing, it leads more often to inept than to economical expression. It is better to rephrase the sentence.

- Our department ~~rates~~ high ~~efficiencywise.~~
  has a   *efficiency rating.*

The -wise suffix is appropriate, however, in instructions that indicate certain space or directional requirements (*lengthwise, clockwise*).

word choice

Mark Twain once said, "The difference between the right word and almost the right word is the difference between 'lightning' and 'lightning bug.'" The most important goal in choosing the right word in technical writing is the preciseness implied by Twain's comment. Vague words and abstract words defeat preciseness because they do not convey the writer's meaning directly and clearly.

> **VAGUE** It was a *productive* meeting.
>
> **PRECISE** The meeting helped both sides understand each other's positions.

In the first sentence, *productive* sounds specific but conveys little. See how the revised sentence says specifically what made the meeting "productive." Although abstract words may at times be appropriate to your topic, using them unnecessarily will make your writing dry and lifeless.

Being aware of the connotations and denotations of words will help you anticipate readers' reaction to the words you choose. Understanding antonyms (*fresh/stale*) and synonyms (*notorious/infamous*) will increase your ability to choose the proper word. Make other usage decisions carefully, especially in technical contexts, such as average/median/mean and female/male.

Although many of the entries throughout this book will help you improve your word choices and avoid problems, the following entries should be particularly helpful:

affectation	26	euphemisms	189
biased language	54	idioms	257
clichés	79	jargon	301
conciseness/wordiness	99	vague words	559

W

A key to choosing the correct and precise word is to keep current in your reading and to be aware of new words in your profession and in the language. In your quest for the right word, remember that there is no substitute for a good <u>dictionary</u>. See also <u>English as a second language</u>.

WEB LINK WISE WORD CHOICES

For electronic exercises on word choice, see *<www.bedfordstmartins.com /alred>* and select *Exercise Central*.

---

word processing and writing

Word-processing software can help your writing in many ways; determine which features will be useful to you. Be aware that working on-screen may focus your attention too narrowly on sentence-level problems and cause you to lose sight of the larger problems of <u>scope</u> or organization. The rapid movement of the text on the screen, together with last-minute editing changes, may allow errors to creep into the text.

DIGITAL TIP GIVING ELECTRONIC FEEDBACK

Adobe Acrobat® and many word-processing packages have sophisticated options for providing feedback on the drafts of collaborators' documents. You can add text or voice annotations within the text, allowing your reader to double-click on an icon to read or hear your comments. In Acrobat, you can also use a drawing tool to input traditional editing marks. Finally, some word processors let you track changes, allowing you to actually change original text and allowing your reader to accept or reject each change. For more on this topic, see *<www.bedfordstmartins.com/alred>* and select *Digital Tips*.

Writer's Checklist: Getting the Most from Word Processing

☑ Avoid the temptation of writing first drafts on the computer *without any planning*. Plan your document carefully, as described in "Five Steps to Successful Writing."

☑ Use the outline feature to brainstorm and organize an initial outline for your topic. As you create the outline, cut and paste to try alternative ways of organizing the information. See <u>outlining</u>.

W

Writer's Checklist: Getting the Most from Word Processing (continued)

☑ Use the search-and-replace command to find and delete wordy phrases such as *that is, there are, the fact that,* and *to be* and unnecessary helping <u>verbs</u> such as *will.* See also <u>conciseness/wordiness</u>.

☑ When writing for <u>readers</u> who are unfamiliar with your topic, use the search command to find technical terms, <u>abbreviations</u>, and other information that may need further explanation.

☑ Use a spell checker and other specialized programs to identify and correct typographical, <u>spelling</u>, grammar, and <u>word-choice</u> errors.

☑ Do not make all of your <u>revisions</u> on the screen. Print a double-spaced copy of your drafts periodically for major revisions and reorganizations.

☑ Always proofread your final copy on paper because catching errors on the screen is difficult. Print an extra copy for your peers to critique before making your final revisions. See also <u>proofreading</u>.

☑ Use the software for effective document design by highlighting major <u>headings</u> and subheadings with bold or italic type and by increasing their size relative to the regular text. Use the copy command to duplicate parallel headings throughout your document, and insert blank lines (hard returns) above and below examples and <u>visuals</u> to set them off from the surrounding text. See also <u>layout and design</u>.

☑ Routinely save to your hard drive and create a backup copy of your documents on separate disks or on the company network.

writing a draft

You are well prepared to write a rough draft when you have established your <u>purpose</u> and <u>readers'</u> needs, defined your <u>scope</u>, completed adequate <u>research</u>, and prepared an outline (whether formal or informal). Writing a draft is simply transcribing and expanding the notes from your outline into <u>paragraphs</u>, without worrying about grammar, refinements of language, or spelling. Refinement will come with <u>revision</u> and <u>proofreading</u>. See "Five Steps to Successful Writing."

Writing and revising are different activities. Do not let worrying about a good opening slow you down. Instead, concentrate on getting your ideas on paper—now is not the time to polish or revise.

Do not wait for inspiration—treat writing a draft as you would any other on-the-job task. The Writer's Checklist that follows offers tactics experienced writers use to get started and keep moving.

W

Writer's Checklist: Writing a Rough Draft

☑ Set up your writing area with the writing tools (paper, pens, complete dictionary, sourcebooks, etc.) you will need to keep going once you get started. Then hang out the "Do Not Disturb" sign.

☑ Use a good outline as a springboard to start and keep going. See also **outlining**.

☑ As you write your draft, keep your readers' needs, expectations, and knowledge of the subject in mind. This will help you write directly to your readers and suggest which ideas need further development.

☑ When you are trying to write quickly and you come to something difficult to explain, try to relate the new concept to something with which the readers are already familiar, as discussed in **figures of speech**.

☑ Start with the section that seems easiest. Your readers will never know that the middle section of the document was the first section you wrote.

☑ Give yourself a set time (ten or fifteen minutes, for example) in which you write continuously, regardless of how good or bad your writing seems to be. But don't stop when you are rolling along easily—if you stop and come back, you may lose momentum.

☑ Give yourself a small reward—a short walk, a soft drink, a brief chat with a friend, an easy task—after you have finished a section.

☑ Reread what you have written when you return to your writing. Seeing what you have already written can return you to a productive frame of mind.

writing for the Web

Writing for the Web requires special attention to organizing and preparing content for Web visitors who must read online. See also **forms design**, **Internet**, and **Web design**.

Write for Rapid Consumption

Because users of the Web expect to get information quickly and efficiently, you need to be clear, concise, and well organized. To keep the attention of your **reader,** your writing style needs to be simple, straightforward, and substantive. Web users seek information: Avoid empty promotional language or **jargon** that says very little about your topic.

EMPTY   The LNK Converter is the best one on the Internet! Buy one today!

W

SUBSTANTIVE   The LNK Converter is a powerful, cost-effective tool. It has been proven 100% effective in <u>industry tests</u> and was voted the #1 Converter of the Year by <u>*Devices Magazine*</u>.

Note that the improved version contains hyperlinks to the sources of the evidence (industry tests and *Devices Magazine*) that support the company's claims about the product.

Use Subheads, Hyperlinks, and Keywords

Divide long text passages into small sections, each containing only one or two related <u>paragraphs</u>. Place boldface <u>headings</u> above each section to set apart a new idea or topic and make them informative. You may also use boldface within the text to emphasize important pieces of information such as due dates, prices, or other essential details.

Because Web users are unlikely to read an online document from beginning to end, avoid directional cues like "as shown in the example below" or "in the graph at the top of this page." Directional phrases like these can be confusing when there is no real reference for "above" and "below" and no "top" or "bottom" of the document. Instead, use internal hyperlinks to connect sections of long documents.

A hyperlinked word usually is underlined or appears in different color from the text so that the user's eye will be drawn to it. Although hyperlinks expand information by offering gateways to other Web pages, they can be very distracting. Rather than embedding hyperlinks throughout your text, combine them into short, well-organized <u>lists</u>. Use these lists throughout your document to break up large blocks of text or place them at the end of the document.

Use keywords and concepts throughout your text so that search engines will point users to your site as in the following passage (keywords are in *italics*).

- The new *Alder* commemorative coin features a portrait of *Wilbur G. Alder,* the founder and president of Alder National Bank. *The coin* can be purchased online at this Web site after December 1, 2003, in honor of the 100th anniversary of the first deposit.

Writer's Checklist: Writing for the Web

☑ Present your information in short paragraphs.

☑ Use short, bulleted lists of hyperlinks to expand your information; if you have few hyperlinks, embed them in the text.

☑ Use informative subheads in boldface to aid skimming and to introduce new or important points.

W

Writer's Checklist: Writing for the Web (continued)

☑   Avoid breaking a single article into a series of separate pages; such in-
terruptions can irritate readers.

☑   Use lists, hyperlinks, and extra white space to present long documents
and to break up dense passages of text.

☑   For documents of more than 500 words, provide a brief summary or
table of contents hyperlinked to each section.

☑   Put keywords that describe your site in the first 50 words so that
search engines will find your site. See **Internet research**.

☑   Avoid humor and jargon that could be easily misunderstood, especially by visitors from other cultures.

☑   Choose language and use a **tone** appropriate to the organization or
product that you represent. See also **style**.

☑   Check for grammatical and typographical errors by carefully **proofreading**.

DIGITAL TIP USING PDF FORMAT

Use HTML for text that introduces or sets the context for information at
your site. If your site contains documents (reports, articles, brochures),
they can be posted in HTML or in PDF (portable document file). Converting the documents to PDF allows you to retain the identical look of the
printed document, except for color. To view PDF documents, visitors to
your site will have to download the Adobe Acrobat Reader software (or
plug-in) if they do not already have it. You should therefore include a link to
the Adobe Acrobat Web site for them to download a free copy of the
reader software. You can add this link to your site Help page, if you have
one. For more on this topic, see *<www.bedfordstmartins.com/alred>* and select *Digital Tips*.

W

Y

"you" viewpoint

The *"you" viewpoint* is a writing technique that places the <u>readers'</u> interest and perspective foremost. It is based on the principle that your readers are naturally more concerned about their own needs than they are about those of the writer or organization. The "you" viewpoint often, but not always, means using the words *you* and *your* rather than *we, our, I,* and *mine*. Consider the following sentence that focuses on the needs of the writer and organization ("we") rather than on those of the reader.

- *We must receive* your receipt with the merchandise before *we can process* your refund.

Even though the sentence uses *your* twice, the words in italics suggest that the <u>point of view</u> centers on the writer's need to receive the receipt in order to process the refund. Consider the following revision, written with the "you" viewpoint.

- So you can receive your refund promptly, please enclose the sales receipt with the returned merchandise.

Because the benefit to the reader is stressed, the writer is more likely to motivate the reader to act. See also <u>persuasion</u>.

The "you" viewpoint, as suggested earlier, means more than simply using *you* and *your*. In some instances, you may even need to avoid using those <u>pronouns</u> to build or maintain goodwill. Notice how the first of the following examples—with *your*—seems to accuse the reader, while the second—without *your*—achieves the goals of the "you" viewpoint with <u>positive writing</u>.

ACCUSATORY   *Your* budget makes no allowance for set-up costs.

POSITIVE   The budget should include an allowance for set-up costs to meet all the concerns of our client.

The "you" viewpoint can be extended beyond the sentence level to include building goodwill, establishing a positive <u>tone</u>, and handling bad news tactfully, as discussed in <u>correspondence</u>.

your / you're

Your is a possessive <u>pronoun</u> (*your* wallet); *you're* is the contraction of *you are* (*you're* late for the meeting). If you tend to make this common typographical error, use the search function for "your" during <u>proofreading</u>.

Y

Index